Lecture Notes in Artificial Intelligence 8577

Subseries of Lecture Notes in Computer Science

Nathalie Hernandez   Robert Jäschke
Madalina Croitoru (Eds.)

Graph-Based Representation and Reasoning

21st International Conference
on Conceptual Structures, ICCS 2014
Iaşi, Romania, July 27-30, 2014
Proceedings

 Springer

Volume Editors

Nathalie Hernandez
Université Toulouse le Mirail
5, Allées Antonio Machado, 31058 Toulouse, France
E-mail: nathalie.hernandez@irit.fr

Robert Jäschke
L3S Research Center
Appelstr. 4, 30167 Hannover, Germany
E-mail: jaeschke@L3S.de

Madalina Croitoru
LIRMM
161 Rue Ada, 34090 Montpellier, France
E-mail: madalina.croitoru@lirmm.fr

ISSN 0302-9743   e-ISSN 1611-3349
ISBN 978-3-319-08388-9   e-ISBN 978-3-319-08389-6
DOI 10.1007/978-3-319-08389-6
Springer Cham Heidelberg New York Dordrecht London

Library of Congress Control Number: 2014941588

LNCS Sublibrary: SL 7 – Artificial Intelligence

Typesetting: Camera-ready by author, data conversion by Scientific Publishing Services, Chennai, India

Printed on acid-free paper

Springer is part of Springer Science+Business Media (www.springer.com)

Preface

This volume contains the proceedings of ICCS 2014, the 21st International Conference on Conceptual Structures held during July 26–29, 2014 in Iaşi, Romania.

The International Conferences on Conceptual Structures[1] have been held annually in Europe, Australia, and North America since 1993. Their focus is on the formal analysis and representation of conceptual knowledge with applications to artificial intelligence, computational linguistics, and related areas of computer science.

The ICCS conferences evolved from a series of 7 annual workshops on conceptual graphs, starting with an informal gathering hosted by John Sowa in 1986. For the 7th conceptual graphs workshop in 1992, the informal workshop notes were upgraded to reviewed and edited proceedings published in the Springer-Verlag series of Lecture Notes in Artificial Intelligence.

Recently, graph-based knowledge representation and reasoning (KRR) paradigms have been receiving more and more attention. The aim of the ICCS 2014 conference is to build upon its long standing expertise in graph-based KRR and focus on providing modeling, formal and application results of graph-based systems. The new Steering Committee, elected in 2013 by the historical members of ICCS, decided to enlarge the themes covered by the conference with the different research domains currently involved in graph-based KRR. The Program Committee has thus been redefined in order to include recognized researchers in fields such as Knowledge Representation and Information Integration, Knowledge Representation and Uncertainty, Knowledge Representation and Alignment, Semantic Web and Web of Data.

The proceedings start with a reprint of the paper "Conceptual Graphs are also Graphs" by Michel Chein and Marie-Laure Mugnier from 1995. Although this work is an important contribution and has been cited frequently, it never got published to an international audience. Therefore, we decided to honor its contribution by including it as a historic paper. The conference also included three invited talks from distinguished speakers whose accompanying extended abstracts are part of this volume: Jonathan Ben-Naim from Université Paul Sabatier, Toulouse, France presented his work on "Argumentation-based Paraconsistent Logics," Jan Ramon from Katholieke Universiteit Leuven, Belgium gave a talk on "Towards a Framework for Learning from Networked Data," and Milos Stojakovic from the University of Novi Sad, Serbia introduced "Games on Graphs."

---

[1] http://conceptualstructures.org/confs.htm

40 research papers were submitted to ICCS 2014 for peer review. Each paper was reviewed by at least three Program Committee members. The top-ranked 17 papers (43%) were selected competitively for this volume. Besides the regular research track, the conference also included an industry track. The aim of this track was to enable the participants to present their current research projects in order to encourage and foster collaboration and joint project proposals. After reviewing, 6 of the 10 submissions to the industry track were accepted and published as short papers.

In 2014 ICCS was held in Iaşi, Romania at the Al. I. Cuza University, the oldest higher education institution in Romania. The university was founded one year after the establishment of the Romanian state in 1860. Iaşi has a long tradition in higher education and has traditionally been one of the leading centers of Romanian social, cultural, academic and artistic life.

The conference included a one-day Workshop on Integration of Heterogeneous dAta Sources and onTologiEs in Life Science and Environment whose proceedings are published separately in the CEUR-WS series.

ICCS 2014 was sponsored by Amazon Romania and BitDefender. We cordially thank the sponsors for their support! We are also thankful to Alexandru Ioan Cuza University for providing the facilities for the conference, the ALUMNI foundation for helping to organize sponsoring, the Faculty of Informatics in Iaşi for its local support, the travel agency Travis for taking care of many arrangements, and EasyChair for enabling a smooth submission and reviewing process.

We wish to thank the Organizing Committee individually: Cornelius Croitoru, Lenuţa Alboaie and Corina Forăscu from Al. I. Cuza University, Alin Patru from Travis, and collectively the Editorial Board and Program Committee members.

May 2014

Nathalie Hernandez
Robert Jäschke
Madalina Croitoru

Organization

General Chair

Madalina Croitoru   University Montpellier 2, France

Program Chairs

Nathalie Hernandez   IRIT, France
Robert Jäschke   L3S Research Center, Germany

Local Chairs

Cornelius Croitoru, Chair
Lenuţa Alboaie, Co-Chair
Corina Forăscu, Sponsorship Chair

Program Committee

Simon Andrews   Sheffield Hallam University, UK
Galia Angelova   Bulgarian Academy of Sciences, Bulgaria
Moulin Bernard   Laval University, Canada
Peggy Cellier   IRISA/INSA Rennes, France
Dan Corbett   Optimodal Technologies, USA
Olivier Corby   Inria, France
Cornelius Croitoru   University "Al. I. Cuza" Iasi, Romania
Madalina Croitoru   LIRMM, France
Frithjof Dau   SAP, Germany
Jérôme David   Inria Rhône-Alpes, France
Aldo De Moor   CommunitySense, The Netherlands
Harry Delugach   University of Alabama in Huntsville, USA
Juliette Dibie-Barthélemy   AgroParisTech, France
Pavlin Dobrev   ProSyst Labs, Bulgaria
Florent Domenach   University of Nicosia, Cyprus
Ged Ellis   Google, Australia
Jérôme Euzenat   Inria & LIG, France
Catherine Faron Zucker   I3S, UNS-CNRS, France
Jerome Fortin   LIRMM, France
Cynthia Vera Glodeanu   Technische Universität Dresden, Germany
Lea Guizol   LIRMM, France

Michaël Thomazo	University Montpellier, France
Francisco Valverde-Albacete	Universidad Carlos III de Madrid, Spain
Srdjan Vesic	Université d'Artois, France
Martin Watmough	Sheffield Hallam University, UK
Karl Erich Wolff	University of Applied Sciences Darmstadt, Germany
Stefan Woltran	Vienna University of Technology, Austria
Guo-Qiang Zhang	Case Western Reserve University, USA

Additional Reviewers

Bliem, Bernhard
Chapman, Peter
Delaney, Aidan
Doerfel, Stephan
Freitas, Andre

Inants, Armen
Martinez, Maria Vanina
Nicholson, Jon
Qasemizadeh, Behrang

Rietveld, Laurens
Schwind, Nicolas
Tellier, Isabelle
Wu, Xia

Sponsoring Institutions

Amazon Romania
BitDefender

Table of Contents

Industrial Papers

Conceptual Graphs Are Also Graphs

Michel Chein and Marie-Laure Mugnier

University of Montpellier 2, France
name@lirmm.fr

Technical Report LIRMM number 95003, v2, 1995

Abstract. The main objective of this paper is to add one more brick in building the CG model as a knowledge representation model *autonomous* from logic. The CG model is not only a graphical representation of logic, it is much more: it is a declarative model encoding knowledge in a mathematical theory, namely *labelled graph theory*, which has efficient computable forms, with a fundamental graph operation on the encodings to do reasoning, *projection*, which is a labelled graph morphism. Main topics of this paper are: a generalized formalism for *simple* CGs; a strong equivalence between CSP (Constraint Satisfaction Problem) and labelled graph morphism. This correspondence allows the transportation of efficient algorithms from one domain to the other, and confirms that projection —or more generally labelled graph morphism— firmly moors CGs to combinatorial algorithmics, which is a cornerstone of computer science. The usual sound and complete first order logic semantics for CGs is still valid for our generalized model. This, plus the ease of doing important reasonings with CGs —for instance plausible reasonings by using some maximal join operations— without, at least for the moment, logical semantics, strengthens our belief that CGs must also be studied and developed independently from logic.

1 Introduction

"Conceptual graphs are significant as the basis for an operational knowledge science and technology encompassing natural language, formal language, visual language, and the wide range of reasoning processes that may be based on them", says Gaines [Gai93]. Works by Ellis, Levinson [Ell93] [Lev94] and others, have strengthen a fourth pillar —besides the three ones emphasized by Gaines: natural language, formal logic, and visual language— of CG theory : *graph theory and graph algorithms*. We share the Peirce group point of view that graph theory and algorithms are at the core of the CG model : this was present in [Sow84], but is not even mentioned by Gaines, and has become marginal in [Sow94a]. As a matter of fact graph theory is interesting in knowledge representation for many reasons. The principal reason is probably that natural language is the most important form of human knowledge representation, and graph models are very useful in linguistics, and specifically in semantics. This is one of the bases of Sowa's work, and we will not develop this point here. In *human/machine*

N. Hernandez et al. (Eds.): ICCS 2014, LNAI 8577, pp. 1–18, 2014.
DOI: 10.1007/978-3-319-08389-6_1, © Springer International Publishing Switzerland 2014

interaction, graphical representations of graphs allow the construction of useful interfaces, in particular convivial graph editors are easy to build (this is the visual language aspect). But, furthermore, a graph is a mathematical object which has "natural" implementations by cells and pointers (cf. [SW86]), which are analogous to the represented graphs. By combining this property with graphical representations, explanation modules may be built, allowing the user to follow computations on a faithful and "readable" image of the model. From a *computational point of view*, efficient algorithms are needed in order to construct usable systems based on CGs, and numerous efficient graph algorithms have been developed these last years.

One purpose of this paper is to add one more brick in building the CG model as a knowledge representation model *autonomous* from logic. The CG model is not (only) a graphical representation of logic, even if it may be used for that. That is to say (without repeating basic principles of knowledge representation as mentioned, for instance, by Sowa in [Sow94a], p.3) the CG model is: a *declarative* model encoding knowledge in a mathematical theory, namely *labelled graph theory*, which has efficient computable forms, with a fundamental graph operation on the encodings to do reasoning, *projection*, which is a labelled graph morphism.

The fact that the simple CG model has *consistent* and *complete* first order logic semantics, plus the ease of doing important reasonings —for instance plausible reasoning by using maximal joins— on CGs, without, at least for the moment, logical semantics, strengthens our belief that CGs must also be studied and developed "independently" from logic. This point of view is illustrated in Figure 1.

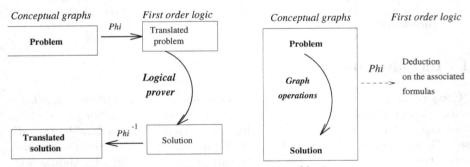

On the left: CGs as a graphical representation of logic.
On the right: CGs as a graph model, with a logical semantics.

Fig. 1. Conceptual graphs and logic

This paper is organized as follows. In section 2 a generalized formalism for the simple CG model is presented. By *simple* graphs we mean *non-nested* graphs, in particular there is no negation. The differences with Sowa's original model [Sow84], studied for instance in [CM92], are the following:

- a CG is not necessarily a connected graph;
- the concept type set is not necessarily a lattice. The concept-types and relation-types both form partial orders —which represent an *A Kind Of* relation;
- the conformity relation is replaced by an *Is A* relation between individual concepts and concept types (a typing of constants).

Generalized definitions for elementary specialization/generalization operations are proposed, and the main results concerning the specialization/generalization relation ($\leq$) remain valid :

- equivalence between $\leq$ and projection,
- existence of a unique irredundant CG in any equivalence class modulo $\leq$,
- furthermore, in the new model, the irredundant CG set has a *lattice* structure.

In our opinion, this slightly generalized model is simpler and more elegant than the previous one. The function Φ introduced by Sowa is studied. Sowa proved [Sow84] that this function may be considered as a *sound* logical semantics for CGs. We proved that this semantics is also *complete* [CM92]. We explain here why it is necessary to consider CGs in a *normal form* (any CG cannot contain more than one concept vertex with a given individual marker) in order to obtain completeness.

In Section 3 a strong equivalence between CSP (Constraint Satisfaction Problem) and graph morphism is established. This correspondence allows the transportation of efficient algorithms from one domain to the other, and confirms that projection —or more generally labelled graphs morphism— firmly moors CGs to combinatorial algorithmics, which is a cornerstone of computer science.

In a short conclusion it is indicated in what directions we are pursuing our research concerning theoretical and algorithmical aspects of the CG model.

2   The Formal Model

2.1   Descriptive Part

A structure, we call a *support*, provides background knowledge on a specific domain application. It represents an ontology of this domain, as the T-Box of terminological systems [Leh92]. It corresponds to the notion of a canon in [Sow84].

A *support* consists of a 5-tuple $S = (T_c,\ T_r,\ \sigma,\ I,\ \tau)$ where:

- T_c is an *ordered* set — not necessarily a lattice — of concept types, with a supremum $\top$ (the universal type) and an infimum $\bot$ (the absurd type),
- T_r is an *ordered* set of relation types, with a partition: $T_r = T_{r_{i_1}} + \dots + T_{r_{i_p}}$ where $T_{r_{i_j}}$ is the set of i_j-ary relations, $i_j \neq 0$; thus, two comparable relations types must have the same arity. The orders on T_c and T_r correspond to an *A Kind Of* relation between types,

- σ is a mapping which associates a signature with each relation type; the signature of a relation specifies the arity and the maximal concept types this relation can link. For any relation type $r \in T_{r_{i_j}}$, $\sigma(r) \in (T_c)^{i_j}$, and $\forall r_1, r_2 \in T_{r_{i_j}}, r_1 < r_2 \Rightarrow \sigma(r_1) \leq \sigma(r_2)$, where the order between signatures is the product order on $(T_c)^{i_j}$ (i.e. each i-ith argument of $\sigma(r_1)$ is less or equal to the i-ith argument of $\sigma(r_2)$),
- I is a set of individual *markers*; in addition there is the generic marker $*$; $I \cup \{*\}$ is provided with an order, such that $*$ is greater than all markers, and any two individual markers are non-comparable,
- τ is a mapping from I to $T_c \setminus \{\bot\}$; if the image of an individual marker m is a concept type t, t is called the type of m. τ corresponds to an *Is A* relation.

A *simple conceptual graph* related to a support S, $G = (R, C, U, label)$ is a bipartite multigraph[1], not necessarily connected. R and C denote the two classes of relation and concept vertices, $C \neq \emptyset$. U is the set of edges, and the edges incident to each relation vertex r are totally ordered —they are numbered from 1 to the arity of r. The i-ith neighbor of r in G is denoted by $G_i(r)$. Every vertex has a label defined by the mapping *label*. If $r \in R$, then $label(r) \in T_r$; if $c \in C$, then $label(c) \in (T_c \setminus \{\bot\}) \times (M \cup \{*\})$; the label of a concept type is a pair $(type(c), marker(c))$. *label* satisfies the constraints given by σ and τ: $\forall r \in R$, $type(G_i(r)) \leq \sigma_i(type(r))$. $\forall c \in C$, if marker$(c) \in I$ then $type(c) = \tau(marker(c))$. Let us point out some aspects of our definitions, which slightly generalize the commonly used definitions of simple conceptual graphs.

- The concept type set is an order with a greatest and a least element, but not necessarily a lattice, as in [Sow84].
- The relation type set is ordered, and relations may be of any arity.
- A simple conceptual graph is not necessarily connected.

Another point is the τ relation, which plays the role of the conformity relation of [Sow84]. We impose that an individual marker always appears with the same type in concept labels[2]. But a more permissive solution could be as follows. Any concept label $l = (t, m)$, where $m \in I$, must satisfy $t \geq \tau(m)$, and not necessarily $t = \tau(m)$. Then, two types are associated with the label l: a "surface" type, t, and a "deep" type $\tau(m)$, this latter being the type used in operations.

---

[1] Graph with possibly several edges between two vertices.

[2] Let us explain why with an example. Let *Jumbo* be a circus elephant, i.e. the marker *Jumbo* conforms to the type *CircusElephant*. Now, suppose there are two graphs $G_1 = $ [Elephant : *Jumbo*]-2-(AGNT)-1-[Eat:$*$] and $G_2 = $ [CircusElephant : $*$]-2-(AGNT)-1-[Eat:$*$], with CircusElephant $\leq$ Elephant in T_c. Since one knows that [Elephant : *Jumbo*] refers to the same individual as [CircusElephant : *Jumbo*], G_2 should be considered as more general than G_1. However it is not possible to project G_2 onto G_1 (see section 2.2). When considering the logical semantics of CGs, one has the formulas: $s_1 = CircusElephant(Jumbo)$ and $s_2 = \forall x\ CircusElephant(x) \rightarrow Elephant(x)$ coming from the logical interpretation of the support (see section 2.3), $f_1 = \exists y Elephant(Jumbo) \wedge Eat(y) \wedge Agent(y, Jumbo)$ and $f_2 = \exists x \exists y CircusElephant(x) \wedge Eat(y) \wedge Agent(y, x)$, respectively associated with G_1 and G_2. $f_1, s_1 \vdash f_2$, what cannot be inferred with the graph operations.

2.2 The Specialization/Generalization Relation between CGs

Elementary Operations. The fundamental notion for reasoning about CGs is the *specialization/generalization* relation on CGs — or *subsumption* relation. It is denoted by $\leq$. This relation can be defined in terms of a sequence of elementary relations, or by a global operation, called *projection*, which is a graph morphism.

The elementary specialization/generalization operations are internal operations on CGs, related to a given support. There are five specialization operations and five inverse operations, called generalization operations. In the list below, each specialization operation is followed by the inverse operation described in italics. All generalization operations are unary. One specialization operation, Disjoint Sum, is binary, the other four are unary. For all operations, the resulting graph is disjoint from the operand graph(s).

Let G be a CG, a CG H can be obtained from G by:

- **Simplify:** delete a duplicate relation vertex of G (i.e. a relation vertex r such that there is another relation vertex with the same type and exactly the same i-ith neighbor, for all i).
  Duplicate: duplicate a relation vertex of G.
- **Relation restrict:** decrease the label of a relation vertex r of G, providing that the new type (say t) conforms to σ: for all i ($1 \leq i \leq arity(r)$), $type(G_i(r)) \leq \sigma_i(t)$.
  Relation augment: increase the label of a relation vertex r of G (no constraint on this augmentation).
- **Concept restrict:** decrease the label of a concept vertex c of G —replace (t, m) by (t', m') with $t' \leq t$ and $m' \leq m$— providing that the new label conforms to τ: if m' is an individual marker, then $t' = \tau(m')$.
  Concept augment: increase the label of a concept vertex c of G —replace (t, m) by (t', m') with $t \leq t'$ and $m' = *$— providing that the new label conforms to σ: for all relation neighbor r of c, let $c = G_i(r)$, $type(c) \leq \sigma_i(r)$.
- **Join:** merge two concept vertices with same label.
  Split: split a concept vertex c into vertices c_1 and c_2, with the same label as c, such that the union of the sets of edges incident to c_1 and c_2 is the edge set of c —one of the two sets may be empty.
  And the unique binary operation:
- **Disjoint sum:** Let G_1 and G_2 be two disjoint CGs; H is obtained from G_1 and G_2 by juxtaposing G_1 and G_2.
  Substract: delete some (not all)[3] connected components of G.

G is a specialization of H ($G \leq H$) if G can be derived from H by specialization operations. Since there is a binary operation, the derivation of H from G may have other sources than H (see [MC93] for a formal definition of a specialization derivation). G is a generalization of H ($G \geq H$) if G can be derived from H by a sequence of generalization operations. Note that if $G \geq H$, then G can be

---

[3] Indeed we do not consider empty graphs.

obtained from H by considering H only (in terms of [MC93], the derivation from G to H is a path, and not a graph with several sources). It is straightforward to prove that $G \leq H$ *iff* $G \geq H$.

Projection. Projection is the fundamental operation on CGs, since it permits the *effective* computation of the $\leq$ relation. This operation is a graph morphism,(see section 3.1), keeping the order on the edges, and allowing restriction of vertex labels. More precisely: A *projection* Π from a conceptual graph $H = (R_H, C_H, U_H, lab_H)$ to a conceptual graph $G = (R_G, C_G, U_G, lab_G)$ is an ordered pair $\Pi=(f,g)$ of mappings, f from R_H to R_G, and g from C_H to C_G, such that

1. $\forall \ rc \in U_H$, $f(r)g(c) \in U_G$, and if $c = H_i(r)$ then $g(c) = G_i(f(r))$ —i.e. it preserves edges and orderings on edges,
2. For each $r \in R_H$, $lab_H(r) \geq lab_G(f(r))$,
3. For each $c \in C_H$, $lab_H(c) \geq lab_G(g(c))$.

In the following, a projection is simply noted Π, with an implicit decomposition into f and g. Sowa [Sow84] proves that, given two conceptual graphs G and H, if $G \leq H$ then there is a projection from H to G. We prove in [CM92] that the reciprocal also holds (in [MC93] the proof is extended to canonical graphs). These proofs are still valid for the generalized model given in this paper.

Theorem 1. *Let G and H be two simple conceptual graphs; $G \leq H$ iff there is a projection from H to G.*

Structure of the $\leq$ Relation. $\leq$ is a reflexive and transitive relation, but, as several times remarked (e.g. [Jac88]), it is not a partial order because the antisymmetry property is not fulfilled (Figure 2). If $G \leq H$ and $H \leq G$, then G and H are said to be equivalent, denoted $G \equiv H$.

A natural question is then to measure the difficulty of coming back to an order. The notion of an *irredundant* graph [CM92] provides a characterization of the quotient order of $\leq$. A graph G is an *irredundant* graph if it has no strict subgraph equivalent to it. Thus G is irredundant *iff* there is no projection from G to one of its strict subgraphs. Otherwise G is *redundant*. Note that

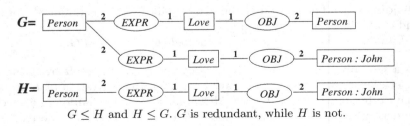

$G \leq H$ and $H \leq G$. G is redundant, while H is not.

Fig. 2. Equivalent graphs

an irredundant graph is not necessarily connected. It appeared later that this notion corresponds to the notion of a *core graph* for non-directed graphs with homorphism operation (e.g. [HN90]).

Proposition 1. [CM92] *Given two canonical graphs G and H, if G ≡ H and H is irredundant, then G has a subgraph isomorphic to H.*

Corollary 1. [CM92] *Two irredundant conceptual graphs are equivalent iff they are isomorphic.*

Proposition 2. [CM92] *An equivalence class contains one and only one irredundant graph, which is the (unique) graph with the smallest vertex number.*

However, this characterization of an equivalence class is not polynomial (see part 3). It is shown in [Mug95] that, if the projection is polynomial for a class of graphs, then the computation of the irredundant form of a graph of this class is polynomial too.

The restriction of $\leq$ to the set of irredundant graphs is not only a partial order, it is a *lattice*.

Theorem 2. *The restriction of $\leq$ to the set of irredundant graphs is a lattice.*

Proof. Let G_1 and G_2 be two irredundant graphs and $G_1 + G_2$ be their disjoint sum. The following property is immediate: for every graph K, $K \leq G_1 + G_2$ *iff* $K \leq G_1$ and $K \leq G_2$. Indeed, a projection from $G_1 + G_2$ to K can be split into two projections, one from G_1 to K and one from G_2 to K. Reciprocally, the union of two projections, from G_1 to K and from G_2 to K yields a projection from $G_1 + G_2$ to K. Now there is a unique irredundant graph equivalent to $G_1 + G_2$, which is thus the greatest lower bound of G_1 and G_2. Thus, the irredundant graph set is an inf-semi-lattice. The irredundant graph set also has a maximum element, the graph $[\top : *]$. It is thus a lattice.   □

Note that this theorem is true even if the vertex labels (or the relation and concept types) are not ordered by a lattice. The only condition is that the concept type set forms a partial order with a greatest element. The theorem does not hold if one only consider connected graphs, even if the vertex labels are lattices.

This lattice structure is important when one wants to find minimal common generalization or maximal common specialization of two graphs. However, although it is easy to compute an equivalent graph to the infimum of two graphs (just compute their disjoint sum), this does not solve the problem of the effective computation of this infimum. The same problem arises with the computation of their supremum.

2.3   First Order Logic Semantics

Sowa [Sow84] proposes to associate with every conceptual graph G a well-formed formula $\Phi(G)$ of the first-order-predicate calculus. Given a support S, with each individual marker is associated a constant and with each relation or concept type is associated a predicate. With each concept vertex of a graph G related to S is associated a *term*: a distinct variable for each generic concept, and the constant corresponding to the marker for each individual concept. With each relation or concept vertex is associated an *atom*: for a concept c, the atom $type(c)(id(c))$, for a relation r, the atom $type(r)(id(c_1), ..., id(c_p))$, where c_i denotes the i-ith neighbor of r and $id(c_i)$ stands for the term associated with c_i. $\Phi(G)$ is the existential closure of the conjunction of the atoms associated with the vertices of G.

With the support is also associated a set of well-formed formulas, say $\Phi(S)$, which corresponds to a logical interpretation of the type orders.

1. for all $t_1, t_2 \in T_c$, if t_2 is covered by t_1[4], then:
   $\forall x t_1(x) \rightarrow t_2(x)$
2. for all $t_{r_1}, t_{r_2} \in T_r$, with same arity p, if t_{r_2} is covered by t_{r_1} then:
   $\forall x_1 ... x_p t_{r_1}(x_1 ... x_p) \rightarrow t_{r_2}(x_1 ... x_p)$
3. $\forall x \top(x)$
4. $\forall x \neg \bot(x)$
5. For every $t_r \in T_r$, with arity p, and $\sigma(t_r) = (t_1 ... t_p)$:
   $\forall x_1 ... x_p t_r(x_1 ... x_p) \rightarrow t_1(x_1) \wedge ... \wedge t_p(x_p)$
6. For every individual marker m, with $\tau(m) = t$:
   $t(m)$

Axioms 1 and 2 are the logical interpretation of the *A Kind Of* links between types. Axioms 3 to 6 are not used in the proof of the completeness theorem, but they are needed for the model-theoretic semantics proposed in [MC94].

It can be directly proven that the generalization rules constitute a *sound* set of inference rules for the set of logical formulas associated with simple CGs (see [Sow84], [CM92]). More precisely:

Theorem 3. *Let $\Phi(S)$ be the set of formulas associated with a support S. Let G and H be simple CGs related to S. If $G \leq H$ then $\Phi(S), \Phi(G) \vdash \Phi(H)$.*

It is more difficult to prove that the generalization rules constitute a *complete* set of inference rules. As far as we know, [CM92] provides the first proof of this result. In order to obtain completeness property, CGs need to be in a specific form, we call a *normal form*: all individual vertices must have different markers. Any graph can be transformed to normal form by joining all vertices with same individual marker. When CGs are not in normal form, some logical deductions can not be simulated with the generalization rules. See for instance Figures 3 and 4: in both figures, formulas $\Phi(G)$ and $\Phi(H)$ are equivalent, but G and H are not equivalent, even non comparable (Figure 4).

---

[4] i.e. $t_2 \leq t_1$ and there does not exist a type t with $t_2 < t < t_1$. We consider here the minimal set of edges in T_C that is necessary to generate all formulas associated to the partial order.

Let us mention that the specialization/generalization rules are not internal rules on the set of CGs in normal form. Even worse: there are graphs in normal form, say G and H, such that $G \leq H$ and there is no specialization derivation from H to G containing only CGs in normal form (Figure 5).

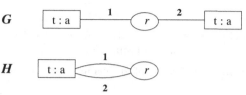

$\Phi(G) = t(a) \wedge t(a) \wedge r(a, a)$. $\Phi(H) = t(a) \wedge r(a, a)$. $\Phi(G)$ and $\Phi(H)$ are equivalent, while G and H are not equivalent $(G > H)$.

Fig. 3. The need for normal form (1)

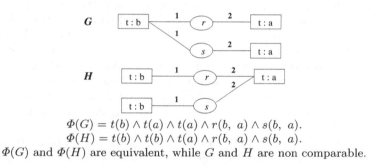

$\Phi(G) = t(b) \wedge t(a) \wedge t(a) \wedge r(b, a) \wedge s(b, a)$.
$\Phi(H) = t(b) \wedge t(b) \wedge t(a) \wedge r(b, a) \wedge s(b, a)$.
$\Phi(G)$ and $\Phi(H)$ are equivalent, while G and H are non comparable.

Fig. 4. The need for normal form (2)

Theorem 4. [CM92], [MC94] *Let $\Phi(S)$ be the set of formulas associated with a support S. Let g and h be well formed formulas associated with G and H in normal form. If $\Phi(S), g \vdash h$ then $G \leq H$.*

3   Algorithmic Considerations

Determining whether there is a projection from a graph to another is an NP-complete problem —we call this problem *Projection*. [CM92] proves that problems *a priori* simpler are also NP-complete: *Equivalence* (given G and H, is G equivalent to H?), *Redundancy* (given G, is there a projection from G to one of its strict subgraphs?). This part is devoted to a further study of the projection from the viewpoint of its relations with two domains : graph theory and Constraint Satisfaction Problems (CSP).

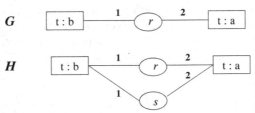

G and H are in normal form, but all derivations from G to H involve a graph which is not in normal form.

Fig. 5. Derivation of graphs in normal form

3.1   Conceptual Graphs and Graphs, Projection and Morphism

From a graph theory viewpoint, projection is a *morphism* —also called homomorphism. A *morphism* from a non-directed graph $G = (V_G, E_G)$ to a non-directed graph $H = (V_H, E_H)$, where V and E respectively denote the vertex set and the edge set, is a mapping $\Pi : V_G \to V_H$, which keeps the edges (i.e. for any edge xy of E_G, $\Pi(x)\Pi(y)$ is an edge of E_H). See Figure 6. When considering *directed* graphs, morphism must keep the edge direction (Figure 7). When there are labels on vertices and/or on edges, there are additional constraints; if the label sets have no specific structure, corresponding vertices (edges) must have the same label. The definition extends naturally to multigraphs.

We call *Morphism* the following generic problem: Given the two graphs G and H (both directed/non-directed, labelled or not, ...), is there a morphism from G to H?

Figure 6 shows an example of a non-directed morphism problem. It may also be interpreted as an instance of a graph vertex coloring problem (Given a graph $G = (V_G, E_G)$ and a positive integer k, is there a coloring of G with k colors, i.e. a mapping from V_G to $\{1, \dots, k\}$, such that no adjacent vertices have the same image (color)?). Here it would be the 3-coloring problem ($k = 3$). More generally, there is a coloring of a graph G with k colors *iff* there is a morphism from G to the k-clique (graph with k vertices and all possible edges except for loops).

Simple conceptual graphs may be seen as particular non-directed graphs, labelled on vertices and edges (furthermore they are bipartite graphs). Conversely, non-directed graphs may be seen as particular conceptual graphs. Indeed, a non-directed graph may be seen, from the morphism operation viewpoint, as a directed graph, without loops, each edge being replaced by two symmetric directed edges (Figure 8). In other words, given two non-directed graphs G and H, there is a non-directed graph morphism from G to H *iff* there is a directed graph morphism from the directed graph associated with G to the directed graph associated with H. A directed graph is itself a specific case of directed labelled graph, with labels on vertices and edges, all vertices (*resp.* edges) having the same label. Finally, a directed graph labelled on vertices and edges is a specific case of conceptual graph: each edge (x, y) is replaced by a binary relation vertex, with x as first neighbor, and y as second neighbor, and same label as the edge (x, y).

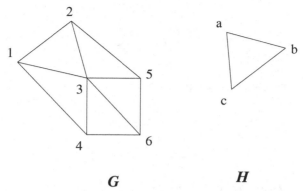

$$G \qquad\qquad H$$

There is no morphism from G to H. But, there are several morphisms from G minus
any edge, to H. For example, if the edge 12 is removed from G,
$$\{(1, a), (3, b), (4, c), (6, a), (5, c), (2, a)\}$$

Fig. 6. Non-directed graph morphism

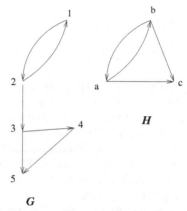

$$H$$

$$G$$

There are two morphisms from G to H: Π_1 and Π_2
$$\Pi_1 = \{(1, a), (2, b), (3, a), (4, b), (5, c)\}$$
$$\Pi_2 = \{(1, b), (2, a), (3, b), (4, a), (5, c)\}$$

Fig. 7. Directed graph morphism

Any polynomial case for Morphism, based upon the *structure* of G or H, provides a polynomial case for Projection. The complexity of Morphism has been particularly studied in the framework of the following problem, called *general coloring problem* or *H-coloring*, where H is a fixed graph, that may be directed or not: Given a graph G, is there a morphism from G to H? This problem is a generalization of the classical coloring problem, since H is not necessarily a clique.

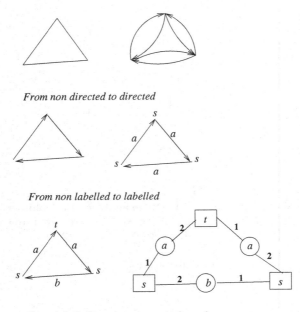

From non directed to directed

From non labelled to labelled

From labelled graph to conceptual graph

Fig. 8. Transformations

The complexity study of *H-coloring* focused on the structure of *H*, i.e. what are exactly the conditions on the structure of *H* for the problem to be polynomial? Concerning non-directed graphs, [HN90] proves that: *H-coloring* is polynomial if *H* is bipartite (or if *H* has a loop or is restricted to one vertex), and NP-complete in all other cases. Concerning directed graphs, [GWW92] and [HNZ93] exhibit some trees *H*, such that *H-coloring* remains NP-complete. Consequently, Morphism is NP-complete, even if *H* is a tree.

But Morphism is polynomial if *G* is a tree (this fact has been shown in several works independently). It has been shown in [HNZ94] that the problem remains polynomial when *G* has bounded treewidth (*i.e.* *G* is a partial k-tree[5], for a fixed *k*). But note that determining for a given graph *G* and a given integer *k*, whether the treewidth of *G* is at most *k* is NP-complete [ACP87]. Concerning conceptual graphs, we give in [MC92]— also [Mug95]— a polynomial algorithm for computing a projection from a "conceptual tree" *G* to a conceptual graph *H*, and for counting the number of such projections. This projection algorithm from a tree to a graph can be exploited in a general projection algorithm (i.e. when *G* has any structure), as a preprocessing part [Mug95].

---

[5] A non-directed graph is a *k-tree* if it can be obtained from a k-clique by repeatedly adding a vertex connected to *k* existing vertices which form a k-clique. It is a *partial k-tree* if it is a spanning subgraph of a k-tree (some edges are removed). In particular, a tree is a 1-tree. The *treewidth* of a graph is he minimum *k* for which the graph is a partial k-tree. For a directed graph, one considers the underlying non-directed graph.

The above mentioned complexity studies do not consider the case of *labelled* graphs, i.e. the effects of some labelling function properties on the complexity. However such a work would be of great interest for conceptual graphs. If the condition on labels for two vertices (or edges) that match is the equality of their label, then the presence of labels may simplify the problem and provide some polynomial cases. For instance, when the labelling function is locally injective (i.e. injective on the neighborhood of each vertex (or if edges are labelled, on the set of edges incident to each vertex), then morphism is polynomial, whatever the structure of G and H may be (see [Kac84] for feature structures and [LB94] for conceptual graphs). But when labels are partially ordered then the condition for matching vertices (or edges) can be "for any vertex c, the label of the image must be *less or equal to* the label of c". Then, morphism becomes a projection, which can make the problem more complex. For example: the problem of determining whether there is an isomorphism between two graphs is not classified but is supposed to be non NP-complete; for labelled graphs, the complexity remains the same if matching vertices necessarily have the same label; but the problem becomes NP-complete when there is a partial order on labels and the label of any vertex must be greater or equal to the label of its image (see the problem "isoprojection" in [CM92]).

3.2   Conceptual Graphs and Constraint Satisfaction Problems

There is a strong equivalence between the Projection problem and the well-known Constraint Satisfaction Problem (CSP) —e.g. [P.89]. The polynomial transformations between these problems, which are detailed below, keep the solution sets (actually an instance of one problem has the same set of solutions as its transformation into an instance of the other problem).

Let us consider a simple definition of a CSP, where constraints are defined in extension and are not typed. A *constraint network* $P = (X, C, D, R)$ involves:

- a set of n variables, $X = \{x_1, \ ... \ , x_n\}$
- a set of constraints, $C = \{C_1, \ ... \ , C_p\} \subset \mathcal{P}(X)$,
  When constraints are binary, i.e. involve two variables, the pair (X, C) can be seen as a non directed graph, whose vertices are the variables, and edges are the constraints. In the general case, (X, C) is an hypergraph.
- a set of variable domains, $D = \{D_1, \ ... \ , D_n\}$
- a set of constraint definitions, $R = \{R_1, \ ... \ , R_p\}$; for every constraint C_i, let $(x_{i_1} \ ... \ x_{i_q})$ be any ordering of C_i variables, then $R_i \subset D_{i_1} \times \ ... \ \times D_{i_q}$.

A *solution* of P is an assignment of values to the n variables such that all the constraints are satisfied, more formally:

a mapping $\mathcal{I} \colon X \to \bigcup D_i \in D$
$$x_i \mapsto a \in D_i,$$
and, for every constraint $C_j = (x_{j_1} \ ... \ x_{j_q})$, $(\mathcal{I}(x_{j_1}) \ ... \ \mathcal{I}(x_{j_q})) \in R_i$.

A *constraint satisfaction problem* is the associated problem: Given a constraint network P, is there a solution to P?

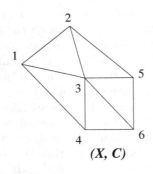

For all constraint Ci={x, y},

$Ri =$	x	y
	a	b
	a	c
	b	a
	b	c
	c	a
	c	b

(X, C)

D1=D2= ... = D6 = {a, b, c}

This constraint network represents an instance of the 3-coloring problem. The constraint graph and the graph to be colored are the same graph. There is no solution to the CSP: cf. Figure 6, which represents the same 3-coloring problem (as an instance of Morphism).

Fig. 9. A constraint network

In short, a solution of a CSP is a mapping from the variable set to the set of values of domains, which satisfies the constraints. In comparison, a morphism (or a projection)is a mapping from one vertex set to another vertex set, which keeps the edges. This formulation summarizes the strong equivalence of CSP and Morphism.

Theorem 5. *The problems* CSP *and directed labelled graph Morphism are polynomially equivalent, and both transformations from one problem to the other satisfy:*
(1) Every solution to one problem is a solution to the other one.
(2) The constraint graph and the graph to be mapped (G) have the same underlying non-directed graph.

This theorem basically comes from [FV93], where the equivalence between both problems is mentioned (but the transformations are not given in this paper). We give here a proof schema. It may be easily adapted for Projection instead of Morphism.

Transformation from Morphism to CSP. Let $G = (V_G, E_G)$ and $H = (V_H, E_H)$ be an instance of Morphism. We first consider G and H as directed antisymmetric graphs, without loops and without labels, we then extend the transformation.

A network $P = (X, C, D, R)$ is built from G and H as follows. (X, C) corresponds to the non-directed graph underlying G, that is: $X = V_G$, and $C = E_G$ with removal of the edge directions. With each variable x_i is associated the domain $D_i = V_H$. With each constraint C_i is associated the definition $R_i = E_H$.

When considering G and H as more complex graphs, variable domains and/or constraint definitions are restricted. E.g. if there are labels on edges, then with each constraint $C_i = (x_j \ x_k)$ is associated the definition $R_i = \{(y \ z) \in E_H \mid label(y, \ z) = label((x_j \ x_k))\}$. If there are loops in G, then for each vertex x_i with a loop, $D_i = \{y \in V_H | (y, \ y) \in E_H\}$, i.e. a loop may be seen as a kind of label.

This transformation is polynomial in the size of the instance of Morphism. It is straightforward to check that there is a morphism from G to H *iff* there is a solution to the CSP, and that points (1) and (2) of the theorem are fulfilled. Figure 10 shows the constraint network obtained from the instance of Morphism of Figure 7.

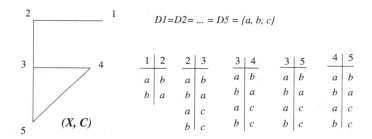

Fig. 10. Transformation from Morphism(Figure 7) to CSP

Transformation from CSP to Morphism. Let us first consider a binary CSP, with the constraint network $P = (X, \ C, \ D, \ R)$. From P, we build $G = (V_G, \ E_G, \ label_G)$ and $H = (V_H, \ E_H, \ label_H)$, an instance of directed labelled graph Morphism (only edges need to be labelled). G is built from the constraint graph: $V_G = X$, $E_G = C$, giving any direction to edges (we take here the order fixed by the constraint definitions). The labels are the "names" of the constraints. Each edge of E_G is labelled by the name of the constraint it comes from. H is built from D and R. V_H is the union of the variable domains, $V_H = \bigcup D_i \in D$. E_H is the disjoint union of the constraint definitions, that is, for each pair $(a \ b)$ of constraint definition R_j, one builds an edge $(a \ b)$ labelled by the name of C_j.

This transformation is polynomial in the size of the constraint network. It is straightforward to check that there is a solution to the CSP *iff* there is a morphism from G to H, and that points (1) and (2) of the theorem are satisfied. Figure 11 shows the graphs obtained from the constraint network of Figure 10.

H actually is a multigraph. It is possible to remove the multi-edges, by subdividing each edge. One obtains a bipartite graph. Labels on edges disappear, and vertices are now labelled: new vertices are labelled by the name of the corresponding constraint, and original vertices are all labelled with the same new value.

When the CSP is not binary, it can be either first transformed into a binary CSP (the transformation is polynomial, see for instance [Ber70]), or the previous transformation can be extended. G and H become non directed bipartite graphs, one vertex class corresponds to constraints, these vertices are labelled by the name of the corresponding constraint, the other class corresponds to variables,

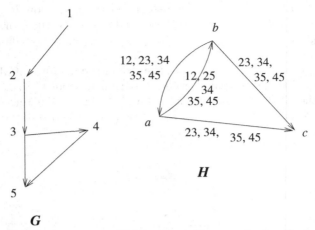

Lists of labels represent multi-edges, e.g. there are 5 edges from a to b.

Fig. 11. Transformation from CSP(Figure 10) to Morphism

and these vertices are labelled with the same value. The edges incident to any "constraint vertex" are totally ordered, what we represent by labelling the edges from 1 to the constraint arity. In short, one obtains a kind of conceptual graph.

These transformations show that Morphism (or Projection) and CSP are polynomially equivalent. It was already known since both are NP-complete problems. But, the interesting points are:

- The graph to be mapped (i.e. G) and the constraint graph have the same non-directed underlying graph. Thus, any polynomial case known for CSP and based on the structure of the constraint graph directly applies to the G of Morphism, and reciprocally. For instance, it is well-known in the CSP community that the problem CSP is polynomial when the constraint graph is acyclic (or is a tree) [Fre82].

- Every solution to the CSP *is* a morphism from G to H. Thus, the algorithmic techniques for determining whether there is a solution, but also enumerating the solutions, can be transferred from one domain to the other.

4   Conclusion

The simple CG model studied in this paper is a firm basis for building a knowledge representation language. We are indeed involved in the construction of a software platform (CORALI) for developing applications in which all kinds of knowledge are represented by CGs, and all reasonings are supported by projection.

But this model is too restrictive for many applications. Extensions are needed. Although it is easy to extend the descriptive features, extensions we are interested in have to respect essential characteristics of the kernel: objects must be sorts of graphs and inferences must be done by graph algorithms, not by graphical translations of logical operations. For instance, an extension to nested CGs —in the sense of [Sow94b]— is proposed in [MC94].

Acknowledgements. We thank Anne Preller for the example of Figure 4, which shows the problems with individual markers in the proof of logical completeness.

References

[ACP87]   Arnborg, S., Corneil, D.G., Proskurovski, A.: Complexity of finding embeddings in a k-tree. SIAM Journal Algebraic Discrete Methods (8), 277–284 (1987)

[Ber70]   Berge, C.: Graphes et hypergraphes. Dunond (1970)

[CM92]   Chein, M., Mugnier, M.L.: Conceptual Graphs: fundamental notions. Revue d'Intelligence Artificielle 6(4) (1992) (in English)

[Ell93]   Ellis, G.: Efficient Retrieval from Hierarchies of Objects using Lattice Operations. In: Mineau, G.W., Moulin, B., Sowa, J.F. (eds.) ICCS 1993. LNCS, vol. 699, pp. 274–293. Springer, Heidelberg (1993)

[Fre82]   Freuder, E.C.: A sufficient condition for backtrack-free search. JACM, 29–45 (1982)

[FV93]   Feder, T., Vardi, M.Y.: Monotone Monadic SNP and Constraint Satisfaction. In: Proceedings of the 25th ACM STOC, pp. 612–622 (1993)

[Gai93]   Gaines, B.R.: Representation, discourse, logic and truth: Situating knowledge technology. In: Mineau, G.W., Moulin, B., Sowa, J.F. (eds.) ICCS 1993. LNCS, vol. 699, pp. 36–63. Springer, Heidelberg (1993)

[GWW92]   Gutjahr, W., Welzl, E., Woeginger, G.: Polynomial graph coloring. Discrete Applied Math. (35), 29–45 (1992)

[HN90]   Hell, P., Nešetřil, J.: On the complexity of H-coloring. J. Combin. Th. (B), 33–42 (1990)

[HNZ93]   Hell, P., Nešetřil, J., Zhu, X.: Complexity of tree homomorphisms (1993) (manuscript)

[HNZ94]   Hell, P., Nešetřil, J., Zhu, X.: Duality and Polynomial Testing of Tree Homomorphisms (1994) (manuscript)

[Jac88]   Jackman, M.K.: Inference and the Conceptual Graph Knowledge Representation Language. In: Moralee, S. (ed.) Research and Development in Expert Systems IV. Cambridge University Press (1988)

[Kac84]   Ait Kaci, H.: A lattice-Theoretic Approach to Computation Based on A Calculus of Partially Ordered Type Structures. PhD Thesis, University of Pennsylvania (1984)

[LB94]   Liquiere, M., Brissac, O.: A class of conceptual graphs with polynomial isoprojection. In: Supplemental Proceedings of the International Conference on Conceptual Structures (1994) (revised version)

[Leh92]   Lehmann, F. (ed.): Semantics Networks in Artificial Intelligence. Pergamon Press (1992)

[Lev94]   Levinson, R.: UDS: A Universal Data Structure. In: Tepfenhart, W.M., Dick, J.P., Sowa, J.F. (eds.) ICCS 1994. LNCS (LNAI), vol. 835, pp. 230–250. Springer, Heidelberg (1994)

[MC92]   Mugnier, M.L., Chein, M.: Polynomial algorithms for projection and matching. In: Pfeiffer, H.D., Nagle, T.E. (eds.) Conceptual Structures: Theory and Implementation. LNCS, vol. 754, pp. 239–251. Springer, Heidelberg (1993)

[MC93]   Mugnier, M.L., Chein, M.: Characterization and Algorithmic Recognition of Canonical Conceptual Graphs. In: Mineau, G.W., Moulin, B., Sowa, J.F. (eds.) ICCS 1993. LNCS, vol. 699, pp. 294–311. Springer, Heidelberg (1993)

[MC94]   Mugnier, M.L., Chein, M.: Représenter des connaissances et raisonner avec des graphes. To Appear in Revue d'Intelligence Artificielle, 52 pages (1994)

[Mug95]   Mugnier, M.L.: On generalization/specialization for conceptual graphs. J. Expt. Theor. Artif. Intell. 7(3), 325–344 (1995)

[P.89]   Meseguer, P.: Constraint satisfaction problems: an overview. AICOM 2(1), 3–17 (1989)

[Sow84]   Sowa, J.F.: Conceptual Structures - Information Processing in Mind and Machine. Addison-Wesley (1984)

[Sow94a]   Sowa, J.F.: Knowledge representation: Logical, philosophical, and computational foundations. Preliminary Edition (August 1994)

[Sow94b]   Sowa, J.F.: Logical Foundations for Representing Object-oriented Systems. Journal of Experimental and Theoretical Artificial Intelligence 5 (1994)

[SW86]   Sowa, J.F., Way, E.C.: Implementing a semantic interpreter using conceptual graphs. IBM Journal of Research and Development 30(1), 57–69 (1986)

Argumentation-Based Paraconsistent Logics

Jonathan Ben-Naim

IRIT – CNRS
118, route de Narbonne
31062, Toulouse Cedex 09, France

Abstract. *Argumentation* is a promising approach for reasoning with inconsistent information. Starting from a knowledge base encoded in a logical language, an argumentation system defines *arguments* and *attacks* between them using the consequence operator associated with the language. Finally, it uses a *semantics* for evaluating the arguments. The plausible conclusions to be drawn from the knowledge base are those supported by "good" arguments.

In this paper, we discuss two families of such systems: the family using extension semantics and the one using ranking semantics. We discuss the outcomes of both families and compare them.

1 Introduction

A paraconsistent logic consists of a language and a consequence operator returning rational conclusions even from inconsistent premises. Possibly, a paraconsistent logic may attach absolute or relative weights to the conclusions.

Note that inconsistency may be present in knowledge bases for mainly three reasons: i) a knowledge base may contain a default rule [10] which has exceptions and the former leads to an opposite conclusion than the latter; ii) in model-based diagnosis [9], the description of the normal behavior of a system may be conflicting with the observations made on this system; iii) an inconsistent knowledge base may result from the union of several consistent knowledge bases pertaining to the same domain [5]. Whatever the source of inconsistency, a paraconsistent logic is needed to deal with it.

As a consequence, there has been much work on constructing and investigating such logics. Two families can be distinguished: those that restore consistency (e.g., [10,11]) and those that tolerate inconsistency without exploding (e.g., [4,6]). One important instance of the first family computes the maximal (for set inclusion) consistent subbases of a knowledge base, then chooses the conclusions that follow from all those subbases. Regarding the second family, a prominent approach considers many valued interpretations with the crucial particularity that they can be models of even inconsistent premises and thus can be used to draw conclusions.

Since early nineties, due to its explanatory power, *argumentation* has become a promising approach for handling inconsistency. An argumentation system consists of *arguments*, *attacks* between them and a *semantics* for evaluating the arguments. The latter are built from a knowledge base encoded in a particular language and using the consequence operator associated with the language. The attacks generally refer to the inconsistency of the base. Finally, the rational conclusions induced by such an argumentation system are those supported by acceptable arguments wrt the semantics.

N. Hernandez et al. (Eds.): ICCS 2014, LNAI 8577, pp. 19–24, 2014.
DOI: 10.1007/978-3-319-08389-6_2, © Springer International Publishing Switzerland 2014

In this paper, we recall the two families of semantics developed in the literature, namely *extension semantics* introduced by Dung in [8] and *ranking semantics* defined more recently in [1]. We discuss and compare the paraconsistent logics induced by each family. We show that logics based on extension semantics return flat conclusions and restore consistency while those based on ranking semantics return ranked conclusions and tolerate inconsistency.

2   Argumentation Systems

Argumentation systems are built on an underlying *monotonic logic*. In this paper, we focus on Tarski's monotonic logics [12]. Indeed, we consider logics $(\mathcal{L}, \text{CN})$ where $\mathcal{L}$ is a set of well-formed *formulas* and CN is a *consequence operator*. It is a function from $2^{\mathcal{L}}$ to $2^{\mathcal{L}}$ which returns the set of formulas that are logical consequences of another set of formulas according to the logic in question. It satisfies the following basic properties:

1. $X \subseteq \text{CN}(X)$ (Expansion)
2. $\text{CN}(\text{CN}(X)) = \text{CN}(X)$ (Idempotence)
3. $\text{CN}(X) = \bigcup_{Y \subseteq_f X} \text{CN}(Y)^1$ (Compactness)
4. $\text{CN}(\{x\}) = \mathcal{L}$ for some $x \in \mathcal{L}$ (Absurdity)
5. $\text{CN}(\emptyset) \neq \mathcal{L}$ (Coherence)

A CN that satisfies the above properties is *monotonic*. The associated notion of *consistency* is defined as follows:

Definition 1 (Consistency). *A set $X \subseteq \mathcal{L}$ is* consistent *wrt a logic $(\mathcal{L}, \text{CN})$ iff $\text{CN}(X) \neq \mathcal{L}$. It is* inconsistent *otherwise.*

Arguments are built from a *knowledge base* $\Sigma \subseteq \mathcal{L}$ as follows:

Definition 2 (Argument). *Let Σ be a knowledge base. An* argument *is a pair (X, x) s.t. 1) $X \subseteq \Sigma$, 2) X is consistent, 3) $x \in \text{CN}(X)$ and X is minimal (for set inclusion) wrt 1), 2) and 3).*

Notations: Supp and Conc denote respectively the *support* X and the *conclusion* x of an argument (X, x). For all $\mathcal{S} \subseteq \Sigma$, Arg($\mathcal{S}$) denotes the set of all arguments that can be built from $\mathcal{S}$ by means of Definition 2, Sub is a function that returns all the sub-arguments of a given argument. For all $\mathcal{E} \subseteq \text{Arg}(\Sigma)$, Base($\mathcal{E}$) = $\bigcup_{a \in \mathcal{E}} \text{Supp}(a)$. Max($\Sigma$) is the set of all maximal (for set inclusion) consistent subbases of Σ.

An argumentation system is defined as follows.

Definition 3. *An* argumentation system *(AS) over a knowledge base Σ is a pair $\mathcal{T} = (\text{Arg}(\Sigma), \mathcal{R})$ where $\mathcal{R} \subseteq \text{Arg}(\Sigma) \times \text{Arg}(\Sigma)$ is an attack relation such that for all $a, b \in \text{Arg}(\Sigma)$, if $(a, b) \in \mathcal{R}$, then $\text{Supp}(a) \cup \text{Supp}(b)$ is inconsistent.*

Note that $(a, b) \in \mathcal{R}$ (or $a\mathcal{R}b$) means that a attacks b. It is also worth mentioning that the set Arg(Σ) may be infinite even when the base Σ is finite.

---

[1] $Y \subseteq_f X$ means that Y is a finite subset of X.

3   Logics Induced by Extension Semantics

The most popular semantics were proposed by Dung in his seminal paper [8]. Those semantics as well as their refinements (e.g. in [3,7]) partition the powerset of the set of arguments into two classes: *extensions* and *non-extensions*. Every extension represents a coherent point of view. We illustrate the kind of paraconsistent logics induced by such semantics on the most popular ones, namely naive, stable and preferred. Before giving the formal definitions of the three semantics, we first introduce two key concepts on which they are based.

Definition 4 (Conflict-freeness–Defence). *Let* $\mathcal{T} = (\mathcal{A}, \mathcal{R})$ *be an argumentation system,* $\mathcal{E} \subseteq \mathcal{A}$ *and* $a \in \mathcal{A}$.

- $\mathcal{E}$ *is* conflict-free *iff* $\nexists a, b \in \mathcal{E}$ *such that* $a\mathcal{R}b$.
- $\mathcal{E}$ *defends an argument* a *iff* $\forall b \in \mathcal{A}$ *such that* $b\mathcal{R}a$, $\exists c \in \mathcal{E}$ *such that* $c\mathcal{R}b$.

Definition 5 (Semantics). *Let* $\mathcal{T} = (\mathcal{A}, \mathcal{R})$ *be an argumentation system and* $\mathcal{E} \subseteq \mathcal{A}$.

- $\mathcal{E}$ *is a* naive *extension iff it is a maximal (w.r.t. set* $\subseteq$) *conflict-free set.*
- $\mathcal{E}$ *is a* preferred *extension iff it is a maximal (w.r.t. set* $\subseteq$) *set that is conflict-free and defends its elements.*
- $\mathcal{E}$ *is a* stable *extension iff it is conflict-free and attacks any argument in* $\mathcal{A} \setminus \mathcal{E}$.

Notations: $\text{Ext}_x(\mathcal{T})$ denotes the set of all extensions of $\mathcal{T}$ under semantics x where $x \in \{n, p, s\}$ and n (resp. p and s) stands for naive (respectively preferred and stable). When we do not need to refer to a particular semantics, we write $\text{Ext}(\mathcal{H})(\mathcal{T})$ for short.

Example 1. *The argumentation system depicted below*

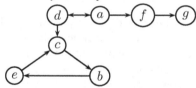

has five naive extensions: $\mathcal{E}_1 = \{a, c, g\}$, $\mathcal{E}_2 = \{d, e, f\}$, $\mathcal{E}_3 = \{b, d, f\}$, $\mathcal{E}_4 = \{a, e, g\}$, $\mathcal{E}_5 = \{a, b, g\}$; *one stable* $\mathcal{E}_3$ *and two preferred extensions* $\mathcal{E}_3$ *and* $\mathcal{E}_6 = \{a, g\}$.

It is worth recalling that stable extensions are naive (respectively preferred) extensions but the converses are not always true. Moreover, an argumentation framework may have no stable extensions.

Let us now define the plausible conclusions that may be drawn from a knowledge base Σ by an argumentation system. The idea is to infer a formula x from Σ iff it is the conclusion of at least one argument in every extension of the system.

Definition 6 (Flat conclusions). *Let* $\mathcal{T} = (\text{Arg}(\Sigma), \mathcal{R})$ *be an argumentation system over a knowledge base* Σ. *The set of* plausible conclusions *of* $\mathcal{T}$ *is*

$$\text{Output}(\mathcal{T}) = \begin{cases} \{x \in \mathcal{L} \mid \forall \mathcal{E} \in \text{Ext}(\mathcal{T}) \exists a \in \mathcal{E} \text{ s.t. } \text{Conc}(a) = x\} & \text{if } \text{Ext}(\mathcal{T}) \neq \emptyset \\ \emptyset & \text{else} \end{cases}$$

In [2] a comprehensive study has been made on the family of logics described in this section. It has been shown that when an argumentation system satisfies two key properties, then there is a full correspondence between the naive extensions of the system and the maximal consistent subbases of the knowledge base. More formally:

Postulates 1 (Closure under sub-arguments – Consistency). *Let* $\mathcal{T} = (\text{Arg}(\Sigma), \mathcal{R})$ *be an argumentation system over a knowledge base* Σ. *For all* $\mathcal{E} \in \text{Ext}(\mathcal{T})$,

- *if* $a \in \mathcal{E}$, *then* $\text{Sub}(a) \subseteq \mathcal{E}$. *We say that* $\mathcal{T}$ *is closed under sub-arguments.*
- $\text{Concs}(\mathcal{E})$ *is consistent. We say that* $\mathcal{T}$ *satisfies consistency.*

Theorem 1. *[2] Let* $\mathcal{T} = (\text{Arg}(\Sigma), \mathcal{R})$ *be an argumentation system over a knowledge base* Σ. *If* $\mathcal{T}$ *satisfies consistency and is closed under sub-arguments (under naive semantics), then:*

- *For all* $\mathcal{E} \in \text{Ext}_n(\mathcal{T})$, $\text{Base}(\mathcal{E}) \in \text{Max}(\Sigma)$.
- *For all* $\mathcal{S} \in \text{Max}(\Sigma)$, $\text{Arg}(\mathcal{S}) \in \text{Ext}_n(\mathcal{T})$.

Let us now characterize the set of inferences that may be drawn from a knowledge base Σ by any argumentation system under naive semantics. It coincides with the set of inferences that are drawn from the maximal consistent subsets of Σ.

Theorem 2. *[2] Let* $\mathcal{T} = (\text{Arg}(\Sigma), \mathcal{R})$ *be an argumentation system over a knowledge base* Σ *such that* $\mathcal{T}$ *satisfies consistency and is closed under sub-arguments (under naive semantics).*

$$\text{Output}(\mathcal{T}) = \bigcap_{\mathcal{S}_i \in \text{Max}(\Sigma)} \text{CN}(\mathcal{S}_i).$$

A similar study has been conducted for stable and preferred semantics. It has been shown that there are two families of attack relations. The first family leads to coherent systems (i.e., their stable extensions coincide with their preferred ones). Furthermore, stable extensions coincide with the naive ones. Such systems collapse then with the above discussed ones. The second family of relations allows choosing only *some* maximal consistent subbases of the knowledge base.

It is worth noticing that the paraconsistent logics defined from argumentation systems that use extension semantics restore consistency and return flat consequences.

4   Logics Induced by Ranking Semantics

Ranking semantics have been introduced in [1] as an alternative approach for evaluating arguments. Their basic idea is to rank arguments from the most to the less acceptable ones, instead of computing extensions. In what follows, we illustrate the approach with *burden-based semantics* (Bbs) introduced in [1]. Bbs assigns a *burden number* to every argument. The heavier the burden of an argument, the weaker its attacks.

Definition 7 (Burden numbers). *Let $\mathcal{T} = (\text{Arg}(\Sigma), \mathcal{R})$ be an argumentation system, $i \in \{0, 1, \ldots\}$, and $a \in \text{Arg}(\Sigma)$. We denote by $\text{Bur}_{\mathcal{T}i}(a)$ the burden number of a in the ith step, i.e.:*

$$\text{Bur}_i(a) = \begin{cases} 1 & \text{if } i = 0; \\ 1 + \sum_{b \in \text{Att}(a)} 1/\text{Bur}_{i-1}(b) & \text{otherwise.} \end{cases}$$

where $\text{Att}(a) = \{b \in \text{Arg}(\Sigma) \mid (b, a) \in \mathcal{R}\}$.

By convention, if $\text{Att}(a) = \emptyset$, then

$$\sum_{b \in \text{Att}(a)} 1/\text{Bur}_{i-1}(b) = 0.$$

Let us illustrate this function in the following example.

Example 2. *Assume the argumentation system depicted in the figure below.*

The burden numbers of each argument are summarized in the table below. Note that these numbers will not change beyond step 2.

Step i	a	b	c
0	1	1	1
1	1	2	2
2	1	2	1.5
⋮	⋮	⋮	⋮

Arguments are compared lexicographically on the basis of their burden numbers as follows:

Definition 8 (Bbs). *The burden-based semantics Bbs transforms any argumentation system $\mathcal{T} = (\text{Arg}(\Sigma), \mathcal{R})$ into the ranking $\text{Bbs}(\mathcal{T})$ on $\text{Arg}(\Sigma)$ such that $\forall a, b \in \text{Arg}(\Sigma)$, $\langle a, b \rangle \in \text{Bbs}(\mathcal{T})$ iff one of the two following cases holds:*

- $\forall i \in \{0, 1, \ldots\}, \text{Bur}_i(a) = \text{Bur}_i(b)$;
- $\exists i \in \{0, 1, \ldots\}, \text{Bur}_i(a) < \text{Bur}_i(b)$ *and* $\forall j \in \{0, 1, \ldots, i-1\}, \text{Bur}_j(a) = \text{Bur}_j(b)$.

Intuitively, $\langle a, b \rangle \in \text{Bbs}(\mathcal{T})$ means that a is *at least as acceptable* as b. Let us see in an example how the semantics works.

Example 2 (Cont). According to Bbs, the argument a is strictly more acceptable than c which is itself strictly more acceptable than b.

The plausible conclusions of an argumentation system that uses ranking semantics are simply those supported by at least one argument. Note that a formula and its negation may both be plausible. This means that the approach tolerates inconsistency. More importantly, the conclusions are ranked from the most to the least plausible ones. A formula is ranked higher than another formula if it is supported by an argument which is more acceptable than any argument supporting the second formula.

Definition 9 (Ranked conclusions). *Let* $\mathcal{T} = (\text{Arg}(\Sigma), \mathcal{R})$ *be an argumentation system over a knowledge base* Σ. *The output of* $\mathcal{T}$ *is the pair* $\text{Output}(\mathcal{T}) = \langle \mathcal{C}, \preceq \rangle$ *such that:*

- $\mathcal{C} = \{\text{Conc}(a) \mid a \in \text{Arg}(\Sigma)\}$
- $x \preceq y$ *iff* $\exists a \in \text{Arg}(\Sigma)$ *such that* $\text{Conc}(a) = x$ *and* $\forall b \in \text{Arg}(\Sigma)$ *such that* $\text{Conc}(b) = y$, $\langle a, b \rangle \in \text{Bbs}(\mathcal{T})$.

Unlike certain well-known inconsistency-tolerating logics (like the the 3 and 4 valued ones [4,6]), the above logics satisfy the following crucial property: if the premises are consistent, the conclusions coincide with those of CN.

5 Conclusion

Argumentation is a natural approach for handling inconsistency. It is more akin to the way humans deal with inconsistency in everyday life. Indeed, generally people construct arguments pros and arguments cons opinions. As a consequence, the results of an argumentation approach are easier to grasp for an end-user. Yet, the approach is efficient since, as discussed in the paper, it is able to capture existing approaches (in the case of extension semantics) or even also to outperform some approaches on certain points (in the case of ranking semantics).

References

1. Amgoud, L., Ben-Naim, J.: Ranking-based semantics for argumentation frameworks. In: Liu, W., Subrahmanian, V.S., Wijsen, J. (eds.) SUM 2013. LNCS (LNAI), vol. 8078, pp. 134–147. Springer, Heidelberg (2013)
2. Amgoud, L., Besnard, P.: Logical limits of abstract argumentation frameworks. Journal of Applied Non-Classical Logics 23(3), 229–267 (2013)
3. Baroni, P., Giacomin, M., Guida, G.: Scc-recursiveness: A general schema for argumentation semantics. Artificial Intelligence Journal 168, 162–210 (2005)
4. Belnap, N.D.: A Useful Four-Valued Logic. In: Dunn, J., Epstein, G. (eds.) Modern Uses of Multiple-Valued Logic, pp. 7–37. Oriel Press (1977)
5. Cholvy, L.: Automated reasoning with merged contradictory information whose reliability depends on topics. In: Froidevaux, C., Kohlas, J. (eds.) ECSQARU 1995. LNCS, vol. 946, pp. 125–132. Springer, Heidelberg (1995)
6. D'Ottaviano, I., da Costa, N.: Sur un problème de Jaśkowski. In: Comptes Rendus de l'Académie des Sciences de Paris, vol. 270, pp. 1349–1353 (1970)
7. Dung, P., Mancarella, P., Toni, F.: Computing ideal skeptical argumentation. Artificial Intelligence Journal 171, 642–674 (2007)
8. Dung, P.M.: On the Acceptability of Arguments and its Fundamental Role in Non-Monotonic Reasoning, Logic Programming and n-Person Games. AIJ 77, 321–357 (1995)
9. Kleer, J.D.: Using crude probability estimates to guide diagnosis. Artificial Intelligence 45, 381–391 (1990)
10. Reiter, R.: A logic for default reasoning. Artificial Intelligence 13(1-2), 81–132 (1980)
11. Rescher, N., Manor, R.: On inference from inconsistent premises. Journal of Theory and Decision 1, 179–219 (1970)
12. Tarski, A.: On Some Fundamental Concepts of Metamathematics. In: Woodger, E.H. (ed.) Logic, Semantics, Metamathematics. Oxford Uni. Press (1956)

Towards a Framework
for Learning from Networked Data

Jan Ramon

Department of Computer Science, KU Leuven
Celestijnenlaan 200A, 3001 Heverlee, Belgium
Jan.Ramon@cs.kuleuven.be

1   Introduction

Over the past decades, one has seen databases of ever increasing size and complexity. While the increasing size is easy to measure in bytes, kilobytes or terabytes, the increase in complexity is more difficult to quantify, however, it has a very deep effect on the theory we use to reason about the data. While in earlier days many researchers reasoned in terms of sets of similarly structured and independent objects, today we are facing large networks of data where everything is connected directly or indirectly to everything else. Examples include social networks, traffic networks, biological networks, administrative networks and economic networks.

These developments have spurred a renewed interest in data storage and knowledge extraction (answers to queries, patterns, models, ...). Three key underlying challenges are the representation of the data and knowledge, managing the computational cost of the problems which we need to solve and the statistical challenge related to the complexity of the data.

In this contribution, I will survey these challenges from a data mining point of view. I will argue that in order to address the current challenges it is valuable to gain a better understanding of fundamental statistical and algorithmic properties of large data networks and to integrate ideas from the many fields of research that are concerned with such networks.

2   Knowledge Representation

More than 15 years ago, many datasets were transactional. They consisted of a set of independent and separate transactions. Examples are a database of transactions describing what an anonymous customer bought in a shop, a database of molecules available to some chemical company, a set of responses to a survey, a database of patient records of a medical doctor, etc.

However, in reality pieces of information are rarely independent. Even for the classical example of transactions containing items bought together in a shop, transactions are related because they featured the same customer, the same shop staff member, the same calendar day, the same discount offer or other more subtle relationships which may have let one transaction influence the behavior

N. Hernandez et al. (Eds.): ICCS 2014, LNAI 8577, pp. 25–30, 2014.
DOI: 10.1007/978-3-319-08389-6_3, © Springer International Publishing Switzerland 2014

of a customer in another transaction. To the other extreme, in many current big databases including social networks, economic networks, biological regulatory networks and traffic networks the relationships are considered the most important part of the information. This shift in the complexity of data has two major implications on the level of knowledge representation, in particular on representing data and representing dependencies.

Representing Data. In the early nineties, one became aware that sets of vectors (the so-called propositional or attribute-value representation) was not anymore sufficient to represent datasets. The field of inductive logic programming (ILP) [11] was the first to be successful in machine learning describing data with the more powerful language of first order logic. In fact, first order logic is so powerful that many problems including deduction are undecidable, and to mitigate the computational intractability, several settings between the propositional one and the first order logic one were explored [4]. After a while, a lot of researchers shifted to graphs to represent data, as they hit a nice balance between expressivity (being as powerful as relational databases or datalog) and computational tractability (in the sense that most things are decidable and the computational complexity of many tasks have been studied in the field of algorithmic graph theory). While transactional graph mining is closely related to the learning from interpretations setting in ILP, the learning from entailment setting was the first one to see instances as elements of a large connected knowledgebase [3]. When the graph mining community moved away from transactional datasets, this idea of a global knowledgebase was revisited under the name of network analysis, a term which is also used in the branch of statistical physics studying graphs representing complex systems [10]. In this representation framework, networks are graphs whose vertices represent objects (or parts thereof) and edges represent relations between them. Depending on what is most convenient, often for theoretical purposes more simpler and abstract settings and for practical applications richer settings, one can use directed or undirected graphs, graphs or hypergraphs, labeled or unlabeled graphs, but usually there are straightforward transformations from the one type to the other type of graph (as illustrated e.g. in [2]).

Representing Dependencies. A second implication of recognizing that instances are part of a single large world is of a more statistical nature. Indeed, instances sharing relations to the same objects in the world may not be independent from a statistical point of view. For instance, friends may share interests, well connected cities may share economic activity and interacting molecules may participate in a shared biological process. In many statistical approaches it is important to represent these dependencies, and the field of statistical relational learning (SRL) [6] has investigated many ways to extend probabilistic models to graph-structured databases. Such models represent explicitly dependence or independence relationships between variables using (hyper)edges, such that again a graph is obtained. Most SRL approaches somehow assume that one can model these statistical dependencies, or at least learn them in a reasonable way from data. This holds in a number of applications, to a large extent also the one discussed below, but as Section 4 will argue sometimes things are more difficult.

A Case Study in Experimental Research. Experimental research in the field of computational biology is a typical domain where the knowledge base integrating domain knowledge, experimental setups and experimental results may get very complex. Here, only a simplified illustration from the domain of protein mass spectrometry [9] is provided.

Mass spectrometry is a technique to detect what molecules are present in a sample (e.g. a blood sample), in the case of protein mass spectrometry one aims at detecting the proteins in such sample. A typical experimental setup involve a pipeline of several treatments (e.g. a typical sequence is digestion, chromatography, ionization, fragmentation, detection). The way each step in such pipeline transforms the sample depends on the characteristics on the instrument and its parameter settings. Recently, there is a growing interest in modeling more precisely each of these transformations (see [5] for an illustration on the digest step). The more accurate are such models, the more accurately one can reason about what was in the sample at the beginning based on the output of the detection step at the end of the pipeline.

In experimental biomedical research, one often uses mass spectrometry to detect whether a particular protein is present in a sample or not. However, the results of such experiments are not independent. Several proteins may be part of a common pathway (chain of chemical reactions in the cell), and hence detecting one protein may be correlated with detecting another protein. The graph representing the interactions between proteins is called the regulatory network. The closer the regulatory network relates two observed proteins, the less statistical evidence it provides. The more unrelated the proteins are, the more we can see them as independent evidence / indications of a particular phenomenon of interest (e.g. a disease we want to diagnose). The better our knowledge of this regulatory network is, the better we can assess how much independent evidence a set of observations provide.

3   The Question of Computational Tractability

A second implication of the increasing complexity of available data is that many tasks which were almost trivial for transactional databases get intractable for networked data. One prototypical example is the problem of pattern matching. The pattern matching operator which is most widely studied in the field of graph mining is subgraph isomorphism. Unfortunately, it is an NP-complete problem to decide whether a pattern is subgraph isomorphic to a database graph. For transactional graph databases this was not very problematic as transactions are usually limited in size. Moreover, for many specific applications, e.g., molecule databases, optimized pattern mining solutions have been developed [7] exploiting the structure of the database graphs. Unfortunately, large data networks don't have an easy to exploit structure and are at the same time orders of magnitude larger. The NP-hardness of the problem suggests that the increase in computational complexity caused by the increasing database sizes is not expected to be compensated by the increasing computing power predicted by the law of Moore.

We therefore face the fundamental problem of making data analysis algorithms scale well with the growing databases. Fortunately, a lot of useful inspiration is provided by recently emerged research lines in theoretical computer science. In particular, despite the fact that the work in theoretical computer science is not necessarily intended for immediate applicability, we recently demonstrated that the above mentioned pattern matching problem can be addressed to a large extent by the use of fixed parameter tractable algorithms [8].

4   Towards a Statistical Framework

A third challenge raised by considering data in networks is of a statistical nature. As explained above, a first step is to explicitly model the dependencies between variables. However, one can argue that even when these dependencies are modeled, the problem of learning a predictive model is still not completely well defined, as we didn't specify complete statistical assumptions.

To see this, let us first recall the most common statistical assumption. In classical statistics, when learning a model on training data and then performing predictions on unseen data, one usually makes the assumption that both the seen and the unseen instances are drawn (independently) from a fixed (but unknown) distribution. If one rolls a dice 1000 times and gets a 2 in 50 of these 1000 cases, then when rolling the dice again, one expects that the result will be a 2 with probability 0.05 because it concerns the same (clearly biased) dice. If the dice would have been replaced by another one, there is no reason to have the same belief.

A similar mechanism is needed for network statistics. However, it is unclear what it means "to be drawn from the same distribution". As an example, consider the following fictitious world. Suppose that 90% of computer users use an operating system called windows and like sunny weather over rainy weather, while 10% of computer users use an operating system called linux and they are nerds not caring about the weather because they are always inside. Designers of operating systems randomly choose some delimiter to separate directory names in paths. linux designers preferred forward slashes while the CEO of the company making windows preferred backslashes. Moreover, assume the CEO of the company making the windows system likes sunny weather. Now, if we take a random computer user, there is a 90% probability that he uses backslashes in pathnames and it is equally probable he likes sunny weather. Now suppose that a tiny change happens: the CEO of the company making windows is replaced by a nerd not interested in the weather, and he prefers to use hash signs as delimiter. Note that everything is still drawn from exactly the same distribution. Now, if we take a random user, will he like sunny weather and/or will he use backslashes? The users using windows didn't change, so they still like sunny weather. However, they are forced now to use hash signs rather than backslashes.

From this example we can see that in situations which are structurally identical from the point of view of data, the result can be quite different due to a difference in the underlying process. In this specific case, we can't distinguish between variables which are functions of the features of individuals and variables which are functions of their connectivity. Of course, this problem would not arise when there would be a large number of providers of operating systems. Unfortunately in the real world many networks have been shown to follow a powerlaw distribution [1], implying there are often a few dominating "hubs" with an exceptionally high connectivity.

Essential to any solution to this problem is to perform a more systematic analysis of learning theory and to specify clearly under which assumptions some prediction (generalization) effort is valid. For instance, in [12] we showed preliminary results providing learning guarantees under the assumption that the connectivity of the objects involved in an observation and the function mapping the features of these objects on the target value are independent.

5 Conclusions

In this contribution, I argued that due to the increasing amount and more importantly the increasing complexity of data, in order to store, process and analyze data we face challenges on the level of knowledge representation, computational costs and statistical inference. Over the past few years, several ideas to address these challenges have been developed, but several open problems remain.

Of particular interest is the statistical challenge. In the past, the majority of efforts were directed at extending data mining algorithms towards graph-based representations and making them computationally feasible. but less attention has been payed to developing a consistent theory of statistics on graphs. It seems plausible that integrating ideas from statistical relational learning theory and random graph theory can further progress the field.

Acknowledgements. This research has been supported by the European Research Council, grant ERC-StG 240186 "MiGraNT: Mining Graphs and Networks, a Theory-based approach".

References

1. Barabási, A.L.: Scale-free networks: A decade and beyond. Science 325(5939), 412–413 (2009)
2. Calders, T., Ramon, J., Van Dyck, D.: All normalized anti-monotonic overlap graph measures are bounded. Data Mining and Knowledge Discovery 23, 503–548 (2011)
3. De Raedt, L.: Logical settings for concept learning. Artificial Intelligence 95, 187–201 (1997)
4. De Raedt, L.: Attribute-value learning versus inductive logic programming: The missing links (extended abstract). In: Page, D. (ed.) ILP 1998. LNCS (LNAI), vol. 1446, pp. 1–8. Springer, Heidelberg (1998)

5. Fannes, T., Vandermarliere, E., Schietgat, L., Degroeve, S., Martens, L., Ramon, J.: Predicting tryptic cleavage from proteomics data using decision tree ensembles. Journal of Proteome Research 12, 2253–2259 (2013)
6. Getoor, L., Taskar, B.: An Introduction to Statistical Relational Learning. MIT Press (2007)
7. Horváth, T., Ramon, J., Wrobel, S.: Frequent subgraph mining in outerplanar graphs. Knowledge Discovery and Data Mining 21(3), 472–508 (2010)
8. Kibriya, A., Ramon, J.: Nearly exact mining of frequent trees in large networks. Data Mining and Knowledge Discovery 27, 478–504 (2013)
9. Martens, L., Laukens, K., Ramon, J., Valkenborg, D.: Inspector: An integrated informatics platform for mass-spectrometry protein assays
10. Newman, M.: Networks: An introduction. Oxford University Press (2010)
11. Nienhuys-Cheng, S.-H., de Wolf, R.: Foundations of Inductive Logic Programming. LNCS (LNAI), vol. 1228. Springer, Heidelberg (1997)
12. Wang, Y., Ramon, J., Guo, Z.-C.: Learning from networked examples in a k-partite graph. In: Proceedings of la Confrence sur l'Apprentissage Automatique, Lille, France, pp. 1–8 (July 2013)

Games on Graphs

Miloš Stojaković*

Department of Mathematics and Informatics, University of Novi Sad, Serbia
milos.stojakovic@dmi.uns.ac.rs
http://www.inf.ethz.ch/personal/smilos/

Abstract. Positional Games is a branch of Combinatorics which focuses on a variety of two player games, ranging from well-known games such as Tic-Tac-Toe and Hex, to purely abstract games played on graphs. The field has experienced quite a growth in recent years, with more than a few applications in related areas.

We aim to introduce the basic notions, approaches and tools, as well as to survey the recent developments, open problems and promising research directions, keeping the main focus on the games played on graphs.

Keywords: positional game, Maker-Breaker, Avoider-Enforcer, probabilistic intuition.

1 Introduction

Positional games are a class of combinatorial two-player games of perfect information, with no chance moves and with players moving sequentially. These properties already distinguish this area of research from its popular relative, Game Theory, which has its roots in Economics. Some of the more prominent positional games include Tic-Tac-Toe, Hex, Bridg-It and the Shannon's switching game.

The basic structure of a positional game is fairly simple. Let X be a finite set and let $\mathcal{F} \subseteq 2^X$ be a family of subsets of X. The set X is called the "board", and the members of $\mathcal{F}$ are referred to as the "winning sets". In the positional game $(X, \mathcal{F})$, two players take turns in claiming previously unclaimed elements of X, until all the elements are claimed. In a more general setting, given positive integers a and b, in the *biased* $(a : b)$ *game* the first player claims a elements per move and the second player claims b elements per move. If $a = b = 1$, the game is called *unbiased*.

As for determining the winner, there are several standard sets of rules, and here we mention three.

- In a *strong game*, the player who is first to claim all elements of a winning set is the winner, and if all elements of the board are claimed and no player has won, the game is a draw. A strategy stealing argument ensures that the

* Partly supported by Ministry of Science and Technological Development, Republic of Serbia, and Provincial Secretariat for Science, Province of Vojvodina.

N. Hernandez et al. (Eds.): ICCS 2014, LNAI 8577, pp. 31–36, 2014.
DOI: 10.1007/978-3-319-08389-6_4, © Springer International Publishing Switzerland 2014

first player can achieve at least a draw, so the two possible outcomes of a game (if played by two perfect players) are: a first player's win, and a draw. Tic-Tac-Toe (a.k.a. Noughts and Crosses, or Xs and Os) is an example of a strong game – the board consists of nine elements (usually drawn as a 3-by-3 grid), with eight winning sets. As most kids would readily tell you, this game is a draw.

Even though these games are quite easy to introduce, they turn out to be notoriously hard to analyze, and hence there are very few results in that area.

- A *Maker-Breaker game* features two players, Maker and Breaker. Maker wins if he claims all elements of a winning set (not necessarily first). Breaker wins otherwise, i.e., if all the elements of the board are claimed and Maker has not won. Hence, one of the players always wins – a draw is not possible. As it turns out, the widely popular game Hex is a Maker-Breaker game, a fact that requires a proof, see [5].
- Finally, in an *Avoider-Enforcer game* players are called Avoider and Enforcer. Here, Enforcer wins if at any point of the game Avoider claims all elements of a winning set. Avoider wins otherwise, i.e., if he manages to avoid claiming a whole winning set to the end of the game. Due to the nature of the game, the winning sets in Avoider-Enforcer games are sometimes referred to as the losing sets.

In what follows we deal with the positional games played on graphs. That means that the board of the game is the *edge set* of a graph, usually the complete graph on n vertices. The winning sets typically are all representatives of a standard graph-theoretic structure. We introduce a few games that stand out when it comes to importance and attention received in the recent years.

The research in this area was initiated by Lehman [13], who studied the *connectivity game*, a generalization of the well-known Shannon switching game, where the winning sets are the edge sets of all spanning trees of the base graph. We denote the connectivity game played on the complete graph by $(E(K_n), C)$. Another important game is the *Hamiltonicity game* $(E(K_n), \mathcal{H})$, where $\mathcal{H}$ consists of the edge sets of all Hamiltonian cycles of K_n. In the *clique game* the winning sets are the edge sets of all the k-cliques, for a fixed integer $k \geq 3$. We denote this game with $(E(K_n), \mathcal{K}_k)$. Note that in this game the size of the winning sets is fixed and does not depend on n, which distinguishes it from the connectivity game and the Hamiltonicity game. A simple Ramsey argument coupled with the strategy stealing argument (see [1] for details) ensures Maker's win if n is large.

Numerous topics on positional games are covered in the monograph of Beck [1]. The new book [10] gives a gentle introduction to the subject, along with a view to the recent developments.

2 Maker-Breaker Games

It is not hard to verify that the connectivity game and the k-clique game are Maker's wins when n is large enough. Showing the same for the Hamiltonicity game requires a one-page argument [4]. This, however, is not the end of the story. A standard approach to even out the odds is introduced by Chvátal and Erdős in [4], giving Breaker more power with the help of a bias.

If an unbiased game $(X, \mathcal{F})$ is a Maker's win, we choose to play the same game with $(1 : b)$ bias, increasing b until Breaker starts winning. Formally, we want to answer the following question: What is the largest integer $b_{\mathcal{F}}$ for which Maker can win the biased $(1 : b_{\mathcal{F}})$ game? This value is called the *threshold bias* of $\mathcal{F}$. The existence of the threshold bias for every game is guaranteed by the so-called *bias monotonicity* of Maker-Breaker games, the fact that a player can only benefit from claiming additional elements at any point of the game.

For the connectivity game, it was shown by Chvátal and Erdős [4] and Gebauer and Szabó [6] that the threshold bias is $b_{\mathcal{C}} = (1 + o(1)) \frac{n}{\log n}$. The result of Krivelevich [11] gives the leading term of the threshold bias for the Hamiltonicity game, $b_{\mathcal{H}} = (1 + o(1)) \frac{n}{\log n}$. In the k-clique game, Bednarska and Łuczak [2] found the order of the threshold bias, $b_{\mathcal{K}_k} = \Theta(n^{\frac{2}{k+1}})$. Determining the leading constant inside the $\Theta(.)$ remains an open problem that appears to be very challenging.

3 Avoider-Enforcer Games

Combinatorial game theory devotes a lot of attention to pairs of two-player games where the way for a player to win in one game becomes the way for him to lose in the other game – while the playing rules in both games are identical, the rule for deciding if the first player won one game is exactly the negation of the same rule for the first player in the other game. We have that setup in corresponding Maker-Breaker and Avoider-Enforcer variants of a positional game. In light of that, an Avoider-Enforcer game is said to be the *misére* version of its Maker-Breaker counterpart.

We already mentioned that in Maker-Breaker games bonus moves do not harm players, if a player is given one or more elements of the board at any point of the game he can only profit from it. Naturally, one wonders if an analogous statement holds for Avoider-Enforcer games. At first sight, it makes sense that a player trying to avoid something cannot be harmed when some of the elements he claimed are "unclaimed". But this turns out not to be true, as the following example shows.

Consider the Avoider-Enforcer $(a : b)$ game played on the board with four elements, and two disjoint winning sets of size two. It is easy to see that for $a = b = 2$ Avoider wins, for $a = 1, b = 2$ the win is Enforcer's, and finally for $a = b = 1$ Avoider is the winner again.

This feature is somewhat disturbing as, to start with, the existence of the threshold bias is not guaranteed. This prompted the authors of [9] to adjust, in a rather natural way, the game rules to ensure bias monotonicity. Under the so-called *monotone rules*, for given bias parameters a and b and a positional game $\mathcal{F}$, in a monotone $(a : b)$ Avoider-Enforcer game $\mathcal{F}$ in each turn Avoider claims *at least* a elements of the board, where Enforcer claims at *least* b elements of the board. These rules can be easily argued to be bias monotone, and thus the threshold bias becomes a well defined notion. We will refer to the original rules, where each player claims *exactly* as many elements as the respective bias suggests, as the *strict* rules. Perhaps somewhat surprisingly, monotone Avoider-Enforces games turn out to be rather different from those played under strict rules, and in quite a few cases known results about strict rules provide a rather misleading clue about the location of the threshold bias for the monotone version.

From now on, each game can be viewed under two different sets of rules – the *strict* game and the *monotone* game. Given a positional game $\mathcal{F}$, for its *strict* version we define the *lower threshold bias* $f_{\mathcal{F}}^-$ to be the largest integer such that Enforcer can win the $(1 : b)$ game $\mathcal{F}$ for every $b \leq f_{\mathcal{F}}^-$, and the *upper threshold bias* $f_{\mathcal{F}}^+$ to be the smallest non-negative integer such that Avoider can win the $(1 : b)$ game $\mathcal{F}$ for every $b > f_{\mathcal{F}}^+$.

If we play the game $\mathcal{F}$ under *monotone* rules, the bias monotonicity implies the existence of the unique threshold bias $f_{\mathcal{F}}^{mon}$ as the non-negative integer for which Enforcer has a winning strategy in the $(1 : b)$ game if and only if $b \leq f_{\mathcal{F}}^{mon}$.

The leading term of the threshold bias for the monotone version of several well-studied positional games with spanning winning sets is given by the following two results. In [9], it was shown that for $b \geq (1 + o(1)) \frac{n}{\ln n}$ Avoider has a winning strategy in the monotone $(1 : b)$ min-degree-1 game, the game in which his goal is to avoid touching all vertices. On the other hand, we have that for $b \leq (1 - o(1)) \frac{n}{\ln n}$ Enforcer has a winning strategy in the monotone $(1 : b)$ Hamiltonicity game, and also in the k-connectivity game, for any fixed k, see [12].

These results give that the leading term of the threshold biases for the mono-tone versions of the connectivity game and the Hamiltonicity game (as well as for some other important games, like the perfect matching game, the min-degree-k game, for $k \geq 1$, the k-edge-connectivity game, for $k \geq 1$, and the k-connectivity game, for $k \geq 1$) is $(1 + o(1)) \frac{n}{\ln n}$. Indeed, each of these graph properties im-plies min-degree-1, and each of them is implied either by Hamiltonicity or k-connectivity. Note that for all these games we have the same threshold bias in the Maker-Breaker version of the game.

Now we switch our attention to the games played under strict rules. For the connectivity game under strict rules we know the exact value of the lower and upper threshold bias, and they are the same, $f_{\mathcal{C}}^- = f_{\mathcal{C}}^+ = \lfloor \frac{n-1}{2} \rfloor$, see [7]. This is one of very few games on graphs for which we have completely tight bounds for the threshold bias, with equal upper and lower threshold biases. Note the substantial difference between these threshold biases and the monotone thresh-old bias for the connectivity game. Much less is known for the Hamiltonicity game, where we just have the lower bound $(1 - o(1)) \frac{n}{\ln n}$ for the lower threshold bias [12]. As for the bounds from above, we have only the obvious. We say that Avoider has a *trivial strategy* when Enforcer's bias is that large that the total number of edges Avoider will claim in the whole game is less than the size of the smallest losing set, so he can win no matter how he plays. It is not clear how far can we expect to get, as for example in the connectivity game a trivial Avoider's strategy turns out to be the optimal one.

As for the k-clique game, as well as for most of the other games in which the winning sets are of constant size, we are quite far from determining the leading term for any of the threshold biases, with only few non-trivial bounds currently available. This gives a whole range of very important open problems.

4   Games on the Random Board

As we have already mentioned, for many standard positional games the outcome of the unbiased Maker-Breaker game played on a (large) complete graph is an obvious Maker's win, and one way to help Breaker gain power is by increasing his bias. An alternative way is the so-called game on the random board, introduced in [16]. Informally speaking, we randomly thin out the board before the game starts, some of the winning sets disappear in that process, Maker's chances drop and Breaker gains momentum.

For a positional game $(X, \mathcal{F})$ and probability p, the *game on the random board* $(X_p, \mathcal{F}_p)$ is a probability space of games, where each $x \in X$ is included in X_p with probability p (independently), and $\mathcal{F}_p = \{W \in \mathcal{F}|\ W \subseteq X_p\}$.

Now even if an unbiased game is an easy Maker's win, as we decrease p the game gets harder for Maker and at some point he is not expected to be able to win anymore. To formalize that, we observe that "being a Maker's win in $\mathcal{F}$" is an increasing graph property. Indeed, no matter what positional game $\mathcal{F}$ we take, addition of board elements does not hurt Maker. Hence, there has to exist a threshold probability $p_\mathcal{F}$ for this property, and we are searching for $p_\mathcal{F}$ such that in the $(1 : 1)$ game $\Pr[\text{Breaker wins } (X_p, \mathcal{F}_p)] \to 1$ for $p \ll p_\mathcal{F}$, and $\Pr[\text{Maker wins } (X_p, \mathcal{F}_p)] \to 1$ for $p \gg p_\mathcal{F}$.

For games played on the edge set of the complete graph K_n, note that the board in now the edge set of the Erdős-Rényi random graph, $G(n, p)$.

The threshold probability for the connectivity game was determined to be $\frac{\log n}{n}$ in [16], and shown to be sharp. As for the Hamiltonicity game, the order of magnitude of the threshold was given in [15]. Using a different approach, it was proven in [8] that the threshold is $\frac{\log n}{n}$ and it is sharp. Finally, as a consequence of a hitting time result, Ben-Shimon et al. [3] closed this question by giving a very precise description of the low order terms of the limiting probability.

The threshold for the triangle game was determined in [16], $p_{K_3} = n^{-\frac{5}{9}}$. The leading term for the threshold probability in the k-clique game for $k \geq 4$ was shown to be $n^{-\frac{2}{k+1}}$ in [14].

Probabilistic Intuition. As it turns out for many standard games on graphs $\mathcal{F}$, the outcome of the game played by "perfect" players is often similar to the game played by "random" players. In other words, the inverse of the threshold bias $b_\mathcal{F}$ in the Maker-Breaker game played on the complete graph is "closely related" to the probability threshold for the appearance of a member of $\mathcal{F}$ in $G(n, p)$. Now we add another related parameter to the picture – the threshold probability $p_\mathcal{F}$ for Maker's win when the game is played on the edge set of $G(n, p)$. As we have seen in case of the connectivity game and the Hamilton cycle game, for both of those games all three mentioned parameters are exactly equal to $\frac{\ln n}{n}$.

In the k-clique game, for $k \geq 4$, the threshold bias is $\Theta(n^{\frac{2}{k+1}})$ and the threshold probability for Maker's win is the inverse (up to the leading constant), $n^{-\frac{2}{k+1}}$, supporting the random graph intuition. But, the threshold probability for the appearance of a k-clique in $G(n,p)$ is not at the same place, it is $n^{-\frac{2}{k-1}}$. And in the triangle game there is even more disagreement, as all three parameters are different – they are, respectively, $n^{\frac{1}{2}}$, $n^{-\frac{5}{9}}$ and n^{-1}.

References

1. Beck, J.: Combinatorial Games: Tic-Tac-Toe Theory. Encyclopedia of Mathematics and Its Applications, vol. 114. Cambridge University Press (2008)
2. Bednarska, M., Luczak, T.: Biased positional games for which random strategies are nearly optimal. Combinatorica 20, 477–488 (2000)
3. Ben-Shimon, S., Ferber, A., Hefetz, D., Krivelevich, M.: Hitting time results for Maker-Breaker games. Random Structures and Algorithms 41, 23–46 (2012)
4. Chvátal, V., Erdős, P.: Biased positional games. Annals of Discrete Mathematics 2, 221–228 (1978)
5. Gale, D.: The game of Hex and the Brouwer fixed-point theorem. The American Mathematical Monthly 86, 818–827 (1979)
6. Gebauer, H., Szabó, T.: Asymptotic random graph intuition for the biased connectivity game. Random Structures and Algorithms 35, 431–443 (2009)
7. Hefetz, D., Krivelevich, M., Szabó, T.: Avoider-Enforcer games. Journal of Combinatorial Theory Series A 114, 840–853 (2007)
8. Hefetz, D., Krivelevich, M., Stojaković, M., Szabó, T.: A sharp threshold for the Hamilton cycle Maker-Breaker game. Random Structures and Algorithms 34, 112–122 (2009)
9. Hefetz, D., Krivelevich, M., Stojaković, M., Szabó, T.: Avoider-Enforcer: The rules of the Game. Journal of Combinatorial Theory Series A 117, 152–163 (2010)
10. Hefetz, D., Krivelevich, M., Stojaković, M., Szabó, T.: Positional Games. Oberwolfach Seminars, vol. 44. Birkhäuser (2014)
11. Krivelevich, M.: The critical bias for the Hamiltonicity game is $(1 + o(1))n/\ln n$. Journal of the American Mathematical Society 24, 125–131 (2011)
12. Krivelevich, M., Szabó, T.: Biased positional games and small hypergraphs with large covers. Electronic Journal of Combinatorics 15, R70 (2008)
13. Lehman, A.: A solution of the Shannon switching game. Journal of the Society for Industrial and Applied Mathematics 12, 687–725 (1964)
14. Müller, T., Stojaković, M.: A threshold for the Maker-Breaker clique game. Random Structures and Algorithms (to appear)
15. Stojaković, M.: Games on Graphs, PhD Thesis. ETH Zürich (2005)
16. Stojaković, M., Szabó, T.: Positional games on random graphs. Random Structures and Algorithms 26, 204–223 (2005)

A Partial-Closure Canonicity Test to Increase the Efficiency of CbO-Type Algorithms

Simon Andrews

Conceptual Structures Research Group
Communication and Computing Research Centre
Faculty of Arts, Computing, Engineering and Sciences
Sheffield Hallam University, Sheffield, UK
s.andrews@shu.ac.uk

Abstract. Computing formal concepts is a fundamental part of Formal Concept Analysis and the design of increasingly efficient algorithms to carry out this task is a continuing strand of FCA research. Most approaches suffer from the repeated computation of the same formal concepts and, initially, algorithms concentrated on efficient searches through already computed results to detect these repeats, until the so-called canonicity test was introduced. The canonicity test meant that it was sufficient to examine the attributes of a computed concept to determine its newness: searching through previously computed concepts was no longer necessary. The employment of this test in Close-by-One type algorithms has proved to be highly effective. The typical CbO approach is to compute a concept and then test its canonicity. This paper describes a more efficient approach, whereby a concept need only be partially computed in order to carry out the test. Only if it passes the test does the computation of the concept need to be completed. This paper presents this 'partial-closure' canonicity test in the In-Close algorithm and compares it to a traditional CbO algorithm to demonstrate the increase in efficiency.

Keywords: Formal Concept Analysis, FCA, FCA algorithms, Computing formal concepts, Canonicity test, Partial-closure canonicity test, Close-by-One, In-Close, CbO.

1 Introduction

The emergence of Formal Concept Analysis (FCA) as a data analysis technique [1–3] has increased the need for algorithms that compute formal concepts quickly. A problem in computing these formal concepts is the large number that can exist in a typical dataset. It is known that the number of concepts can be exponential in the size of the input context and the problem of determining this number is #P-complete [4]. Furthermore, computation of formal concepts typically involves repeated computation of the same concept [5], which is usually unwanted. Older algorithms relied on ever more efficient search techniques to

N. Hernandez et al. (Eds.): ICCS 2014, LNAI 8577, pp. 37–50, 2014.
DOI: 10.1007/978-3-319-08389-6_5, © Springer International Publishing Switzerland 2014

find repeated concepts, but these algorithms were superseded by the discovery of the so-called 'canonicty test', whereby the attributes in a concept could be examined to determine its newness in the computation [6]. This test has given rise to a number of algorithms based on the Close-by-One (CbO) algorithm [6] and the focus of research has been to develop improvements of the original CbO algorithm. This paper presents one such improvement in the form of a more efficient canonicity test: the partial-closure canonicity test.

The next two sections of the paper briefly describe formal concepts and the main issues involved in their efficient computation, showing how testing a computed concept's canonicity is an efficient means of detecting the computation of repeated results. Section 4 shows a CbO algorithm incorporating a canonicity test that involves the complete closure of a concept intent before testing. Section 5 presents a partial-closure canonicity test that avoids the need for complete closure before testing and in Section 6 the partial-closure test is realised in the In-CloseI algorithm. Section 7 compares and evaluates the efficiency of the algorithms by carrying out 'level playing field' implementations and experiments on a variety of data sets. Another recent advance in CbO-type algorithms, that allows the inheritance of failed canonicity tests in the recursion, is included in the evaluations. Finally, the paper makes its conclusions and suggestions for further work in Section 8.

2   Formal Concepts

A description of formal concepts [7] begins with a set of objects X and a set of attributes Y. X and Y, along with a binary relation $I \subseteq X \times Y$ is called the *formal context*. If $x \in X$ and $y \in Y$ then xIy says that object x has attribute y. For a set of objects $A \subseteq X$, a closure operator $^\uparrow$ is defined to obtain the set of attributes common to the objects in A as follows:

$$A^\uparrow := \{\, y \in Y \mid \forall x \in A : xIy \,\}. \tag{1}$$

Similarly, for a set of attributes $B \subseteq Y$, the $^\downarrow$ operator is defined to obtain the set of objects common to the attributes in B as follows:

$$B^\downarrow := \{\, x \in X \mid \forall y \in B : xIy \,\}. \tag{2}$$

(A, B) is called a formal concept *iff* $A^\uparrow = B$ and $B^\downarrow = A$. (A, B) is a pair of two closed sets. A is called the extent and B the intent of the formal concept (A, B). The closure operator $(^\uparrow)$ can be thought of as a *full* or *complete* closure operator, which will help differentiate it from the *partial*-closure operator defined later in this paper.

A formal context (X, Y, I) is typically represented as a cross table, with crosses indicating the relation I between objects (in rows) and attributes (in columns). The following is a simple example of a formal context:

	0	1	2	3	4
a	×			×	×
b		×	×	×	×
c	×		×		
d		×	×		×

Formal concepts in a cross table can be visualised as maximal rectangles of crosses, where the rows and columns in the rectangle are not necessarily contiguous. The formal concepts in the example context are:

$$C_1 = (\{a, b, c, d\}, \emptyset) \quad C_6 = (\{b\}, \{1, 2, 3, 4\})$$
$$C_2 = (\{a, c\}, \{0\}) \quad C_7 = (\{b, d\}, \{1, 2, 4\})$$
$$C_3 = (\emptyset, \{0, 1, 2, 3, 4\}) \quad C_8 = (\{b, c, d\}, \{2\})$$
$$C_4 = (\{c\}, \{0, 2\}) \quad C_9 = (\{a, b\}, \{3, 4\})$$
$$C_5 = (\{a\}, \{0, 3, 4\}) \quad C_{10} = (\{a, b, d\}, \{4\})$$

3   Computation of Formal Concepts

A formal concept can be computed by applying the $^\downarrow$ operator to a set of attributes to obtain (close) its extent, and then applying the $^\uparrow$ operator to the extent to obtain (close) the intent. For example, taking an arbitrary set of attributes, say $\{1, 2\}$, from the the context above, $\{1, 2\}^\downarrow = \{b, d\}$ and $\{b, d\}^\uparrow = \{1, 2, 4\}$. $(\{b, d\}, \{1, 2, 4\})$ is concept C_7 in the list above. If this procedure is applied to every possible combination of attributes, then all the concepts in the context will be computed.

Thus, if there are n attributes in a formal context there are, potentially, 2^n concepts. It is the exponential nature of the problem that provides the computational challenge, compounded by the fact that the same concept can be computed more than once - in the worst case, exponentially many times. For example, in the context above, the concept C_7 will be computed six times because the extent $\{b, d\}$ can be obtained from the closure of six different combinations of attributes: $\{b, d\} = \{1\}^\downarrow = \{1, 2\}^\downarrow = \{1, 2, 3\}^\downarrow = \{1, 4\}^\downarrow = \{2, 4\}^\downarrow = \{4\}^\downarrow$.

Thus determining that a concept has already been computed is a key halting condition of the algorithm since each concept should be calculated only once. The algorithm `ComputeConceptsOnce`, below, illustrates this approach by intersecting a current extent A with successive attribute closures, to form a set of objects C from which an intent B is closed. The resulting concept (C, B) is tested for newness and, if it passes the test, the new concept is processed in some way (stored for example) and the algorithm called again, passing the new extent and the next attribute to the next level of recursion. In the following, an attribute set with n elements is denoted by $Y = \{0, , n - 1\}$. The algorithm is invoked with an extent and a starting attribute, initially $(X, 0)$.

`ComputeConceptsOnce(A, y)`

---

for $j \leftarrow y$ **upto** $n - 1$ **do**
    $C \leftarrow A \cap \{j\}^\downarrow$
    $B \leftarrow C^\uparrow$
    if `NewConcept`(C, B) **then**
        `ProcessConcept`(C, B)
        `ComputeConceptsOnce`$(C, j + 1)$

Early algorithms concentrated on finding repeats by efficiently searching the previously generated concepts. Lindig's algorithm [8], and others like it, use a search tree to quickly find repeated results. Others use a hash function where the cardinality of results is used to divide them into groups, thus narrowing the search [9]. The problem with these approaches is that even an efficient search becomes expensive when there is a large number of repeated concepts.

By iterating the attributes $0, ..., n-1$ recursively, an algorithm will generate intents in their lexicographic order. For example, the concepts in Section 2, above, are listed in the lexicographic order of their intents: $\emptyset < \{0\} < \{0, 1, 2, 3, 4\} < \{0, 2\} < \{0, 3, 4\} < \{1, 2, 3, 4\}$, and so on. During such an algorithm, when the closure operator is applied to an extent, the intent computed is thus canonical if its lexicographic position is after the lexicographic position of the previous canonical intent. If the last canonical intent was $\{0, 3, 4\}$ and the next closure produced the intent $\{0, 2\}$, it would not be canonical: the corresponding concept will have been generated earlier in the computation. It is this canonicity, and the testing thereof, that has become fundamental in detecting the repeated computation of a concept. Kuznetsov specified this canonicty test in Close-by-One (CbO), an algorithm with a time complexity of $O(|X||Y|^2|L|)$, where L is the set of concepts [6, 10, 11].

CbO has thus formed the basis for several variants and improvements [12–16] and it has been shown, in numerous tests, that these CbO-type algorithms are significantly faster than previous algorithms [11–19], outperforming algorithms including Chein [20], Norris [21], Next-Closure [22], Bordat [23], Godin [9], Nourine [24] and Add-Intent [25]. All CbO variants have the same time complexity, although it would be written $O(|Y||X|^2|L|)$ if the main cycle of the algorithm is over attributes instead of objects.

4   CbO Algorithm [6, 10, 12]

In the CbO algorithm [6, 10, 12], the canonicity test for a new concept is carried out by comparing a newly closed intent, D, with the last canonical intent that was closed, B. If they agree in all attributes up to the current attribute, j, then the new concept is canonical. If, however, there is an attribute in D that is not in B and that comes before j then the concept is not canonical (it will have been computed earlier). Thus a new concept is canonical if:

$$B \cap Y_j = D \cap Y_j \tag{3}$$

Where Y_j is the set of attributes up to but not including j:

$$Y_j := \{y \in Y \mid y < j\} \tag{4}$$

The correctness of the original CbO algorithm is given in [6], and proof of the canonicity test has been shown in [12]. The algorithm is written below and is called `ComputeConceptsFrom`. The procedure is invoked with the initial concept $(A, B) = (X, X^{\uparrow})$ and attribute $y = 0$.

CbO

ComputeConceptsFrom$((A, B), y)$

---

1 ProcessConcept$((A, B))$
2 **for** $j \leftarrow y$ **upto** $n - 1$ **do**
3 | **if** $j \notin B$ **then**
4 | | $C \leftarrow A \cap \{j\}^{\downarrow}$
5 | | $D \leftarrow C^{\uparrow}$
6 | | **if** $B \cap Y_j = D \cap Y_j$ **then**
7 | | | ComputeConceptsFrom$((C, D), j + 1)$

A line-by-line explanation of the algorithm is as follows:

Line 1 - Pass concept (A, B) to notional procedure ProcessConcept to process it in some way (for example, storing it in a set of concepts).

Line 2 - Iterate across the context, from attribute y up to attribute $n - 1$.

Line 3 - Test if the next attribute is in the current intent, B. If it is, skip it to avoid computing the same concept again.

Line 4 - Otherwise, form an extent, C, by intersecting the current extent, A, with the extent of the next attribute.

Line 5 - From the extent, fully close an intent, D.

Line 6 - Perform the canonicity test by checking that attributes in B and D agree up to the current attribute. If they do then the concept (C, D), is a new one so:

Line 7 - Recursively compute concepts from the new one, starting from the next attribute in the context.

CbO Call Tree – Figure 1 shows the CbO call tree for the simple context in Section 2. A rounded box represents the computation of a concept and a square box represents the computation of a repeated concept (that then fails the canonicity test). The lower part of a box shows the intersections carried out in the computation of the corresponding concept: an empty-square arrow represents the intersection carried out in Line 4 and the filled-circle arrows are the intersections implicit in the closure operation in Line 5. In each case the number pointed to represents the attribute extent involved. Note that each concept, new or repeat, requires a full closure of the intent. The notation $< C_x, j >$ represents the invocation of the next level in the form of a concept and initial attribute. The lines connecting the boxes are numbered with the current attribute in the cycle $(y, ..., n - 1)$ and are thus the same as those involved in the Line 4 intersections.

There are a total of 15 closures in the tree, with five intersections required for each closure, plus 14 extent intersections, making a total of 89 intersections for CbO.

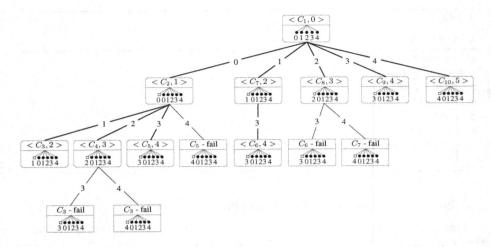

Fig. 1. CbO Call Tree

5   A Partial-Closure Canonicity Test

In the CbO algorithm, above, a concept intent is fully closed before the canonicity test is applied to it. However, it is sufficient to close an intent only up to the current attribute j to determine its canonicity. Remember that the canonicity test in CbO compares the previous intent with the new intent to see that they agree in all attributes *up to the current attribute*. Attributes after the current one have no bearing on the canonicity.

Thus in designing a partial-closure canonicity test we define, from 1 and 4, a *partial-closure* operator $\uparrow_j$ as a modification of the original closure operator:

$$A^{\uparrow_j} := \{\, y \in Y_j \mid \forall x \in A : xIy \,\}. \tag{5}$$

The partial-closure canonicity test will thus determine if the attributes in the intent B, up to j, agree with the attributes in the intent closed using C, up to j, and can be defined from 5 and 3 as:

$$B \cap Y_j = C^{\uparrow_j} \tag{6}$$

If there is an attribute in $C^{\uparrow_j}$ that is in B before j then the concept is not canonical and is a repeat.

The efficiency is that a partial-closure, $C^{\uparrow_j}$, is clearly less expensive to compute than a complete closure, $C^\uparrow$. The cost of the complete closure is the number of intersections equal to the number of attributes, n, whereas the cost of the partial-closure is always $< n$.

In fact, the implementation of the partial closure test can be even more efficient because the closure can be halted *as soon as* an attribute is found that is in B before j, so the cost of the closure will actually be $< j$ for each test failure.

Of course, if the canonicity test is passed, the new concept will then need to have the closure of its intent completed.

6   In-CloseI Algorithm with Partial-Closure Test

This section gives an updated version of the algorithm given in [13] that now incorporates the partial-closure test. The key difference to the CbO algorithm in Section 4, above, is that the canonicity test is applied *before* an intent is fully closed. The closure of an intent that passes the test is completed at the next level of recursion, by adding the current attribute j to the intent B whenever the current extent A is found: i.e. whenever $A \cap \{j\}^{\downarrow} = A$. This test of equality can be implemented with almost zero additional cost, given that the intersection $A \cap \{j\}^{\downarrow}$ is carried out in any case. Whenever a new concept is detected, the current, partial, intent is passed to the next level. The closure of the intent is completed when the main cycle at the next level is completed.

The algorithm, given as **In-CloseI** below, is invoked with the initial pair $(A, B) = (X, \emptyset)$ and initial attribute $y = 0$.

In-CloseI

ComputeConceptsFrom$((A, B), y)$

---

1   **for** $j \leftarrow y$ **upto** $n - 1$ **do**
2   $\quad$ $C \leftarrow A \cap \{j\}^{\downarrow}$
3   $\quad$ **if** $A = C$ **then**
4   $\quad\quad$ $B \leftarrow B \cup \{j\}$
5   $\quad$ **else**
6   $\quad\quad$ **if** $B = C^{\uparrow_j}$ **then**
7   $\quad\quad\quad$ $D \leftarrow B \cup \{j\}$
8   $\quad\quad\quad$ ComputeConceptsFrom$((C, D), j + 1)$

9   ProcessConcept$((A, B))$

Line 1 - Iterate across the context, from starting attribute y up to attribute $n - 1$.

Line 2 - Form an extent, C, by intersecting the current extent, A, with the extent of the next attribute.

Line 3 - If the extent formed, C, equals the given extent, A, then

Line 4 - ...add the current attribute j to the intent being closed, B.

Line 6 - Otherwise the new partial-closure canonicity test is applied. A small simplification to the canonicity test can be made because B is being completed incrementally with j. In other words, at the time of the test, $B = B_j = B \cap Y_j$. Therefore, in the canonicity test, $B \cap Y_j$ can be replaced with B. So if the attributes in B agree with those in the partial-closure $C^{\uparrow_j}$ the extent must be a new one so...

Line 7 - Create a new partial intent D that inherits the attributes of B, plus the current attribute j.

Line 8 - Pass the new extent C, the partial intent D and the next location $j + 1$ to the next level so that concepts from there can be computed and so that the closure of D can be completed.

Line 9 - Pass concept (A, B) to notional procedure `ProcessConcept` to process it in some way (for example, storing it in a set of concepts). Note that in In-CloseI this happens at the end of the procedure, once the main cycle has completed the closure of the intent, B.

Correctness of In-CloseI – CbO and In-CloseI use the same cycle to form the same extents, C. The only pertinent difference up to this point is the skipping of attributes in CbO when $j \in B$ (Line 3). However, this test is only to avoid forming an extent that will fail the canonicity test anyway. Thus to show that In-CloseI is correct it is sufficient to show that the partial-closure canonicity test is equivalent to the original test, in other words, given the same extent C, show that the tests are equivalent:

$$(B = C^{\uparrow_j}) \equiv (B \cap Y_j = D \cap Y_j)$$

As previously stated, in In-CloseI the intent B is incrementally closed up to the current attribute j, thus $B = B \cap Y_j$, so it is sufficient to show that:

$$C^{\uparrow_j} \equiv D \cap Y_j$$

Replacing D with $C^{\uparrow}$, from Line 5 of CbO:

$$C^{\uparrow_j} \equiv C^{\uparrow} \cap Y_j$$

Thus, from (1) and (5):

$$\{\, y \in Y_j \mid \forall x \in C : xIy \,\} \equiv \{\, y \in Y \mid \forall x \in C : xIy \,\} \cap Y_j.$$

So on the left side are all the attributes up to j that are related to the extent and on the right are all the attributes that are related to the extent, intersected with all the attributes up to j. Both sides are thus clearly equivalent.

In-CloseI Call Tree – Figure 2 shows the In-CloseI call tree for the simple context in Section 2. As in the CbO call tree, a rounded box represents the computation of a new concept and a square box represents the computation of a concept that fails the canonicity test. The notation $< C_x, j >$ is the invocation of the next level in the form of an extent, a partial intent and an initial attribute. The lower part of a box shows the intersections carried out in the computation of the corresponding concept: an empty-square arrow represents the intersections carried out by Line 2 and the filled-circle arrows are the intersections implicit in the partial closure operation in Line 6. Thus the In-CloseI call tree shows fewer closure intersections than CbO. If the canonicity test is passed the intent is fully closed at the next level, requiring some additional extent intersections. Altogether there are 59 intersections in the tree, 30 fewer than for CbO.

Table 1 summarises the performance of each algorithm using the simple context from Section 2. The table counts and compares the number of full and partial-closures and the number of extent intersections, $A \cap \{j\}^{\downarrow}$. Because closure itself involves repeated intersections of an extent with extents of attributes, it is convenient to use the intersection as a measure of performance. For a full closure, $C^{\uparrow}$, there are n intersections (in the example, $n = 5$) and for a partial-closure, $C^{\uparrow j}$, there are $j - 1$ intersections. To perform the evaluation and create the call-trees, a paper run of each algorithm was carried out, line by line. These are too lengthy to be presented here but are available as an on-line appendix [26].

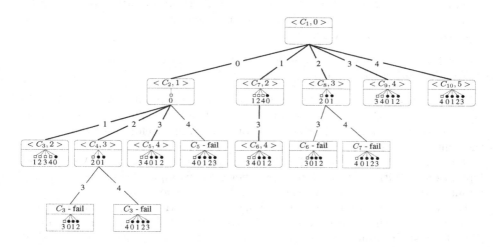

Fig. 2. In-Close Call Tree

Table 1. Comparison of closures and intersections for the simple context example

Algorithm	Full closures (intersections)	Partial closures (intersections)	Extent intersections: $A \cap \{j\}^{\downarrow}$	Total intersections
CbO	15 (75)		14	89
In-CloseI		14 (37)	22	59

7   Performance Evaluation

7.1   Incorporating the Partial-Closure Test into FCbO

Another significant advance in CbO-type algorithms was made with the FCbO algorithm [15, 16]. FCbO is an enhancement of CbO where a failed canonicity test is inherited by the next level of recursion. Thus an attribute that has caused a previous failure can be skipped in subsequent levels, avoiding unnecessary closures. To provide a performance evaluation that included this feature it was implemented in In-CloseI to produce In-CloseII.

7.2   Implementations

Implementations were carried out using C++ and a series of tests were performed using contexts created from real data sets, artificial data sets and randomised data sets. The experiments were carried out using a standard Windows PC with an Intel E4600 2.39GHz processor and 3GB of RAM. The times for the programs include data pre-processing, such as sorting, but exclude administrative aspects, such as data file input.

To create a level playing field for testing, the algorithms were implemented with the same two optimisations. Although it would have been possible to create un-optimised implementations, times for real data sets would be prohibitively slow and comparison with times presented elsewhere would have been unhelpful. The two optimisations used were

- Sorting context columns in order of density (lower density columns first)
- Implementing the context as a bit-array

The practice of column-sorting to improve concept computation is well known [5]. By doing so, there are fewer canonicity test failures. This is because there is less chance of finding A in the extent of an attribute before attribute j since the context is less dense before attribute j.

The use of bit-arrays is well known in computation, allowing a SIMD (Single Operation - Multiple Data) approach, where, for example, 32 context rows can be intersected simultaneously using standard 32-bit operations [12].

For the In-Close variants, the full savings of the partial-closure canonicity test were realised in the implementations: for the test to fail it is only necessary to find the first attribute that is not canonical. Thus the partial-closure can be halted as soon as such an attribute is found.

The inherited canonicity failure required the creation of a two-dimensional array to implement the failed intents that are passed to the next level of recursion. Although the use of pointers reduces the need for copying arrays in memory, the updating and accessing of the arrays gives rise to some additional complexity in the computation.

The notional **ProcessConcept** procedure was implemented simply by storing the computed concepts. Extents were stored as 'end-to-end' lists of integers in a one-dimensional array to make it possible to store them in the memory available. Intents were stored in a two-dimensional bit-array, each being stored as n bits. The use of bits and the fact that the number of attributes is typically much smaller than objects, makes the memory requirements tractable. For testing $j \notin B$, the Boolean nature of the bit-array version of intents is an efficient structure, with the test implemented simply as if not(B[j]).

For carrying out the extent intersection, $C \leftarrow A \cap \{j\}^{\downarrow}$, it was a simple case of testing the bit-position in column j of the context for each integer in A. For the In-Close variants, the size of C (produced as a by-product of the extent intersection) was used to test the equality of C and A. In effect the test becomes $if \ |A| = |C|$, incurring no additional overhead during the completion of partial-closures.

For the closure $C^{\uparrow}$ and partial-closure $C^{\uparrow_j}$, a bit-wise Boolean *and* operator was used to 'parse' 32 columns of the context at a time, using the integers in A to identify the rows to test. For the In-Close variants, as soon as C was found in a column before j and not in B, the partial-closure was halted.

7.3 Data Set Experiments

A series of experiments were carried out to compare the performance of the algorithm implementations using a variety of real, artificial and randomised data sets. These provided a wide range of size and density of formal context to test the implementations under a variety of conditions. Three of the real data sets are from the UCI Machine Learning Repository [27]: *Mushroom*, *Adult* and *Internet Ads*. A fourth data set, *Student*, is a set of results from a student questionnaire used to obtain course feedback at Sheffield Hallam University, UK in 2010. The results are given in Table 2.

Table 2. Real data set results (timings in seconds)

	Mushroom	Adult	Internet Ads	Student
$\lvert G \rvert \times \lvert M \rvert$	$8,124 \times 125$	$32,561 \times 124$	$3,279 \times 1,565$	587×145
Density	17.36%	11.29%	0.97%	24.50%
#Concepts	226,921	1,388,469	16,570	2,276,0243
CbO	0.66	3.06	0.56	32.68
In-CloseI	0.40	1.65	0.12	11.42
FCbO	0.35	2.06	0.21	17.20
In-CloseII	**0.29**	**1.62**	**0.10**	**9.38**

Artificial data sets were used that, although randomised, the randomisation was constrained by properties of real data sets, such as many-valued attributes having a fixed number of possible values. The results of the artificial data set experiments are given in Table 3.

Table 3. Artificial data set results (timings in seconds)

	M7X10G120K	M10X30G120K	T10I4D100K
$\lvert G \rvert \times \lvert M \rvert$	$120,000 \times 70$	$120,000 \times 300$	$100,000 \times 1,000$
Density	10.00%	3.33%	1.01%
#Concepts	1,166,343	4,570,498	2,347,376
CbO	2.51	31.26	49.45
In-CloseI	**1.26**	18.95	16.02
FCbO	1.67	22.33	29.41
In-CloseII	1.39	**10.42**	**11.04**

Three series of random data experiments were carried out, testing the affect of changes in the number of attributes, context density, and number of objects. The results are shown in Figure 3.

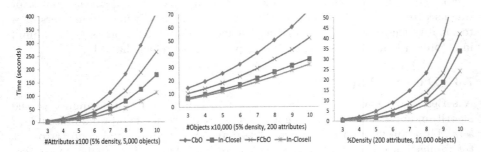

Fig. 3. Comparison of performance with varying number of attributes, density and objects

8   Conclusions and Further Work

The results of the performance experiments suggest that the partial-closure canonicity test significantly improves the efficiency of CbO-type algorithms. Not only was In-CloseI faster than the basic CbO algorithm, in most cases it was faster than FCbO. Although FCbO can significantly reduce the number of closures carried out [16] this appears to be outweighed by the savings in intersections made by partial-closure.

The results also show that further efficiency is possible by combining advances in CbO: combining the inherited canonicity test failure of FCbO with the partial closure canonicity test of In-CloseI to create In-CloseII. In most cases In-CloseII was faster than In-CloseI, but it is interesting that for the objects series of randomised experiments the saving was relatively small and for the artificial data set M7X10G120K, In-CloseI was actually faster. It is quite possible that the complexity of the inherited failure feature is resulting in some significant overheads in the computation, but further work is required to confirm this.

There is also scope for comparing concept mining algorithms with frequent itemset miners such as LCM [28]. Although they carry out different tasks, there is enough similarity for one to learn from the other.

Future work is also required in increasing performance through parallel processing. Work has been carried out to develop parallel versions of CbO (PCbO) and FCbO (PFCbO) [12,15] but not as yet for the In-Close variants.

Implementations of In-Close and FCbO are available open-source at *Source-Forge* [29,30].

References

1. Andrews, S., Orphanides, C.: Analysis of large data sets using formal concept lattices. In: [31], pp. 104–115
2. Tanabata, T., Sawase, K., Nobuhara, H., Bede, B.: Interactive data mining for image databases based on fca. Journal of Advanced Computational Intelligence and Intelligent Informatics 14, 303–308 (2010)

3. Kaytoue, M., Duplessis, S., Kuznetsov, S.O., Napoli, A.: Two FCA-based methods for mining gene expression data. In: Ferré, S., Rudolph, S. (eds.) ICFCA 2009. LNCS (LNAI), vol. 5548, pp. 251–266. Springer, Heidelberg (2009)
4. Kuznetsov, S.O.: On computing the size of a lattice and related decision problems. Order 18, 313–321 (2001)
5. Carpineto, C., Romano, G.: Concept Data Analysis: Theory and Applications. J. Wiley (2004)
6. Kuznetsov, S.O.: Mathematical aspects of concept analysis. Mathematical Science 80, 1654–1698 (1996)
7. Ganter, B., Wille, R.: Formal Concept Analysis: Mathematical Foundations. Springer (1998)
8. Lindig, C.: Fast concept analysis. In: Working with Conceptual Structures: Contributions to ICCS 2000, pp. 152–161. Shaker Verlag, Aachen (2000)
9. Godin, R., Missaoui, R., Alaoui, H.: Incremental concept formation algorithms based on Galois lattices. Computational Intelligence 11, 246–267 (1995)
10. Kuznetsov, S.O.: Learning of simple conceptual graphs from positive and negative examples. In: Żytkow, J.M., Rauch, J. (eds.) PKDD 1999. LNCS (LNAI), vol. 1704, pp. 384–391. Springer, Heidelberg (1999)
11. Kuznetsov, S., Obiedkov, S.: Comparing performance of algorithms for generating concept lattices. Journal of Experimental and Theoretical Artificial Intelligence 14, 189–216 (2002)
12. Krajca, P., Outrata, J., Vychodil, V.: Parallel recursive algorithm for FCA. In: Belohavlek, R., Kuznetsov, S. (eds.) Proceedings of Concept Lattices and their Applications (2008)
13. Andrews, S.: In-close, a fast algorithm for computing formal concepts. In: Rudolph, S., Dau, F., Kuznetsov, S.O. (eds.) ICCS 2009. CEUR WS, vol. 483 (2009), http://sunsite.informatik.rwth-aachen.de/Publications/CEUR-WS/Vol-483/
14. Andrews, S.: In-close2, a high performance formal concept miner. In: Andrews, S., Polovina, S., Hill, R., Akhgar, B. (eds.) ICCS 2011. LNCS (LNAI), vol. 6828, pp. 50–62. Springer, Heidelberg (2011)
15. Krajca, P., Vychodil, V., Outrata, J.: Advances in algorithms based on CbO. In: [31], pp. 325–337
16. Outrata, J., Vychodil, V.: Fast algorithm for computing fixpoints of Galois connections induced by object-attribute relational data. Inf. Sci. 185, 114–127 (2012)
17. Strok, F., Neznanov, A.: Comparing and analyzing the computational complexity of fca algorithms. In: Proceedings of the 2010 Annual Research Conference of the South African Institute of Computer Scientists and Information Technologists, pp. 417–420 (2010)
18. Kirchberg, M., Leonardi, E., Tan, Y.S., Link, S., Ko, R.K.L., Lee, B.S.: Formal concept discovery in semantic web data. In: Domenach, F., Ignatov, D.I., Poelmans, J. (eds.) ICFCA 2012. LNCS (LNAI), vol. 7278, pp. 164–179. Springer, Heidelberg (2012)
19. Borchman, D.: A generalized next-closure algorithm - enumerating semilattice elements from a generating set. In: Szathmary, L., Priss, U. (eds.) Proceedings of Concept Lattices and thie Applications (CLA 2012), pp. 9–20. Universidad de Malaga (2012)
20. Chein, M.: Algorithme de recherche des sous-matrices premieres dune matrice. Bull. Math. Soc. Sci. Math. R.S. Roumanie 13, 21–25 (1969)
21. Norris, E.M.: Maximal rectangular relations. In: Karpinski, M. (ed.) FCT 1977. LNCS, vol. 56, pp. 476–481. Springer, Heidelberg (1977)

22. Ganter, B.: Two basic algorithms in concept analysis. FB4-Preprint 831. TH Darmstadt (1984)
23. Bordat, J.P.: Calcul pratique du treillis de Galois dune correspondance. Math. Sci. Hum. 96, 31–47 (1986)
24. Nourine, L., Raynaud, O.: A fast algorithm for building lattices. Information Procesing Letters 71, 199–204 (1999)
25. van der Merwe, D., Obiedkov, S., Kourie, D.: Addintent: A new incremental algorithm for constructing concept lattices. In: Eklund, P. (ed.) ICFCA 2004. LNCS (LNAI), vol. 2961, pp. 372–385. Springer, Heidelberg (2004)
26. Andrews, S.: Appendix to a partial-closure canonicity test to increase the efficiency of CbO-type algorithms (2013),
   https://dl.dropboxusercontent.com/u/3318140/partialclosureappendix.pdf
27. Frank, A., Asuncion, A.: UCI machine learning repository (2010),
   http://archive.ics.uci.edu/ml
28. Uno, T., Kiyomi, M., Arimura, H.: Lcm ver. 3: Collaboration of array, bitmap and prefix tree for frequent itemset mining. In: Proceedings of the 1st International Workshop on Open Source Data Mining: Frequent Pattern Mining Implementations, pp. 77–86. ACM (2005)
29. Andrews, S.: In-Close program (2013),
   http://sourceforge.net/projects/inclose/
30. Krajca, P., Outrata, J., Vychodil, V.: FCbO program (2012),
   http://fcalgs.sourceforge.net/
31. Kryszkiewicz, M., Obiedkov, S. (eds.): Proceeding of 7th International Conference on Concept Lattices and Their Applications, CLA 2010. University of Sevilla, Seville (2010)

On Conceptual Graphs and Explanation of Query Answering under Inconsistency

Abdallah Arioua, Nouredine Tamani, and Madalina Croitoru

University of Montpellier II, France
{arioua,tamani,croitoru}@lirmm.fr

Abstract. Conceptual Graphs are a powerful visual knowledge representation language. In this paper we are interested in the use of Conceptual Graphs in the setting of Ontology Based Data Access, and, more specifically, in reasoning in the presence of inconsistency. We present different explanation heuristics of query answering under inconsistency and show how they can be implemented under the Conceptual Graphs editor COGUI.

1 Introduction

We place ourselves in a Rule-based Data Access (RBDA) setting that investigates how to query multiple data sources defined over the same ontology represented using a rule based language. The RBDA is a specific case of the Ontology Based Data Access (OBDA) setting. In RBDA we assume that the ontology is encoded using rules. The growing number of distinct data sources defined under the same ontology makes OBDA an important and timely problem to address. The input to the problem is a set of facts, an ontology and a conjunctive query. We aim to find if there is an answer to the query in the facts (eventually enriched by the ontology).

More precisely, the RBDA problem stated in reference to the classical *forward chaining* scheme is the following: "Can we find an answer to the query Q in a database F' that is built from F by adding atoms that can be logically deduced from F and the rule based ontology $\mathcal{R}$?"

In certain cases, the integration of factual information from various data sources may lead to inconsistency. A solution is then to construct maximal (with respect to set inclusion) consistent subsets of $\mathcal{F}$ called *repairs* [6,20]. Once the repairs are computed, there are different ways to combine them in order to obtain an answer for the query.

In this paper we address the RBDA problem from a Conceptual Graphs perspective. Conceptual Graphs are powerful visual formalism for representing a subset of First Order Logic covered by the RBDA setting.

We make explicit these links and focus on the case where we want to perform query answering in presence of inconsistency. We present query answering explanation strategies inspired from the link between the ODBA inconsistent-tolerant semantics and argumentation acceptance semantics[11]. Our work is inspired by the argumentation explanation power [14,25,28].

N. Hernandez et al. (Eds.): ICCS 2014, LNAI 8577, pp. 51–64, 2014.
DOI: 10.1007/978-3-319-08389-6_6, ⓒ Springer International Publishing Switzerland 2014

2   Related Work

There are two major approaches in order to represent an ontology for the OBDA problem and namely Description Logics (such as $\mathcal{EL}$([2]) and DL-Lite [9] families) and rule based languages (such as the Datalog$^+$ [8] language, a generalization of Datalog that allows for existentially quantified variables in the head of the rules). When using rules for representing the ontology we would denote the OBDA problem under the name of RBDA. Despite Datalog$^+$ undecidability when answering conjunctive queries, there exist decidable fragments of Datalog$^+$ which are studied in the literature [5]. These fragments generalize the above mentioned Description Logics families.

Here we follow the second method: representing the ontology via rules. We give a general rule based setting knowledge representation language equivalent to the Datalog$^+$ language and show how this language is equivalent to Conceptual Graphs with rules and negative constraints.

Within this language we are mainly interested in studying the question of "why an *inconsistent* KB entails a certain query α under an inconsistency-tolerant semantics". Indeed, many works focused on the following questions: "Why a concept C is subsumed (non-subsumed) by D" or "Why the KB is unsatisfiable and incoherent"? The need for explanation-aware methods stems from the desire to seek a comprehensive means that facilitates maintenance and repairing of inconsistent knowledge bases as well as understanding the underlying mechanism for reasoning services. In the field of databases there has been work on explaining answer and non-answer returned by database systems [1,24,23,16,15] using causality and responsibility or using a cooperative architecture to provide a cooperative answer for query failing.

In the area of DLs, the question was mainly about explaining either reasoning (subsumption and non-subsumption) or unsatisfiability and incoherence. In a seminal paper McGuinness et al. [22,7] addressed the problem of explaining subsumption and non-subsumption in a coherent and satisfiable DL knowledge base using *formal proofs as explanation* based on a complete and sound deduction system for a fragment of Description Logics, while other proposals [27,26,3,4] have used *Axiom pinpointing* and *Concept pinpointing* as explanation to highlight contradictions within an unsatisfiable and incoherent DL KB.

Another proposal [19,18] is the so-called *justification-oriented proofs* in which the authors proposed a *proof-like* explanation without the need for deduction rules. The explanation then is presented as an acyclic proof graph that relates axioms and lemmas. Another work [12] in the same context proposes a resolution-based framework in which the explanation is constructed from a refutation graph.

3   Logical Language

We consider a (potentially inconsistent) knowledge base composed of the following:

- A set $\mathcal{F}$ of facts that correspond to existentially closed conjunctions of atoms. The atoms can contain n-ary predicates. The following facts are borrowed from [21]: $F_1 : directs(John, d_1)$, $F_2 : directs(Tom, d_1)$, $F_3 : directs(Tom, d_2)$, $F_4 : supervises(Tom, John)$, $F_5 : works\_in(John, d_1)$, $F_6 : works\_in(Tom, d_1)$.

- A set of negative constraints which represent the negation of a fact. Alternatively negative constraints can be seen as rules with the absurd conclusion. Negative constraints can also be n-ary. For example, $N_1 = \forall x, y, z\, (supervises(x, y) \wedge work\_in(x, z) \wedge directs(y, z)) \rightarrow \perp$ and
$N_2 = \forall x, y\, supervises(x, y) \wedge manager(y) \rightarrow \perp$ are negative constraints.
- An ontology composed of a set of rules that represent general implicit knowledge and that can introduce new variables in their head (conclusion). Please note that these variables, in turn, can trigger new rule application and cause the undecidability of the language in the general case. Different rule application strategies (chase), including the skolemized chase, are studied in the literature. For example
$R_1 = \forall x \forall d\, works\_in(x, d) \rightarrow emp(x)$
$R_2 = \forall x \forall d\, directs(x, d) \rightarrow emp(x)$
$R_3 = \forall x \forall d\, directs(x, d) \wedge works\_in(x, d) \rightarrow manager(x)$
$R_4 = \forall x\, emp(x) \rightarrow \exists y (office(y) \wedge uses(x, y))$

A rule is *applicable* to set of facts $\mathcal{F}$ if and only if the set entails the hypothesis of the rule. If rule R is applicable to the set F, the application of R on F produces a new set of facts obtained from the initial set with additional information from the rule conclusion. We then say that the new set is an *immediate derivation* of F by R denoted by $R(F)$. For example $R_1(F_5) = works\_in(John, d_1) \wedge emp(John)$.

Let F be a set of facts and let $\mathcal{R}$ be a set of rules. A set F_n is called an $\mathcal{R}$-*derivation* of F if there is a sequence of sets (*derivation sequence*) $(F_0, F_1, \ldots, F_n)$ such that: (i) $F_0 \subseteq F$, (ii) F_0 is $\mathcal{R}$-consistent, (iii) for every $i \in \{1, \ldots, n-1\}$, it holds that F_i is an immediate derivation of F_{i-1}.

Given a set $\{F_0, \ldots, F_k\}$ and a set of rules $\mathcal{R}$, the closure of $\{F_0, \ldots, F_k\}$ with respect to $\mathcal{R}$, denoted $\mathtt{Cl}_\mathcal{R}(\{F_0, \ldots, F_k\})$, is defined as the smallest set (with respect to $\subseteq$) which contains $\{F_0, \ldots, F_k\}$, and is closed for $\mathcal{R}$-derivation (that is, for every $\mathcal{R}$-derivation F_n of $\{F_0, \ldots, F_k\}$, we have $F_n \subseteq \mathtt{Cl}_\mathcal{R}(\{F_0, \ldots, F_k\})$). Finally, we say that a set $\mathcal{F}$ and a set of rules $\mathcal{R}$ *entail* a fact G (and we write $\mathcal{F}, \mathcal{R} \models G$) iff the closure of the facts by all the rules entails F (i.e. if $\mathtt{Cl}_\mathcal{R}(\mathcal{F}) \models G$).

Given a set of facts $\{F_1, \ldots, F_k\}$, and a set of rules $\mathcal{R}$, the set of facts is called $\mathcal{R}$-*inconsistent* if and only if there exists a constraint $N = \neg \mathcal{F}$ such that $\mathtt{Cl}_\mathcal{R}(\{F_1, \ldots, F_k\}) \models \mathcal{F}$. A set of facts is said to be $\mathcal{R}$-*consistent* if and only if it is not $\mathcal{R}$-inconsistent.

A knowledge base $\mathcal{K} = (\mathcal{F}, \mathcal{R}, \mathcal{N})$, composed of a set of facts (denoted by $\mathcal{F}$), a set of rules (denoted by $\mathcal{R}$) and a set of negative constraints (denoted by $\mathcal{N}$), is said to be *consistent* if and only if $\mathcal{F}$ is $\mathcal{R}$-consistent. A knowledge base is *inconsistent* if and only if it is not consistent.

The above facts $\{F_1, ..., F_6\}$ are $\mathcal{R}$-inconsistent with $\mathcal{R} = \{R_1, ..., R_4\}$ since $\{F_1, F_4, F_6\}$ activate together N_1. Moreover, R_3 can be applied on F_1 and F_5 delivering the new fact $manager(John)$ which put together with F_4 activate N_2.

3.1 Conceptual Graphs Representation

Conceptual Graphs are a visual, logic-based knowledge representation knowledge representation formalism. They encode a part of the ontological knowledge in a structure called *support*. The support consists of a number of taxonomies of the main concepts (unary predicates) and relations (binary or more predicates) used to describe the world. Note that these taxonomies correspond to certain rules in Datalog. More complex rules (for instance representing transitivity or symmetry of relations) or rules that introduce existential variables in the conclusion are represented using Conceptual Graphs rules. Finally, negative constraints represent rules with the conclusion the absurd operator (or, logically equivalent, negation of facts).

The factual information is described using a bipartite graph in which the two classes of the partition are the concepts, and the relations respectively.

We recall the definition of support and fact following [10]. We consider here a simplified version of a support $S = (T_C, T_R, \mathcal{I})$, where: $(T_C, \leq)$ is a finite partially ordered set of *concept types*; $(T_R, \leq)$ is a partially ordered set of *relation types*, with a specified arity; $\mathcal{I}$ is a set of *individual markers*. A (simple) CG is a triple CG= $[S, G, \lambda]$, where:

- S is a support;
- $G = (V_C, V_R, E)$ is an ordered bipartite graph ; $V = V_C \cup V_R$ is the node set of G, V_C is a finite nonempty set of *concept nodes*, V_R is a finite set of *relation nodes*; E is the set of edges $\{v_r, v_c\}$ where the edges incident to each relation node are ordered and this ordering is represented by a positive integer label attached to the edge; if the edge $\{v_r, v_c\}$ is labeled i in this ordering then v_c is the i-neighbor of v_r and is denoted by $N_G^i(v_r)$;
- $\lambda : V \to S$ is a labeling function; if $v \in V_C$ then $\lambda(v) = (type_v, ref_v)$ where $type_v \in T_C$ and $ref_v \in \mathcal{I} \cup \{*\}$; if $r \in V_R$ then $\lambda(r) \in T_R$.

We denote a conceptual graph CG= $[S, G, \lambda]$ by G, keeping support and labeling implicit. The order on $\lambda(v)$ preserves the (pair-wise extended) order on T_C (T_R), considers $\mathcal{I}$ elements mutually incomparable, and $* \geq i$ for each $i \in \mathcal{I}$. Usually, CGs are provided with a logical semantics via the function Φ, which associates to each CG a FOL formula (Sowa (1984)). If S is a support, a constant is associated to each individual marker, a unary predicate to each concept type and a n-ary predicate to each n-ary relation type. We assume that the name for each constant or predicate is the same as the corresponding element of the support. The partial orders specified in S are translated in a set of formulae $\Phi(S)$ by the following rules: if $t_1, t_2 \in T_C$ such that $t_1 \leq t_2$, then $\forall x(t_2(x) \to t_1(x))$ is added to $\Phi(S)$; if $t_1, t_2 \in T_R$, have arity k and $t_1 \leq t_2$, then $\forall x_1 \forall x_2 \ldots \forall x_k(t_2(x_1, x_2, \ldots, x_k) \to t_1(x_1, x_2, \ldots, x_k))$ is added to $\Phi(S)$.

If CG= $[S, G, \lambda]$ is a conceptual graph then a formula $\Phi(CG)$ is constructed as follows. To each concept vertex $v \in V_C$ a term a_v and a formula $\phi(v)$ are associated: if $\lambda(v) = (type_v, *)$ then $a_v = x_v$ (a logical variable) and if $\lambda(v) = (type_v, i_v)$, then $a_v = i_v$ (a logical constant); in both cases, $\phi(v) = type_v(a_v)$. To each relation vertex $r \in V_R$, with $\lambda(r) = type_r$ and $deg_G(r) = k$, the formula associated is $\phi(r) = type_r(a_{N_G^1(r)}, \ldots, a_{N_G^k(r)})$.

$\Phi(CG)$ is the existential closure of the conjunction of all formulas associated with the vertices of the graph. That is, if $V_C(*) = \{v_{i_1}, \ldots, v_{i_p}\}$ is the set of all concept vertices having generic markers, then $\Phi(CG)= \exists v_1 \ldots \exists v_p(\wedge_{v \in V_C \cup V_R} \phi(v))$.

If (G, λ_G) and (H, λ_H) are two CGs (defined on the same support S) then $G \geq H$ (G subsumes H) if there is a *projection* from G to H. A projection is a mapping

π from the vertices set of G to the vertices set of H, which maps concept vertices of G into concept vertices of H, relation vertices of G into relation vertices of H, preserves adjacency (if the concept vertex v in V_C^G is the ith neighbour of relation vertex $r \in V_R^G$ then $\pi(v)$ is the ith neighbour of $\pi(r)$) and furthermore $\lambda_G(x) \geq \lambda_H(\pi(x))$ for each vertex x of G. If $G \geq H$ then $\Phi(S), \Phi(H) \models \Phi(G)$ (*soundness*). *Completeness* (if $\Phi(S), \Phi(H) \models \Phi(G)$ then $G \geq H$) only holds if the graph H is in *normal form*, i.e. if each individual marker appears at most once in concept node labels.

A CG rule $(Hyp, Conc)$ expresses implicit knowledge of the form "if *hypothesis* then *conclusion*", where hypothesis and conclusion are both basic graphs. This knowledge can be made explicit by applying the rule to a specific fact: intuitively, when the hypothesis graph is found in a fact, then the conclusion graph can be added to this fact. There is a one to one correspondence between some concept nodes in the hypothesis with concept nodes in the conclusion. Two nodes in correspondence refer to the same entity. These nodes are said to be *connection nodes*. A rule can be represented by a bicolored graph or by a pair of two CGs (represented, for instance, on the right and respectively left hand side of the screen).

A rule R can be applied to a CG H if there is a homomorphism from its hypothesis to H. Applying R to H according to such a homomorphism π consists of "attaching" to H the conclusion of R by merging each connection node in the conclusion with the image by π of the corresponding connection node in the hypothesis. When a knowledge base contains a set of facts (say $\mathcal{F}$) and a set of rules (say $\mathcal{R}$), the query mechanism has to take implicit knowledge coded in rules into account. The knowledge base answers a query Q if a CG F' can be derived from $\mathcal{F}$ using the rules of $\mathcal{R}$ such that Q maps to F'.

We note that using the $(\mathcal{F}, \mathcal{R}, \mathcal{N})$ Datalog$^+$ notation or the rule based Conceptual Graphs with negative constraints has the same logical expressivity. However, the added value of using Conceptual Graphs comes from the visual depiction of the knowledge. This aspect is shown next, where the previous example knowledge base is depicted using COGUI, a Conceptual Graphs editor developed by the LIRMM, University of Montpellier 2.

3.2   COGUI CG Editor

All figures depict graphs drawn using the conceptual graph editor Cogui[1]. CoGui is a Conceptual Graphs editor. Please note that Cogui is also fully integrated with the conceptual graph engine Cogitant[2] to perform reasoning on the above mentioned graphs.

Let us consider again the knowledge base previously considered:

- $\mathcal{F}$: $F_1 : directs(John, d_1)$, $F_2 : directs(Tom, d_1)$, $F_3 : directs(Tom, d_2)$, $F_4 : supervises(Tom, John)$, $F_5 : works\_in(John, d_1)$, $F_6 : works\_in(Tom, d_1)$.
- $N_1 = \forall x, y, z\ (supervises(x, y) \wedge work\_in(x, z) \wedge directs(y, z)) \rightarrow \bot$ and $N_2 = \forall x, y\ supervises(x, y) \wedge manager(y) \rightarrow \bot$.
- The set of rules: $R_1 = \forall x \forall d\ works\_in(x, d) \rightarrow emp(x)$

---

[1] http://www.lirmm.fr/cogui/
[2] http://cogitant.sourceforge.net/

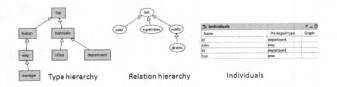

Type hierarchy   Relation hierarchy   Individuals

Fig. 1. Visualisation of a support (vocabulary) using CoGui

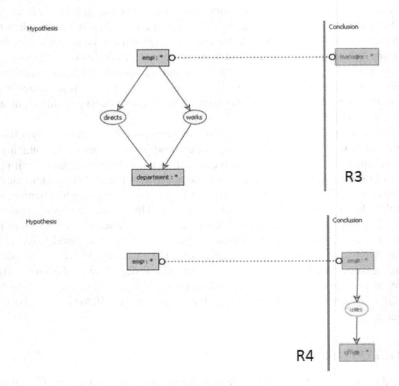

Fig. 2. Visualisation of two rules using CoGui

$R_2 = \forall x \forall d\, directs(x, d) \rightarrow emp(x)$
$R_3 = \forall x \forall d\, directs(x, d) \wedge works\_in(x, d) \rightarrow manager(x)$
$R_4 = \forall x\, emp(x) \rightarrow \exists y (office(y) \wedge uses(x, y))$

Figure 1 presents the concept type hierarchy, the relation type hierarchy and the list of individuals. Please note that the rule hierarchy encodes the rules R_1 and R_2.

Rules R_3 and R_4 respectively are depicted in Figure 2. The negative constraints N_1 and N_2 are depicted in Figure 3.

Finally, the set of facts is represented in Figure 4.

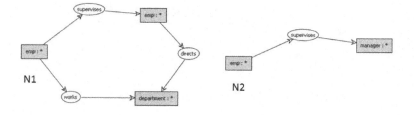

Fig. 3. Visualisation of negative constraints using CoGui

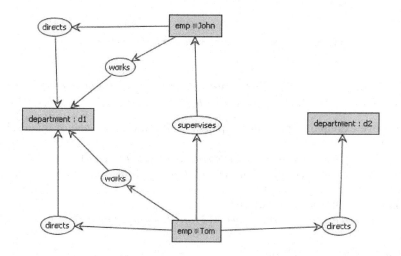

Fig. 4. Visualisation of factual knowledge using CoGui

4   Dealing with Inconsistency

We recall the definition of inconsistency. Given a set of facts $\{F_1, \dots, F_k\}$, and a set of rules $\mathcal{R}$, the set of facts is called $\mathcal{R}$-*inconsistent* if and only if there exists a constraint $N = \neg \mathcal{F}$ such that $\mathtt{Cl}_{\mathcal{R}}(\{F_1, \dots, F_k\}) \models \mathcal{F}$.

In Figure 5 we can see that there is a negative constraint entailed by the facts enriched by the rules. The image of the negative constraint by homomorphism is represented in red (if color is available) or darker shade of grey (greyscale).

Like in classical logic everything can be entailed from an inconsistent knowledge base. Different semantics have been introduced in order to allow query answering in the presence of inconsistency. Here we only focus on the ICR (**I**ntersection of **C**losed **R**epair) semantics defined as follows:

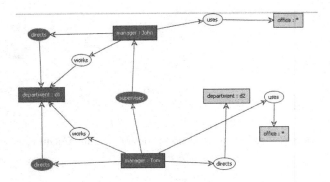

Fig. 5. Visualisation of factual knowledge using CoGui

Definition 1. *Let* $\mathcal{K} = (\mathcal{F}, \mathcal{R}, \mathcal{N})$ *be a knowledge base and let* α *be a query. Then* α *is* **ICR-entailed** *from* $\mathcal{K}$, *written* $\mathcal{K} \models_{ICR} \alpha$ *if and only if* $\bigcap_{A' \in \mathcal{R}epair(\mathcal{K})} Cl_{\mathcal{R}}(A') \models \alpha$.

In the above example, we obtain 6 repairs. The following are one of them (closed under set of rules):
$A_1 = \{directs(John, d_1), directs(Tom, d_1), directs(Tom, d_2),$
$supervises(Tom, John), emp(John), emp(Tom), \exists y_1(office(y_1) \land uses(Tom, y_1)),$
$\exists y_2(office(y_2) \land uses(John, y_2))\}$
The intersection of the closed repairs is:
$\bigcap_{Cl_{\mathcal{R}}(A)} = \{directs(Tom, d_1), directs(Tom, d_2), emp(Tom),$
$\exists y \, uses(Tom, y), \exists y office(y)\}$.

Another possibility to deal with an inconsistent knowledge base in the OBDA setting is to define an instantiation [11] of Dung's abstract argumentation theory [13]. An argumentation framework is composed of a set of arguments and a binary relation defined over arguments, the attack.

Definition 2 (Argument). *[11] An argument* A *in a knowledge base* $\mathcal{K} = (\mathcal{F}, \mathcal{R}, \mathcal{N})$ *is a tuple* $A = (F_0, \ldots, F_n)$ *where:*

- $(F_0, \ldots, F_{n-1})$ *is a derivation sequence w.r.t* $\mathcal{K}$.
- F_n *is an atom, a conjunction of atoms, the existential closure of an atom or the existential closure of a conjunction of atoms such that* $F_{n-1} \models F_n$.

We can extract from each argument its sub-arguments.

Definition 3 (Sub-argument). *Let* $\mathcal{K} = (\mathcal{F}, \mathcal{R}, \mathcal{N})$ *be a knowledge base and* $A = (F_0, F_1, \ldots, F_n)$ *be an argument.* $A' = (F_0, \ldots, F_k)$ *with* $k \in \{0, \ldots, n-1\}$ *is a sub-argument of* A *iff (i)* $A' = (F_0, \ldots, F_k)$ *is an argument and (ii)* $F_k \in F_{k+1}$.

Let $A = (F_0, \ldots, F_n)$ be an argument, then $\text{Supp}(A) = F_0$ and $\text{Conc}(A) = F_n$. Let $S \subseteq \mathcal{F}$ a set of facts, $Arg(S)$ is defined as the set of all arguments A such that $\text{Supp}(A) \subseteq S$.

An argument corresponds to a rule derivation. Therefore we can use the Cogui editor in order to depict arguments (via the depiction of rule derivations). In Figure 6 an example of a derivation is depicted. The added information by the rule is visible due to the changed color (pink in color, darker shade of grey on grey scale).

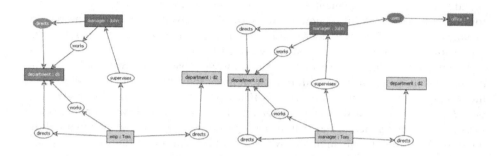

Fig. 6. Visualisation of a rule derivation using CoGui

Definition 4 (Attack). *[11] Let $\mathcal{K} = (\mathcal{F}, \mathcal{R}, \mathcal{N})$ be a knowledge base and let $a, b \in \mathcal{A}$. The argument a attacks b, denoted by $(a, b) \in Att$, iff there exists $\varphi \in \mathrm{Supp}(b)$ such that the set $\{Conc(a), \varphi\}$ is $\mathcal{R}$-inconsistent.*

Definition 5 (Argumentation framework). *[11] Let $\mathcal{K} = (\mathcal{F}, \mathcal{R}, \mathcal{N})$ be a knowledge base, the corresponding argumentation framework $\mathcal{AF}_K$ is a pair $(\mathcal{A} = \mathrm{Arg}(\mathcal{F}), Att)$ where $\mathcal{A}$ is the set of arguments that can be built from $\mathcal{F}$ and Att is the attack relation. Let $\mathcal{E} \subseteq \mathcal{A}$ and $a \in \mathcal{A}$.*

- *We say that $\mathcal{E}$ is conflict free iff there exists no arguments $a, b \in \mathcal{E}$ such that $(a, b) \in Att$.*
- *$\mathcal{E}$ defends a iff for every argument $b \in \mathcal{A}$, if we have $(b, a) \in Att$ then there exists $c \in \mathcal{E}$ such that $(c, b) \in Att$.*
- *$\mathcal{E}$ is admissible iff it is conflict free and defends all its arguments.*
- *$\mathcal{E}$ is a preferred extension iff it is maximal (with respect to set inclusion) admissible set.*
- *$\mathcal{E}$ is a stable extension iff it is conflict-free and $\forall a \in \mathcal{A} \backslash \mathcal{E}$, there exists an argument $b \in \mathcal{E}$ such that $(b, a) \in Att$.*
- *$\mathcal{E}$ is a grounded extension iff $\mathcal{E}$ is a minimal (for set inclusion) complete extension.*

We denote by $\mathrm{Ext}(\mathcal{AF}_K)$ the set of extensions of $\mathcal{AF}_K$. We use the abbreviations p, s, and g for respectively preferred, stable and grounded semantics. An argument is skeptically accepted if it is in all extensions, credulously accepted if it is in at least one extension and rejected if it is not in any extension.

The following results are then showed by [11]:

Theorem 1. *[11] Let $\mathcal{K} = (\mathcal{F}, \mathcal{R}, \mathcal{N})$ be a knowledge base, let $\mathcal{AF}_K$ be the corresponding argumentation framework, α be a query, and $x \in \{s, p\}$ be stable or preferred semantics. Then $\mathcal{K} \models_{ICR} \alpha$ iff α sceptically accepted under semantics x.*

5 Argumentative Explanation

In this section we define two different heuristics for explanation of inconsistency tolerant semantics. Since these heuristics work under inconsistent knowledge bases the Cogui editor is not yet adapted to implement them. We note that explanations correspond to the notion of argument, thus, the Cogui visual power could be easily adapted for our case. Moreover, in section 5.1 we show the equivalence between one type of explanation and a visual rule depiction in Cogui. This could be a starting point for the explanation of queries under inconsistency using Cogui.

When handling inconsistent ontological knowledge bases we are interested in the *explanation* of query answers conforming to a given semantics. More precisely we are interested in explaining why a query α is ICR-entailed by an inconsistent knowledge base $\mathcal{K}$. By explanation we mean a *structure* that has to incorporate minimal set of facts (w.r.t $\subseteq$) and general rules that, if put together, will lead to the entailment of the query α. According to this intuition (which coincides with the definition of [17]) and the link between inconsistent ontological knowledge bases and logic-based argumentation framework, a first candidate of explanation is an argument. However, an argument as defined in definition 2 can be cumbersome and difficult to understand, because the information of how these derivations have been achieved and how they lead to the conclusion are missing. Therefore we propose a refined explanation that incorporates rules as a crucial component.

Definition 6 (Explanation). *Let* $\mathcal{K} = (\mathcal{F}, \mathcal{R}, \mathcal{N})$ *be an inconsistent knowledge base, let* α *be query and let* $\mathcal{K} \vDash_{ICR} \alpha$. *An explanation for* α *in* $\mathcal{K}$ *is a 3-tuple* $E = (A, G, C)$ *composed of three finite sets of formulae such that: (1)* $A \subseteq \mathcal{F}, G \subseteq \mathcal{R}$, *(2)* $C \vDash \alpha$, *(3)* $\mathtt{Cl}_G(A) \nvDash \bot$ *(consistency), (4) For every formula* β *in* A, $\mathtt{Cl}_G(A - \beta) \nvDash C$ *(minimality). Such that* $\mathtt{Cl}_G$ *represents the closure w.r.t to the set of rules* G.

We denote by $\mathcal{EXP}$ the universe of explanations and by $\mathcal{EXP}_\alpha$ the set of all explanations for α. We denote the sets A, G and C as antecedents, general laws and conclusions respectively. Here the definition specifies three important components for explaining query α. First, the set A of antecedent conditions which is a minimal subset of facts that entails the query α. Second, the set of general laws G (from now on, rules) that produce the query α, the reason for integrating rules is that the user is often interested in knowing how we achieved the query. Finally, the third component is the conclusion C (the answer for the query α). The definition also imposes a central concept, namely *explanation consistency*.

An explanation can be computed directly from $\mathcal{K}$ or from an argumentation framework using the following mapping $\mathbb{A}$.

Definition 7 (Mapping $\mathbb{A}$). *Given an inconsistent knowledge base* $\mathcal{K} = (\mathcal{F}, \mathcal{R}, \mathcal{N})$ *and* $\mathcal{AF}_K = (\mathcal{A}, \mathtt{Att})$ *the corresponding argumentative framework. The mapping* $\mathbb{A}$ *is a total function defined on* $\mathcal{A} \longrightarrow \mathcal{EXP}$ *as* $E = \mathbb{A}((F_0, ..., F_n))$ *with:*

- *The set of antecedent conditions* $A = F_0$,
- *The set of rules* $G \subseteq \mathcal{R}$, *such that* $\forall r \in \mathcal{R}, r \in G$ *iff for all* F_i *in* x *the rule* r *is applicable to* F_i.
- *The conclusion* $C = F_n$ *iff* $\mathtt{Cl}_G(A) \vDash F_n$.

Proposition 1 (Bijection of $\mathbb{A}$). *For any argument $a \in \mathcal{A}$, the mapping $\mathbb{A}$ is a bijection.*

The proposition follows from the definition of the mapping because for every argument we can construct one explanation. Since the mapping is a bijection, we call the argument $x_e = \mathbb{A}^{-1}(e)$ the corresponding argument of an explanation e. We say the argument x_e supports the explanation e. The following proposition states that there is always an explanation for an ICR-entailed query.

Proposition 2 (Existence of explanation). *For every query α such that $\mathcal{K} \models_{ICR} \alpha$, the set $\mathcal{EXP}_\alpha$ is not empty.*

Proof 1 *On the one hand, if $\mathcal{K} \models_{ICR} \alpha$ then the query α is sceptically accepted. That means $\forall \mathcal{E} \in \text{Ext}(\mathcal{AF}_K), \mathcal{E} \models \alpha$. Hence there is an argument $a \in \mathcal{E}$ such that $Cons(a) \models \alpha$. On the other hand, using the mapping $\mathbb{A}$ we have $e = \mathbb{A}(a)$ is an explanation for α, namely $e \in \mathcal{EXP}_\alpha$. Consequently $\mathcal{EXP}_\alpha \neq \emptyset$*

Example 1 (Corresponding Argument). Let us explain $\alpha = \exists x\, emp(x)$. We can build the following argument for α:
$$a_\alpha^+ = (\{works\_in(Tom, d_1)\}, \{works\_in(Tom, d_1), emp(Tom)\}, emp(Tom)),$$
and the delivered explanation is:
$$E_\alpha = (\{directs(Tom, d_1)\}, \{\forall x \forall d\, works\_in(x, d) \rightarrow emp(x)\}, emp(Tom)).$$

There could be cases where the user wants to know how the set of *rules and facts* interact in order to explain a query α. Put differently, a user-invoked explanation that makes explicit any relation between the facts and the rules which lead to α. Notice that, this type of user-invoked explanation is called *deepened* explanation and it should not confounded with a proof-like explanation, because we are considering an inconsistent and incomplete settings. For this reason the explanation below has not yet been implemented as a stand alone plugin for Cogui (Cogui only deals with querying consistent knowledge).

5.1 Deepened Explanation (d-explanation)

Definition 8 (d-explanation). *Let $\mathcal{K} = (\mathcal{F}, \mathcal{R}, \mathcal{N})$ be an inconsistent knowledge base, let α be a query and let $\mathcal{K} \models_{ICR} \alpha$. Then, the finite sequence of tuples $d = \langle t_1, t_2, ..., t_n \rangle$ is a d-explanation for α iff:*

1. *For every tuple $t_i = (a_i, r_i) \in d$ such that $i \in \{1, ..., n\}$, it holds that $a_i \subseteq \text{Cl}_\mathcal{R}(\mathcal{F})$ and $r_i \in \mathcal{R}$.*
2. *For every tuple $t_i = (a_i, r_i) \in d$ such that $i \in \{2, ..., n\}$ we have $a_i = a_i' \cup a_i''$ where (i) $r_{i-1}(a_{i-1}) \models a_i'$, (ii) $a_i'' \subseteq \text{Cl}_\mathcal{R}(\mathcal{F})$ and (iii) r_i is applicable to a_i. Note that if $i = 1$ then $a_i' = \emptyset$.*
3. *The tuple (a_n, r_n) entails α (i.e $r_n(a_n) \models \alpha$).*
4. *$\text{Cl}_\mathcal{R}(\cup_{i=0}^n a_i) \not\models \bot$ (consistency).*

We denote by $\mathcal{D}$ the universe of d-explanations and by $\mathcal{D}_\alpha$ the set of all d-explanations for a query α.

The intuition about the d-explanation d is as follows: tuples in d represent $\langle fact, applicable\, rule \rangle$, and the sequence of tuples represents the order by which we achieve the answer of the query. Think of it as a chain where each a_i has a link with the previous a_{i-1} through the rule r_{i-1}. This is similar to the notion of derivation depicted in Figure 6.

Example 2 (Deepened explanation). The deepened explanation associated to α is the same as E and doesn't provide more information. Let us consider the explanation of $\alpha_2 = \exists x \, office(x)$. A possible argument for α_2 is:

$a_{\alpha_2}^+ = (\{works\_in(Tom, d_1)\}, \{works\_in(Tom, d_1), emp(Tom)\},$
$\{works\_in(Tom, d_1), emp(Tom), \exists y(office(y) \wedge uses(Tom, y))\},$
$\exists y(office(y) \wedge uses(Tom, y))).$

So $E_{\alpha_2} = (\{works\_in(Tom, d_1)\}, \{\forall x \forall d \, directs(x, d) \rightarrow emp(x), \forall x \, emp(x) \rightarrow \exists y(office(y) \wedge uses(x, y))\}, \exists y office(y)).$

$D_E = (\langle works\_in(Tom, d_1), \forall x \forall d \, directs(x, d) \rightarrow emp(x)\rangle,$
$\langle emp(Tom), \forall x \, emp(x) \rightarrow \exists y(office(y) \wedge uses(x, y))\rangle).$

There is a bijection between an explanation e and a d-explanation d represented here by the following mapping.

Definition 9 (Mapping $\mathbb{D}$). *Let $\mathcal{K} = (\mathcal{F}, \mathcal{R}, \mathcal{N})$ be an inconsistent knowledge base, α be a query, $e = (A, G, C) \in \mathcal{EXP}$ be an explanation for α and $d = \langle t_1, t_2, ..., t_n \rangle \in \mathcal{D}$ be a d-explanation for α. The mapping $\mathbb{D}$ is total function $\mathbb{D} : \mathcal{EXP} \longrightarrow \mathcal{D}, e \longrightarrow d$ defined as follows:*

1. *For every tuple $t_i = (a_i, r_i)$ such that $i \in \{1, ..., n\}$, it holds that $r_i \in G$.*
2. *For every tuple $t_i = (a_i, r_i)$ in D such that $i \in \{2, ..., n\}$ we have $a_i = a_i' \cup a_i''$ where $r_{i-1}(a_{i-1}) \vDash a_i'$, $a_i'' \subseteq A$ and r_i is applicable to a_i. Note that if $i = 1$ then $a_i = A$ and e_i is applicable to a_i.*
3. *The tuple (a_n, r_n) entails α (i.e $r_n(a_n) \vDash \alpha$) and $C \vDash \alpha$.*

Since the mapping is a bijection the existence of the inverse function is guaranteed. Thereby we consider the mapping $\mathbb{D}(e)$ as deepening the explanation e and the inverse mapping $\mathbb{D}^{-1}(d)$ as simplifying the d-explanation d. The advantage of such a mapping is that it gives the users the freedom to shift from an explanation to another which complies better with their level of understanding and their experiences. Also it guarantees that every explanation can be deepened. As done before, we also define the corresponding argument x_d for a d-explanation d, as the corresponding argument x_e for an explanation $e = \mathbb{D}^{-1}(d)$. This can be achieved by the following composition of function: $x_d = (\mathbb{D} \circ \mathbb{A})^{-1}(d)$.

6 Conclusion

In this paper we have presented an argumentative approach for explaining user query answers in a particular setting, Namely, an inconsistent ontological knowledge base where inconsistency is handled by inconsistency-tolerant semantics (ICR) and it is issued from the set of facts. In this paper we have exploited the relation between ontological knowledge base and logical argumentation framework to establish different levels of explanation ranging from an explanation based on the notion of argument to a user-invoked explanation called deepened explanation. We have also shown the relation between every type of explanation using a one-to-one correspondence which gives the user the possibility to deepen (or simplify) the explanation in hand. Future works aims at studying the proposed explanation in the context of other inconsistency-tolerant semantics. We are currently working on a Cogui based plug-in that only deals with reasoning under inconsistency and the above mentioned semantics.

Acknowledgments. A. Arioua and M. Croitoru have been supported by the French National Agency of Research within Dur-Dur project.

References

1. Arora, T., Ramakrishnan, R., Roth, W.G., Seshadri, P., Srivastava, D.: Explaining program execution in deductive systems. In: Ceri, S., Tanaka, K., Tsur, S. (eds.) DOOD 1993. LNCS, vol. 760, pp. 101–119. Springer, Heidelberg (1993)
2. Baader, F., Brandt, S., Lutz, C.: Pushing the el envelope. In: Proc. of IJCAI 2005 (2005)
3. Baader, F., Peñaloza, R., Suntisrivaraporn, B.: Pinpointing in the description logic $\mathcal{EL}^+$. In: Hertzberg, J., Beetz, M., Englert, R. (eds.) KI 2007. LNCS (LNAI), vol. 4667, pp. 52–67. Springer, Heidelberg (2007)
4. Baader, F., Suntisrivaraporn, B.: Debugging snomed ct using axiom pinpointing in the description logic $\mathcal{EL}^+$. In: KR-MED, vol. 410 (2008)
5. Baget, J.-F., Mugnier, M.-L., Rudolph, S., Thomazo, M.: Walking the complexity lines for generalized guarded existential rules. In: Proceedings of the 22nd International Joint Conference on Artificial Intelligence (IJCAI 2011), pp. 712–717 (2011)
6. Bienvenu, M.: On the complexity of consistent query answering in the presence of simple ontologies. In: Proc. of AAAI (2012)
7. Borgida, A., Franconi, E., Horrocks, I., McGuinness, D.L., Patel-Schneider, P.F.: Explaining ALC subsumption. In: Lambrix, P., Borgida, A., Lenzerini, M., Möller, R., Patel-Schneider, P.F. (eds.) Proceedings of the 1999 International Workshop on Description Logics 1999, Linköping, Sweden, vol. 22 (July 1999)
8. Calì, A., Gottlob, G., Lukasiewicz, T.: A general datalog-based framework for tractable query answering over ontologies. In: Proceedings of the Twenty-Eigth ACM SIGMOD-SIGACT-SIGART Symposium on Principles of Database Systems, pp. 77–86. ACM (2009)
9. Calvanese, D., De Giacomo, G., Lembo, D., Lenzerini, M., Rosati, R.: Tractable reasoning and efficient query answering in description logics: The dl-lite family. J. Autom. Reasoning 39(3), 385–429 (2007)
10. Chein, M., Mugnier, M.-L.: Graph-based Knowledge Representation and Reasoning— Computational Foundations of Conceptual Graphs. Advanced Information and Knowledge Processing. Springer (2009)
11. Croitoru, M., Vesic, S.: What can argumentation do for inconsistent ontology query answering? In: Liu, W., Subrahmanian, V.S., Wijsen, J. (eds.) SUM 2013. LNCS (LNAI), vol. 8078, pp. 15–29. Springer, Heidelberg (2013)
12. Deng, X., Haarslev, V., Shiri, N.: A framework for explaining reasoning in description logics. In: Proceedings of the AAAI Fall Symposium on Explanation-Aware Computing, pp. 189–204. AAAI Press (2005)
13. Dung, P.M.: On the acceptability of arguments and its fundamental role in nonmonotonic reasoning, logic programming and n-persons games. Artificial Intelligence 77(2), 321–357 (1995)

14. Garcıa, A., Chesnevar, C.I., Rotstein, N.D., Simari, G.R.: An abstract presentation of dialectical explanations in defeasible argumentation. In: ArgNMR 2007, pp. 17–32 (2007)
15. Godfrey, P.: Minimization in cooperative response to failing database queries. International Journal of Cooperative Information Systems 06(02), 95–149 (1997)
16. Godfrey, P., Minker, J., Novik, L.: An architecture for a cooperative database system. In: Litwin, W., Risch, T. (eds.) ADB 1994. LNCS, vol. 819, pp. 3–24. Springer, Heidelberg (1994)
17. Hempel, C.G., Oppenheim, P.: Studies in the logic of explanation. Philosophy of Science 15(2), 135–175 (1948)
18. Horridge, M., Parsia, B., Sattler, U.: Explaining inconsistencies in owl ontologies. In: Godo, L., Pugliese, A. (eds.) SUM 2009. LNCS (LNAI), vol. 5785, pp. 124–137. Springer, Heidelberg (2009)
19. Horridge, M., Parsia, B., Sattler, U.: Justification oriented proofs in OWL. In: Patel-Schneider, P.F., Pan, Y., Hitzler, P., Mika, P., Zhang, L., Pan, J.Z., Horrocks, I., Glimm, B. (eds.) ISWC 2010, Part I. LNCS, vol. 6496, pp. 354–369. Springer, Heidelberg (2010)
20. Lembo, D., Lenzerini, M., Rosati, R., Ruzzi, M., Savo, D.F.: Inconsistency-tolerant semantics for description logics. In: Hitzler, P., Lukasiewicz, T. (eds.) RR 2010. LNCS, vol. 6333, pp. 103–117. Springer, Heidelberg (2010)
21. Lukasiewicz, T., Martinez, M.V., Simari, G.I., et al.: Inconsistency handling in datalog+/-ontologies. In: ECAI, pp. 558–563 (2012)
22. McGuinness, D.L., Borgida, A.T.: Explaining subsumption in description logics. In: Proceedings of the 14th International Joint Conference on Artificial Intelligence, IJCAI 1995, vol. 1, pp. 816–821. Morgan Kaufmann Publishers Inc., San Francisco (1995)
23. Meliou, A., Gatterbauer, W., Halpern, J.Y., Koch, C., Moore, K.F., Suciu, D.: Causality in databases. IEEE Data Eng. Bull. 33(3), 59–67 (2010)
24. Meliou, A., Gatterbauer, W., Moore, K.F., Suciu, D.: Why so? or why no? functional causality for explaining query answers. In: Proceedings of the Fourth International VLDB Workshop on Management of Uncertain Data (MUD 2010) in Conjunction with VLDB 2010, Singapore, September 13 (2010)
25. Modgil, S., Caminada, M.: Proof theories and algorithms for abstract argumentation frameworks. In: Argumentation in Artificial Intelligence, pp. 105–129. Springer (2009)
26. Schlobach, S.: Debugging and semantic clarification by pinpointing. In: Gómez-Pérez, A., Euzenat, J. (eds.) ESWC 2005. LNCS, vol. 3532, pp. 226–240. Springer, Heidelberg (2005)
27. Schlobach, S., Cornet, R.: Non-standard reasoning services for the debugging of description logic terminologies. In: Proceedings of the 18th International Joint Conference on Artificial Intelligence, IJCAI 2003, pp. 355–360. Morgan Kaufmann Publishers Inc., San Francisco (2003)
28. Seselja, D., Strasser, C.: Abstract argumentation and explanation applied to scientific debates. Synthese 190(12), 2195–2217 (2013)

Defining Key Semantics for the RDF Datasets: Experiments and Evaluations

Manuel Atencia[2,4], Michel Chein[1,4], Madalina Croitoru[1,4], Jérôme David[2,4],
Michel Leclère[1,4], Nathalie Pernelle[3], Fatiha Saïs[3], Francois Scharffe[1],
and Danai Symeonidou[3]

[1] Univ. Montpellier 2, LIRMM, France
{michel.chein,madalina.croitoru,michel.leclere,
francois.scharffe}@lirmm.fr
[2] Univ. Grenoble Alpes, LIG, France
{manuel.atencia,jerome.david}@inria.fr
[3] Univ. Paris Sud, LRI, France
{nathalie.pernelle,fatiha.sais}@lri.fr
[4] Inria, France

Abstract. Many techniques were recently proposed to automate the linkage of RDF datasets. Predicate selection is the step of the linkage process that consists in selecting the smallest set of relevant predicates needed to enable instance comparison. We call keys this set of predicates that is analogous to the notion of keys in relational databases. We explain formally the different assumptions behind two existing key semantics. We then evaluate experimentally the keys by studying how discovered keys could help dataset interlinking or cleaning. We discuss the experimental results and show that the two different semantics lead to comparable results on the studied datasets.

1 Introduction

Linked data facilitates the implementation of applications that reuse data distributed on the web. To ensure the interoperability between applications, data issued by different providers has to be interlinked, i.e., the same entity in different datasets must be identified. Many approaches were recently proposed to detect identity links between entities described in RDF datasets (see [5],[4] for a survey). In most of these approaches, linking rules specify conditions that must hold true for two entities to be linked. Some of the approaches are fully automatic and exploit datasets to learn expressive rules: data transformations, similarity measures, thresholds and the aggregation function [12,8,11]. These rules are specific to the vocabulary used in the data sets. Other approaches exploit the axioms that are declared in an ontology such as keys or (inverse) functional properties [19,6]. Indeed, keys figure largely in the Semantic Web (SW), especially since the inclusion in OWL 2 of HasKey construct, which allows to include in an ontology an axiom stating that a set of data or object properties is a key of a particular class.

Key axioms can be exploited for different purposes. They can be used as logical rules to deduce identity links with a high precision rate [18]. They can also guide the construction of more complex linking rules for which elementary similarity measures

N. Hernandez et al. (Eds.): ICCS 2014, LNAI 8577, pp. 65–78, 2014.
DOI: 10.1007/978-3-319-08389-6_7, © Springer International Publishing Switzerland 2014

or aggregation functions are chosen by user experts [22,19]. Finally, these keys can also be exploited to detect entity pairs that do not need to be compared, as it is done with blocking methods for relational data [9,3].

Additionally, different approaches have been recently proposed which attempt to automatically induce keys from RDF datasets, and then exploit discovered keys for datasets cleansing and interlinking [15,2]. Nevertheless, these keys may have different semantics. In [15], each pair of class instances that share at least one value for each property of the key should be considered as referring to the same entity. In [2], the authors consider the fact that each pair of instances should coincide for all the property values to be considered as referring to the same entity. This last semantics can be interesting when one can guarantee that local completeness assumption is fulfilled. This paper contributes to formalize different notions of a key in the context of RDF dataset cleansing and interlinking, and to show if these different notions can be useful for dataset cleansing and interlinking.

In section 2, we present the different notions of keys currently proposed in the literature and show how these rules can be applied to an example. Then in section 3 we exploit an unifying logic framework to compare the various key notions and how these keys can be discovered from RDF datasets. In section 4, we give the experimental results of these different key notions for both problems of cleansing and interlinking and discuss the differences between the notions of keys theoretically introduced in previous sections. Finally, we present related work in section 5 and give our conclusion and some future plans in section 6.

2 Problem Statement

Keys can be exploited for duplicate detection and data interlinking (i.e., *owl:sameAs* link discovery) of RDF datasets. In general, if a set of properties is declared to be a key for a class, then not satisfying the key within a dataset may be due to errors or unknown duplicates, whereas individuals from different datasets which do not satisfy the key can be seen as candidates for interlinking. Indeed, each pair of instances that share the same values for all the properties of a key should be considered as referring to the same entity.

Different definitions of a key can be considered in the semantic web. It is the case of the two key extraction methods [15] and [2]. For the first approach, two individuals have to share at least one value for each property of a key to be equal, while, with the second one, they have to share all the values.

In the following, we define two notions of keys: S-key and F-key. S-key roughly corresponds to the hasKey axiom of OWL2 and the notion of a key used by [15]. In [2], a tradeoff between S-key and F-key is considered.

Definition 1 (S-key). *The* S-*key* $\{p_1, \ldots, p_n\}$ *for a class expression* C *is the rule defined as follows:*

$$\forall x \forall y \forall z_1 \ldots z_n (C[x] \wedge C[y] \wedge \bigwedge_{i=1}^{n} (p_i(x, z_i) \wedge p_i(y, z_i))) \rightarrow x = y$$

where $C[.]$ *is a unary lambda formula having exactly one free variable.* $C[.]$ *is still named the target expression class of the key.*

Declaring that the set $\{p_1, \ldots, p_n\}$ *is a S-key for an atomic class* C *is denoted by* $S\text{-}key(C, (p_1, \ldots, p_n))$.

The definition of the hasKey axiom given in OWL 2 (c.f. Section 9.5 of [17] and Section 2.3.5 of [16]) enforces the considered instances to be named (*i.e.* they have to be URIs or literals, but not blank nodes). Hence an OWL 2 key can be defined with the following rule:

$$\forall x \forall y \forall z_1 \ldots z_n (C[x] \wedge C[y] \wedge Const(x) \wedge Const(y) \wedge$$
$$\bigwedge_{i=1}^{n} (p_i(x, z_i) \wedge p_i(y, z_i) \wedge Const(z_i)) \ \rightarrow \ x = y)$$

where $Const$ is a built-in unary predicate which is *true* for a constant and *false* otherwise.

Since KD2R [15] approach discovers S-keys without considering blank nodes, it follows the above OWL 2 key definition.

S-key and OWL 2 hasKey do not require that the two instances x and y of C coincide on all values of the key properties to be equal: it suffices to have at least one pair of values that coincide for all p_i to decide that x and y refer to the same entity. However, in case of not functional properties that represent a full list of items (e.g., a list of authors of a given paper, a list of actors of a given movie) it can be more meaningful to consider the fact that the instances should coincide for all the property values. We call this second semantics *Forall-key* and we give its formal definition in the following.

Definition 2 (F-key). *The F-key for a class* C *is the rule defined as follows:*

$$\forall x \forall y (C[x] \wedge C[y] \wedge \bigwedge_{i=1}^{n} (\forall z_i (p_i(y, z_i) \rightarrow p_i(x, z_i)) \wedge$$
$$(\forall w_i (p_i(x, w_i) \rightarrow p_i(y, w_i))) \ \rightarrow \ (x = y))$$

Declaring that the set $\{p_1, \ldots, p_n\}$ *as a F-key for an atomic class* C *is denoted by* $F\text{-}key(C, (p_1, \ldots, p_n))$.

Atencia et al. [2] propose an approach that automatically discovers F-keys by considering only class instances for which all the key properties are instantiated. This variant of F-key is called SF-key.

Illustrative Example. Figure 1 and Figure 2 help to compare the two notions of a key described above in a scenario of datasets cleansing. First, consider the RDF graph $G1$ shown in Figure 1. If the datatype property myLab:hasEMail is declared to be a S-key of the class myLab:Researcher then the two researchers myLab:ThomasDupond and myLab:TomDupond must be the same, since they share "thomas.dupond@mylab.org"[1]. If the datatype property myLab:hasEMail is declared to be a F-key of the class myLab:Researcher then we cannot infer that these two researchers are the same. In this

---

[1] This is solved by adding myLab:ThomasDupond owl:sameAs myLab:TomDupond to the graph.

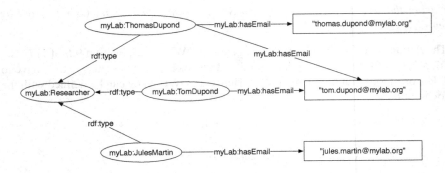

Fig. 1. A RDF graph $G1$

case declaring myLab:hasEMail as a S-key is more appropriate. Now, consider the graph $G2$ depicted in Figure 2. It seems not to be appropriate to declare that the object property myLab:isAuthor is a S-key for myLab:Researcher. Indeed, this would lead us to infer that myLab:TomDupond and myLab:JulesMartin are the same just because they have coauthored http://papersdb.org/conf/145. On the other hand, it is unlikely that different researchers are authors of exactly the same publications. If we declare that the object property myLab:isAuthor as a F-key for myLab:Researcher then we can infer only that the two researchers myLab:ThomasDupond and myLab:TomDupond are the same person. Indeed, using this F-key would not lead us to equate myLab:TomDupond and myLab:JulesMartin because the latter is not an author of http://papersdb.org/conf/26.

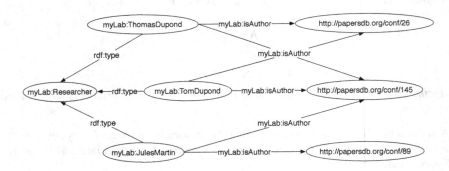

Fig. 2. A RDF graph $G2$

In this paper, we raise the problem of a comparative study of these two different semantics in a practical view (see Section 4).

3 Semantic Comparison of S-keys and F-keys

The two key discovery algorithms assessed in Section 4 are described using set-based notions and properties corresponding to the interpretations satisfying (i.e. to models of) S-keys and F-keys. In this section, we define these notions and we briefly present the relationships between S-keys and F-keys. For simplicity reasons, we consider C as a unary predicate.

3.1 S-key and F-key Models

Let C be a set, p be a binary relation over C and $c, c' \in C$, $P(c)$ denotes the set of elements in C related to c by P, i.e. $P(c) = \{u \mid (c, u) \in P\}$.

Definition 3. *A FOL interpretation I with domain Δ^I of the predicates occurring in a S-key, or a F-key, is given by:*

- *$C^I \subseteq \Delta^I$ the interpretation of the class C;*
- *$p_i^I \subseteq \Delta^I \times \Delta^I$ the interpretation of the predicates p_i;*

Proposition 1. *An interpretation I is a model of the S-key $(C, (p_1, \ldots, p_n))$ iff for any c and c' in C^I such that for any $i = 1, \ldots, k$, $p_i^I(c) \cap p_i^I(c') \neq \emptyset$, one has $(c = c')$.*

Proposition 2. *An interpretation I is a model of the F-key $(C, (p_1, \ldots, p_n))$ iff for any c and c' in C^I such that for any $i = 1, \ldots, k$, $p_i^I(c) = p_i^I(c')$, one has $(c = c')$.*

Relationships between S-keys and F-keys are given by the following proposition. The first property states that the S-key notion is more restrictive than the F-key notion when one considers interpretations in which for any p_i^I there is at most one element of C^I with no value. In relational databases it corresponds to the case where in each column there is at most one null value. The second property states the opposite when one considers interpretations in which any p_i^I is functional. In relational databases it corresponds to the case where there is no multiple values. Finally, the last property states that these notions are equivalent when the interpretation of any p_i, i.e. p_i^I, is a total function. In the relational databases it corresponds to the case where there is neither column with two null values nor multiple values.

Proposition 3. *Let S-K and F-K be respectively the S-key and the F-key associated with $(C, (p_1, \ldots, p_n))$.*

1. *For any interpretation I such as for any $i = 1, \ldots, n$ there is at most one element $c \in C^I$ with $p_i^I(c) = \emptyset$, one has: if I is a model of S-K then I is a model of F-K (i.e. if $I \models S$-K then $I \models F$-K).*
2. *For any interpretation I such as for any element $c \in C^I$ and any property p_i one has $card(p_i^I(c)) \leq 1$, then if I is a model of F-K then I is a model of S-K (i.e. if $I \models F$-K then $I \models S$-K).*
3. *For any interpretation I such as for any element $c \in C^I$ and any property p_i one has $card(p_i^I(c)) = 1$, then I is a model of S-K iff I is a model of F-K (i.e. $I \models F$-K iff $I \models S$-K).*

3.2   Key Discovery

In certain cases the keys are not asserted by domain expert and need to be obtained from the knowledge base. In this section we give the principle of discovering S-keys and F-keys on RDF datasets. Theoretically, a key K that is discovered on a dataset ensure the logical satisfiabilty of this considered RDF dataset enriched by K. To exploit this theoretical notion, it is needed that all the (not) sameAs of the dataset are expressed. We consider here the problem of key discovery on RDF datasets without considering blank nodes (i.e, with $Const$ atoms). Furthermore, we assume that the RDF datasets have been already saturated using the OWL entailment rules [14].

Definition 4 (Fact). *A fact relative to a class* C *over a vocabulary V is a conjunction of positive atoms built with V.*

Definition 5 (Extended Fact). *An extended fact is composed of a fact plus possibly negated equality atoms.*

To check the satisfiability of the fact enriched by a key we consider that all the equality and non-equality atoms are declared.

Definition 6 (Completion of a fact). *An extended fact* $\mathcal{F}$ *relative to a class* C *is complete if for any pair of terms* (t, t'), *not necessarily distinct, instances of* C, *it contains either* $t = t'$ *or* $t \neq t'$. *Let* $\mathcal{F}$ *be a fact, we denote* $\mathcal{F}^C$ *the complete fact obtained by adding* $t \neq t'$ *atoms for each pair of terms* (t, t') *such that* $t = t'$ *is not occurring in* $\mathcal{F}$.

It is straightforward to see that: *A fact* $\mathcal{F}$ *is consistent iff there is no terms* t, t' *such that atoms* $t = t'$ *and* $t \neq t'$ *both occur in* $\mathcal{F}$. Note that a complete fact is consistent.

Proposition 4. *Let* $\mathcal{F}$ *be a fact relative to a class* C *and let* S *be a* S-key $(C, (p_1, ..., p_n)$. S *is a* S-key for C *in* $\mathcal{F}$ *iff* $(\mathcal{F}^C, S)$ *is satisfiable.*

Proposition 5. *Let* $\mathcal{F}$ *be a fact relative to a class* C *and let* F *be a* F-key $(C, (p_1, ..., p_n)$. F *is a* F-key for C *in* $\mathcal{F}$ *iff* $(\mathcal{F}^C, F)$ *is satisfiable.*

In practice, knowledge bases often contain erroneous data and/or unknown duplicates. Thus discovering keys strictly on the basis of key definitions leads to miss some useful keys. That is why, [2] introduced the notion of pseudo-keys by allowing discovered keys to have some exceptions. Of course, adding these pseudo-keys to the knowledge base leads to its unsatisfiability.

A pseudo-key can be characterized by a discriminability measure which allows to assess the quality of such discovered keys. We propose here a definition of discriminability that can be used to discover S-keys and F-keys with exceptions.

Definition 7 (Discriminability). *The discriminability of a key is the ratio between the size of the maximal subset of instances on which the discovered key is a key and the number of instances of the class in* $\mathcal{F}$.[2]

---

[2] In the case of S- and SF-keys, discriminability is defined only the subset of instances having at least a value for each property of the key.

4   Experimental Results

This section experimentally studies how S-key, F-key and its SF-key variant behave on data cleansing and data interlinking scenarii. In the first experiment, we discovered keys on a benchmark about movies to evaluate the potential of each semantic of key for discovering links between datasets. In a second experiment we discovered keys with exceptions on a subset of DBPedia dataset in order to evaluate the potential of each type of key for discovering duplicates in a dataset. We have used KD2R [15] and [2] to discover the keys.

4.1   Data-Interlinking Experiments

The first experiment presents the performance of the different semantic of key for a data interlinking task. The second one studies the growth resistance of these different notions of a key.

IIMB Dataset. IIMB (ISLab Instance Matching Benchmark) is a set of interlinking tasks used by the instance matching track of the Ontology Alignment Evaluation Initiative (OAEI 2011 & 2012)[3]. An initial dataset that describes movies (films, actors, directors, etc.) has been extracted from the web (file 0). Various kinds of transformations, including value transformations, structural and logical transformations, were applied to this initial dataset to generate a set of 80 different test cases. For each test case, the reference alignments are given (i.e. sameAs links between individuals of the generated test case and the ones of the initial dataset).

Qualitative Evaluation. We evaluate the keys here using the first test case (file 1) in which the modifications only concern data property values (typographical errors, lexical variations). We have focused this experiment on the class $Film$. Each of the two files contains 1228 descriptions of pairwise distinct film instances. These instances are described using six object properties and two datatype properties. The object properties are: $featuring$ that describes all the people involved in, $starring\_in$ that describes the main actors, $directed\_by$, $estimated\_budget$, $filmed\_in$ (i.e. the languages) and $shot\_in$ (i.e. the places where the movie was filmed). The two datatype properties are $name$ and $article$ (a textual detailed description of the film). From all these properties, $featuring$, $starring\_in$, $directed\_by$, $filmed\_in$ and $shot\_in$ are properties that can be multi-valued.

We have discovered S-keys, SF-keys and F-keys on the set of film instances described in file 0. Since we know that the unique name assumption is fulfilled, we have built the extended fact by completing the fact corresponding to the file 0 with all the inequality atoms between all the pairs of class instances. The applied similarity measures are string equality after stop words elimination for data properties and equality for object properties. For the sake of simplicity, we omit the similarity measures in the following.

---

[3] http://oaei.ontologymatching.org/2011/

The three discovered S-keys are the following:
$\{(name, directed\_by, filmed\_in\}, \{article\}, \{estimated\_budget\}.$

The seven F-keys are the following:
$\{name, directed\_by, filmed\_in\}, \{shot\_in, article, starring\_in\}, \{name, article, directed\_by\}, \{article, featuring\}, \{name, featuring\}, \{name, starring\_in\}, \{article, directed\_by, starring\_in\}.$

Finally, we have discovered five SF-keys:
$\{name, directed\_by, filmed\_in\}, \{article\}, \{estimated\_budget\}, \{name, featuring\}, \{name, starring\_in\}.$

For this dataset, every S-key is also an SF-key. Because of the absence of some property values and because of the multi-valuation (see Proposition 3), the set of F-keys is not comparable to the sets of S-keys. Furthermore, some SF-keys are not F-keys. For example, the property $estimated\_budget$ is not a F-key, because there are at least two film instances for which the value of this property is not given.

To evaluate the quality of the discovered keys, they have been applied to infer mappings between film instances of the file 0 and of the file 1. More precisely, we have measured the quality of each set of keys independently from the quality of the possible links that can be found for other class instances. With this goal, we have exploited the correct links appearing in the reference alignments to compare object property values (i.e. complete set of sim_C for all C different from $Film$). In table 1, we present recall, precision and F-measure for each type of keys.

Table 1. Recall, Precision and F-measure for the class film

keys	Recall	Precision	F-Measure
S-keys	27.86%	100%	43.57%
F-keys	22.15%	100%	36.27%
SF-keys	27.86%	100%	43.57%

We notice that the results of S-keys and SF-keys are the same. Indeed, all the mappings that are found by the two additional SF-keys are included in the set of mappings obtained using the three shared keys. Furthermore, the shared keys generate the same links either because the involved properties are monovalued ($estimated\_budget$, $article$), or because the involved multi-valued object properties have the same values in both files for the same film. The recall of F-keys is slightly lower in particular because of the absence of the key $\{estimated\_budget\}$.

Some SF-keys cannot be S-keys and may have a high recall. For example, it is not sufficient to know that two films share one main actor to link them. On the other hand, when the whole sets of actors are the same, they can be linked with a good precision. Moreover, if we consider only instances having at least two values for the property $starring\_in$ (98% of the instances), this property is discovered as an SF-key and allows to find links with a precision of 96.3% and a recall of 97.7%.

Growth Resistance Evaluation. We have evaluated how the quality of each type of keys evolve when they are discovered in smaller parts of the dataset. Thus, we have randomly split the file 0 in four parts. Each part contains the complete description of a subset of the Film instances. We have then discovered the keys in a file that contains only 25% of the data, 50% of the data, and finally 75% of the data. Then we have computed the recall, precision and f-measure that are obtained for each type of keys. The larger the dataset, the more specific are the keys. Also, for all types of keys, precision increases and recall decreases with dataset' size. To obtain a good precision, S-keys need to be discovered in at least 50% of this dataset, while F-keys and SF-keys obtain a rather good precision when the keys are learnt using only 25% of the dataset. Furthermore, some F-keys and some SF-keys have also a very high recall when they are learnt on a subpart of the dataset even if the precision is not 100%. Indeed, new RDF descriptions are introduced that prevent the system from discovering keys that can be very relevant. So, it seems particularly suitable to allow to have exceptions when these two kinds of keys are discovered.

Table 2. Recall, Precision and F-measure for the S-keys, SF-keys & F-keys

S-keys	Recall	Precision	F-Measure		SF-keys	Recall	Precision	F-Measure
25%	27.85%	77.55%	40.98%		25%	100%	94.1%	96.96%
50%	27.85%	99.42%	43.51%		50%	100%	99.03%	99.51%
75%	27.85%	99.42%	43.51%		75%	27.85%	99.42%	43.51%
100%	27.85%	100%	43.56%		100%	27.85%	100%	43.56%

F-keys	Recall	Precision	F-Measure
25%	100%	94.1%	96.96%
50%	100%	99.03%	99.51%
75%	22.14%	99.27%	36.21%
100%	22.14%	100%	36.27%

4.2 Dataset Cleansing Experiment

This experiment aims at studying the differences between S-keys, SF-keys and F-keys on the application consisting in identifying duplicate pairs inside a dataset. The protocol is to learn pseudo keys using a given discriminability threshold less than 1. All pseudo keys will be viewed as key and their exception pairs as deduced similarities. Deduced similarities will be evaluated allowing us to compute precision and recall.

DBPedia Persons Dataset. DBPedia Persons[4] is a subset of DBPedia describing persons (date and place of birth etc.) extracted from the English and German Wikipedia and described using the FOAF vocabulary. This dataset contains 966,460 instances and makes use of 8 properties: foaf:name, foaf:surname, foaf:givenName, dc:description, dbpedia:birthPlace, dbpedia:deathPlace, dbpedia:birthDate, dbpedia:deathDate.

---

[4] For these experiments, we use the version 3.8 of DBPedia.

Experimental Protocol. The goal of this experiment is to evaluate and compare the similarities that can be deduced for each kind of keys: S-keys, F-keys and SF-keys. To that extent, we propose two evaluations. The first one aims at assessing the quality of the similarity deduced by discovered keys. The second evaluation consists in comparing the resistance of discovered keys to dataset alterations.

The first evaluation consists in computing, for each kind of key, the set of pseudo keys on a dataset D for a given discriminability threshold σ_d, then deducing similarities using the discovered keys and finally evaluating the set of deduced similarities. Usually such an evaluation is performed by computing precision and recall. Recall requires a reference set of similarity to be provided. The size of DBPedia prohibits us to manually build such a reference. Then we can only provide a relative recall evaluation. For each kind of key $x \in \{S-key, F-key, SF-key\}$, a set of similarities E_x is computed. All discovered similarities will be evaluated manually allowing to identify true positives TP_x and false positives FP_x. The precision of keys x is $P(x) = |TP_x|/|E_x|$ and its relative recall is $R(x) = |TP_x|/|TP_{S-key} \cup TP_{F-key} \cup TP_{SF-key}|$.

For the second evaluation, we randomly remove $x\%$ triples from the dataset and we compute keys on it. We repeated this procedure for x varying from 0 to 90 with a step of 10%. The ideal evaluation would be to measure precision and recall foreach case. However, it is not straightforward to manually check all exceptions on each case. Thus, we restrict ourselves to compare extracted keys: What is the proportion of keys extracted from the original dataset which are equals to and more general than those extracted from an altered dataset.

Qualitative Evaluation. Quality evaluation has been performed on DBPedia Persons dataset. For the key discovery, we choose a discriminability threshold equal to 99, 90% as it generates an reasonable amount of similarities to be checked. For threshold of 99%, there were several thousand generated similarities.

There is no result for F-keys because on the DBPedia dataset properties are not instantiated for each instance. In average properties are defined for 60% of 1 million of instances then it implies a lot of exceptions to the key. For that kind of dataset, F-keys are clearly not suitable.

Table 3 shows results obtained for S-keys and SF-keys. For this dataset, SF-keys get slightly better results since they get a better relative recall. Comparatively, the difference between similarities deduced by S-keys and SF-keys is marginal since they share

Table 3. Precision, Relative recall and F-measure on DBPedia Person

	Precision	Relative Recall	F-measure
E_{S-key}	63%	86%	73%
E_{SF-key}	65%	96%	78%
$E_{S-key} \cup E_{SF-key}$	59%	100%	74%
$E_{S-key} \cap E_{SF-key}$	73%	82%	77%
$E_{SF-key} - E_{S-key}$	40%	14%	21%
$E_{S-key} - E_{SF-key}$	15%	4%	6%

respectively 83% and 76% of deduced similarities. The precision of similarities shared by the two approaches is much higher than those discovered by only one approach.

In conclusion, these first results do not show any major difference between the similarities discovered by S-keys and SF-keys. Marginally, both approaches generate their own good similarities but the precision values of these exclusive parts are very low.

Since the precision of both approaches is not perfect, we propose two analysis at the key level. The goal of this analysis is to see if one approach is more suitable than the other for filtering similarities in function of the keys generating them. The first one aims at studying if false-positives are specific to some keys. To that extent, Figure 3 provides the distributions of discovered keys in function of precision and recall for both SF-keys and S-keys. This chart shows that the keys having the highest recall tend to have good precision, but it does not validate the hypothesis that false-positive are mainly deduced from particular keys. This trend is observed both for SF-keys and S-keys. The second analysis aims at showing similarities reaching a certain level of consensus among both types of keys positively influence precision and recall. Figure 5 shows that, for both approaches, precision increases significantly and quickly when the minimal consensus increase. Then it becomes stable from 5 keys on. Recall and F-Measure follow almost the inverse trend. It clearly demonstrates that consensus is not sufficient for selecting both only and all good similarities.

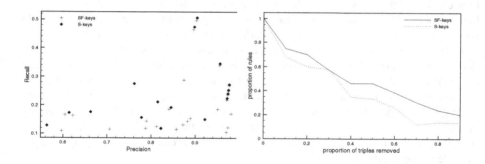

Fig. 3. Distribution of keys in function of the precision and recall

Fig. 4. Evolution of the proportion of keys equals to or generalized by initial keys in function of the quantity of triples removed

To conclude, this qualitative comparison of SF-key and S-key shows that both approaches are able to discover good similarities among a real dataset. When we try to filter only and the most of true positives the two approaches have the same behavior. The most suitable criteria is to only consider similarities which are common to both approaches.

Growth-Resistance Evaluation. Figure 4 shows the loss of generality of the initial set of keys in regard to the keys discovered when triples are randomly removed. S-keys curve decreases faster than those of ADS, but the difference between the two approches remains low.

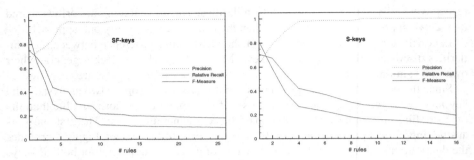

Fig. 5. Evolution of precision, relative recall and F-measure for the deduced similarities in function of the minimal number of keys from which they are deduced

5 Existing Work

The problem of key discovery from RDF datasets is similar to the key discovery problem in relational databases. In both cases, the key discovery problem is a sub-problem of Functional Dependencies (FDs) discovery. A FD states that the value of one attribute is uniquely determined by the values of some other attributes. In relational data keys or FDs can be used in tasks related to data integration, anomaly detection, query formulation, query optimization, or indexing. Discovering FDs in RDF data differs from the relational case as null values and multi-valued properties are to be considered.

In the semantic web, data linking approaches exploit S-key semantics that can be declared using OWL2 when they are restricted to named entities. Indeed, these keys can be used for different purposes. Blocking methods aim at using approximate S-keys to reduce the number of instance pairs that have to be compared by a data linking tool ([10],[20]). In [20], discriminating data type properties (i.e approximate keys) are discovered from a data set. Then, only the instance pairs that have similar literal values for such discriminating properties are selected. These properties are chosen using unsupervised learning techniques.

Other approaches use approximate S-keys to infer possible identity links. In [21], the authors discover (inverse) functional properties from data sources where the unique name assumption (UNA) hypothesis is fulfilled (i.e. non composite S-keys). The functionality degree of a property is computed to generate probable identity links. More precisely, for one instance, the local functionality degree of a property is the number of distinct values (or instances) that are the object of the property when the considered instance is the subject. The functionality degree of one property is the harmonic mean of the local functionality degrees across all the instances; the inverse functionality degree is defined analogously. In the context of Open Linked Data, [13] have proposed a supervised approach to learn (inverse) functional properties on a set of linked data (i.e. non composite S-keys). Other approaches aim to enrich the ontology and/or use these S-keys to generate identity links between pairs of instances that can be propagated to other pairs of instances ([19,1]). Such approaches, are called collective or global approaches of data linking. For example, if the approach can find that two paintings are the same, then their museums can be linked and this link will lead to generate identity links between the cities where the museums are located in. KD2R [15] aims to discover S-keys that are correct with regard to a set of data sources. The approach does not need training data and exploits data sources where the unique name assumption hypothesis is fulfilled. One important feature of KD2R is that it can discover composite keys.

In [2], the authors developed an approach based on TANE [7] algorithm to discover pseudo-keys for which a few instances may have the same values for the properties of a key. In this work, the discovered keys are particular F-keys for which only class instances for which all the key properties are instantiated are considered (i.e SF-keys).

The notion of keys employed by the above papers varies depending on the approach taken. Nevertheless, even if some of them exploit approximated or non composite keys, they are all based on the semantics of F-keys, SF-keys or S-keys.

6   Conclusion

Keys are useful in the web of data for interlinking and cleansing tasks. We have shown that different notions of a key coexist in the semantic web. This paper provides experimental evaluations of the different semantics of key for both interlinking and cleansing scenarios. Results show that learning F-keys from data is not suitable when properties are not almost fully instanciated. The SF-key variant allows to fix this problem by relaxing the equality constraint when instances have no value for a property. In term of data interlinking or data cleansing , S-keys and SF-keys have almost the same relevance in term of precision and recall. SF-keys and F keys seems to be more robust than S-keys to instance removal.

Each semantics of key has its own advantages and we think that it could be interesting to define and discover hybrid keys (i.e. keys composed of F properties and S properties) when data knowledge and/or ontology axioms can be used to decide how properties can be handled.

Acknowledgement. This work has been partially supported by the ANR projects Qualinca (12-CORD-0012) and Lindicle (12-IS02-0002).

References

1. Arasu, A., Ré, C., Suciu, D.: Large-scale deduplication with constraints using dedupalog. In: ICDE, pp. 952–963 (2009)
2. Atencia, M., David, J., Scharffe, F.: Keys and pseudo-keys detection for web datasets cleansing and interlinking. In: ten Teije, A., Völker, J., Handschuh, S., Stuckenschmidt, H., d'Acquin, M., Nikolov, A., Aussenac-Gilles, N., Hernandez, N. (eds.) EKAW 2012. LNCS (LNAI), vol. 7603, pp. 144–153. Springer, Heidelberg (2012)
3. Baxter, R., Christen, P., Churches, T.: A comparison of fast blocking methods for record linkage. In: KDD 2003 Workshops, pp. 25–27 (2003)
4. Elmagarmid, A.K., Ipeirotis, P.G., Verykios, V.S.: Duplicate record detection: A survey. IEEE Transactions on Knowledge and Data Engineering 19, 1–16 (2007)
5. Ferrara, A., Nikolov, A., Scharffe, F.: Data linking for the semantic web. Int. J. Semantic Web Inf. Syst. 7(3), 46–76 (2011)

6. Hu, W., Chen, J., Qu, Y.: A self-training approach for resolving object coreference on the semantic web. In: WWW, pp. 87–96 (2011)
7. Huhtala, Y., Kärkkäinen, J., Porkka, P., Toivonen, H.: Tane: An efficient algorithm for discovering functional and approximate dependencies. The Computer Journal 42(2), 100–111 (1999)
8. Isele, R., Bizer, C.: Learning expressive linkage rules using genetic programming. PVLDB 5(11), 1638–1649 (2012)
9. Isele, R., Jentzsch, A., Bizer, C.: Efficient multidimensional blocking for link discovery without losing recall. In: Proceedings of the 14th International Workshop on the Web and Databases (WebDB), Greece (2011)
10. Michelson, M., Knoblock, C.A.: Learning blocking schemes for record linkage. In: AAAI, pp. 440–445 (2006)
11. Ngonga Ngomo, A.-C., Lyko, K.: EAGLE: Efficient active learning of link specifications using genetic programming. In: Simperl, E., Cimiano, P., Polleres, A., Corcho, O., Presutti, V. (eds.) ESWC 2012. LNCS, vol. 7295, pp. 149–163. Springer, Heidelberg (2012)
12. Nikolov, A., d'Aquin, M., Motta, E.: Unsupervised learning of link discovery configuration. In: Simperl, E., Cimiano, P., Polleres, A., Corcho, O., Presutti, V. (eds.) ESWC 2012. LNCS, vol. 7295, pp. 119–133. Springer, Heidelberg (2012)
13. Nikolov, A., Motta, E.: Data linking: Capturing and utilising implicit schema-level relations. In: Proceedings of Linked Data on the Web Workshop at 19th International World Wide Web Conference (WWW 2010) (2010)
14. Patel-Schneider, P.F., Hayes, P., Horrocks, I.: OWL Web Ontology Language Semantics and Abstract Syntax Section 5. RDF-Compatible Model-Theoretic Semantics. Technical report, W3C (December 2004)
15. Pernelle, N., Sais, F., Symeonidou, D.: An automatic key discovery approach for data linking. Web Semantics: Science, Services and Agents on the World Wide Web (2013)
16. W. Recommendation. Owl 2 web ontology language: Direct semantics. In: Motik, B., Patel-Schneider, P.F., Cuenca Grau, B. (eds.) W3C (October 27, 2009), http://www.w3.org/TR/owl2-direct-semantics/
17. W. Recommendation. Owl 2 web ontology language: Structural specification and functional-style syntax. In: Motik, B., Patel-Schneider, P.F., Parsia, B. (eds.) W3C (October 27, 2009), http://www.w3.org/TR/owl2-syntax/
18. Saïs, F., Pernelle, N., Rousset, M.-C.: L2r: A logical method for reference reconciliation. In: Proceedings of the Twenty-Second AAAI Conference on Artificial Intelligence, Vancouver, British Columbia, Canada, pp. 329–334 (2007)
19. Saïs, F., Pernelle, N., Rousset, M.-C.: Combining a logical and a numerical method for data reconciliation. In: Spaccapietra, S. (ed.) Journal on Data Semantics XII. LNCS, vol. 5480, pp. 66–94. Springer, Heidelberg (2009)
20. Song, D., Heflin, J.: Automatically generating data linkages using a domain-independent candidate selection approach. In: Aroyo, L., Welty, C., Alani, H., Taylor, J., Bernstein, A., Kagal, L., Noy, N., Blomqvist, E. (eds.) ISWC 2011, Part I. LNCS, vol. 7031, pp. 649–664. Springer, Heidelberg (2011)
21. Suchanek, F.M., Abiteboul, S., Senellart, P.: Paris: Probabilistic alignment of relations, instances, and schema. The Proceedings of the VLDB Endowment (PVLDB) 5(3), 157–168 (2011)
22. Volz, J., Bizer, C., Gaedke, M., Kobilarov, G.: Discovering and maintaining links on the web of data. In: Bernstein, A., Karger, D.R., Heath, T., Feigenbaum, L., Maynard, D., Motta, E., Thirunarayan, K. (eds.) ISWC 2009. LNCS, vol. 5823, pp. 650–665. Springer, Heidelberg (2009)

A Framework for Qualitative Representation and Reasoning about Spatiotemporal Patterns

Foued Barouni and Bernard Moulin

Department of Computer Science and Software Engineering
Laval University
Quebec City (Quebec) G1V 0A6, Canada
Foued.barouni.1@ulaval.ca, Bernard.moulin@ift.ulaval.ca

Abstract. A new generation of data acquisition systems has emerged in recent years through the development of communication technologies. These systems make use of geographically distributed devices that generate large amounts of data. Several models of spatiotemporal patterns have been proposed to help users take advantage of such data. However, most of current approaches rely on query languages which are not easily manipulated by end-users. Because of their limited expressiveness, such approaches do not allow for reasoning about spatiotemporal phenomena that are so common in the real world. In this context, we propose a new definition of a spatiotemporal pattern. We use conceptual graphs to allow for a qualitative representation of patterns and we integrate contextual information in the definition of patterns to improve the reasoning ability of software agents. We propose an approach to help users manage spatiotemporal situations and illustrate it with a case study in the domain of power utilities.

1 Introduction

The emergence of internet applications and efficient communication systems led to the deployment of large scale software solutions for data acquisition and environment monitoring using geographically distributed devices and computers. Moreover, the notion of computer has changed over the past decade. There is a clear shift from the big tower desktop computer to much smaller devices having the equivalent hardware capabilities (CPU, memory etc.) to any typical computer. These devices run continuously and report their observations to centralized systems in various formats, generating very large databases and leaving to end-users the task to assess the observations and draw conclusions on their own. Actually, users are interested in identifying typical configurations occurring in their environments. We call an agent's 'environment' all the elements (people, things, devices, software, places, events, etc.) that constitute the surroundings of that agent and provide conditions for its activities, development and growth (adapted from the definition of environment available in Simple English [20]). Users reason about these configurations in order to make decisions. Such typical configurations are called dynamic spatiotemporal situations of interest which are referred to as patterns [2]. Empirical research works indicate that human psychology

N. Hernandez et al. (Eds.): ICCS 2014, LNAI 8577, pp. 79–92, 2014.
DOI: 10.1007/978-3-319-08389-6_8, © Springer International Publishing Switzerland 2014

is structured to find meaning in patterns [1]. Therefore, several models of spatiotemporal patterns have been proposed to help users take advantage of large amounts of data. However, most of current monitoring, event processing and pattern detection approaches rely on query languages which are not easily manipulated by end-users. Because of their limited expressiveness, such approaches do not allow for reasoning about spatiotemporal phenomena that are so common in the real world. There is a clear gap between what users want to see (situations of interest) and what current approaches and tools are proposing in terms of pattern detection and display (see Section 3 for an overview of approaches proposed in the literature). There is also a trend to associate software agents with monitoring systems to help users detect and analyze patterns. Hence, software agents also need to reason about situations of interests in a way that complies with how users perceive and interpret phenomena. Consequently, there is a need for a knowledge representation and reasoning formalism to represent situations of interest in a computer tractable form and to support human experts' reasoning. Russell and Norvig [14] described how knowledge representation formalisms translate human's perception of the world (facts) to sentences that can be stored within software agents (Fig. 1). In the spatiotemporal research area, a situation of interest (i.e. a fact) can be represented by a spatiotemporal pattern (i.e. a sentence). Indeed, spatiotemporal patterns relevant to a specific application domain depend on how human experts perceive the domain environment. Hence, it is useful to make use of a cognitive model of human perception/interpretation of the environment such as Freksa's model of human mental representation of the spatial environment [15]. Taking advantage of its cognitive orientation we propose to adapt Freksa's approach to the pattern modeling domain. We are looking for an enhanced pattern model that allows for using qualitative techniques and contextual information which are key elements in human cognition [25], and which are used by domain experts when detecting and interpreting patterns. Moreover, such an enhanced pattern model should also deal with dynamic spatiotemporal phenomena which can be detected using the new generation of acquisition systems. To the best of our knowledge, current approaches and tools used to process spatiotemporal patterns do not meet these requirements for an enhanced pattern model.

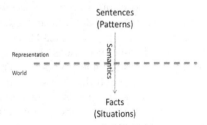

Fig. 1. Representation of situations of interest using patterns (adapted from [14])

To address the above issues, we propose a novel approach and an enhanced pattern model to represent and reason about spatiotemporal situations of interest. Patterns are defined from a cognitive perspective inspired by Freksa's model. They are enhanced with contextual information that (software and human) agents may use during their reasoning process in a dynamic environment. We use the conceptual graph formalism [29] to support the qualitative representation and to enhance the software agents' reasoning capabilities. Fig. 2 depicts an overview of the proposed solution where spatiotemporal patterns are defined using events and states which are generated by sensors and qualified with respect to spatial objects. Software agents use contextual information and pattern instances to make decision. The rest of the paper is organized as follows: Section 2 gives an overview of a case study that is used to illustrate our approach. Section 3 introduces the formal definition of dynamic spatiotemporal phenomena. Section 4 presents the definition of a spatiotemporal pattern. In Section 5, we extend the pattern model using contextual information. Section 6 discusses some related works on spatiotemporal pattern modeling and concludes the paper with some research perspectives.

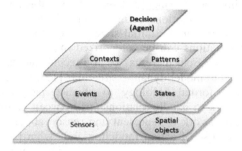

Fig. 2. Overview of the proposed solution

2   Case Study

To illustrate certain aspects of our formalism, we use a case study in the domain of power utilities. An increasing pressure from customers (residential, commercial and industrial) and media is put on electric distribution companies to supply reliable information about outages in their distribution networks. An Outage Management System (OMS) is a computer system used by operators of electric distribution systems to assist in the restoration of power [20]. Using an OMS mainly aims at reducing the outage time in the distribution network for power utilities and at keeping customers updated about the real time status of outage in the network. To this end, an OMS supports a variety of capabilities. It needs to predict the location of electric devices such as switches and breakers that opened during a failure. It also needs to identify the faulty electrical devices such as transformers and cables that caused the outage. Furthermore, it can help prioritizing restoration efforts using several criteria such as the location of emergency facilities, the crew availability, the crew expertise and the outage size. An OMS is usually designed to integrate information from several software applications such as geographic information systems (GIS), sensors and customer relation management systems. These systems are used by the OMS to achieve two main goals: keeping customers informed about the status of an outage; and dispatching the crews taking into account several criteria. Most of the existing commercial OMS do not have the capability to automatically identify spatiotemporal patterns using multiple data sources. The pattern model that we propose in the next section can be used by software agents to automatically detect typical situations of interest and patterns, and to present them to the OMS operators.

3   From Situations to Dynamic Spatiotemporal Patterns

A situation is a finite configuration of objects that are characterized by properties and that are related to each other [29]. When a situation evolves in time and is defined with regards to spatial attributes and entities characterizing the environment, it is called a spatial-temporal situation (STS for short). Desclés [18] proposed to distinguish two types of situations: static and dynamic situations. Static situations remain stable during a given time interval and no change on the subject described by the situation is observed. Dynamic situations define changes in time and space references. These changes will be acting on the initial state of the situation and they result in a final state. Dynamic situations can be categorized into events or processes. In the qualitative spatial reasoning field there has been a recent interest in developing concurrent-based approaches to model dynamic situations [12]. For example, Grenon and Smith [17] proposed a formal ontology to characterize geospatial events and processes and a framework to handle relationships between geospatial entities. Haddad [19] proposed a framework to represent and reason about change and dynamic STS in the context of course of action modeling and "what if" analysis. Usually, a change should attract a user's attention when monitoring a process/situation. Therefore, we claim that a pattern model should integrate a definition of dynamic spatiotemporal situation.

Patterns are often thought of as generic recurring solutions to specific problems [16]. Let us consider Erwig's definition of a pattern [2]: *"Regular structures in space and time, in particular, repeating structures, are often called patterns. Patterns that describe changes in space and time are referred to as spatiotemporal patterns"*. We extend this definition to consider dynamic spatiotemporal phenomena and situations of interest by adding the statement: "Patterns that describe one or a set of dynamic spatiotemporal situations are referred to as dynamic spatiotemporal patterns." Moreover, we aim at defining spatiotemporal patterns to represent typical situations of interest that are meaningful to human beings during their problem solving activities. As a starting point, we choose Haddad's STS model for several reasons. First, a spatiotemporal situation is defined in this model using both states and events whereas most current pattern detection approaches only take into account events. In this way we are able to model change, which is an important characteristic of dynamic environments. Second, Haddad's model uses the Conceptual Graph formalism, which makes it expressive and human readable and meets one of our requirements for a new pattern model. Here we briefly introduce the main elements of the proposed model.

A *state* is a spatiotemporal situation describing stability and things that do not change [18]. An *event* is a spatiotemporal situation describing something that happens in reality [9]. An event occurs in a static background and may or may not change a state. To represent a change, Haddad [19] proposed a generic formalism where an event is specified as a spatiotemporal situation in which temporal relations are used to link two states [13]. Fig. 3 presents a graphical representation of an event linking two states. The state referred by *#cablestate02* describes the state of a power cable referred by *#QuebecMontreal* during the period July 17th to 24th 2013. The cable status is normal. An event type *Break_Event* occurred after this state at distance 125 Km and resulted in a new state where the cable status changed to *broken*.

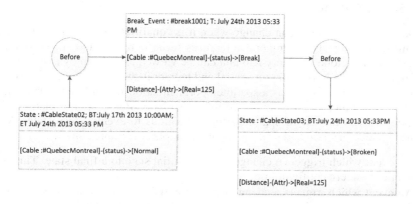

Fig. 3. An event and state representations example

4   Spatiotemporal Pattern

In this section we introduce our spatiotemporal pattern model. Using temporal and spatial relations to link spatiotemporal situations can result to different pattern types. Predefined pattern types have been already proposed as for example the OGC initiative [10] which defines four types of patterns: simple patterns, repetitive patterns, complex patterns and timer patterns. We will only consider complex and simple patterns in the following subsections. The other pattern types can be derived thanks to the flexibility of our formalism and to the expressiveness of the associated representation language.

4.1   Spatial Relations and Spatial Object

Spatial relations are used to link situations and generate different types of patterns. Generally, spatial relations are categorized into three families: directional/orientation relation, distance relation and topological relation. According to [24] spatial relations are usually binary relations. Considering topological relations, one of the most used formalisms is the Region Connection Calculus RCC-8. In our model, we consider RCC-5 which is a simplified version of RCC-8 and provides the 5 following topological relations: *disjoint, overlaps, inside, contains* and *equals*. Proximity relations are used to qualify distances. This qualification takes into consideration, besides Euclidian distance, other parameters such as user preference, contextual information and other spatial and semantic proprieties of the domain. Examples of proximity relations are *near* and *far from*. Directional/Orientation relations describe the direction of one spatial object with respect to another one [24]. Examples of directional relations are *right of* and *in front of*. A spatial object can represent a natural geographic entity (mountain, hill) or a constructed geographic entity (building, road, etc.) [19]. More details about spatial object classification can be found in [17].

4.2   Simple Pattern

The meaning of a situation can change when it is "qualified" with respect to the surrounding spatial objects or temporal references using temporal and spatial relations. For example, a cyclone for a city habitant can be viewed as a storm by a village habitant if both habitants have different spatial and temporal characteristics (i.e. attributes) [23]. Hence, a simple pattern is a spatiotemporal situation which represents a change in space and/or time: it can be qualified using spatial and/or temporal relations. This definition restricts the use of spatiotemporal situations to events, since states do not represent change in space and time. Here are examples of simple patterns:

- An event which triggers a change from an initial state to a final state. This is the simplest pattern configuration.
- An event which triggers a change from an initial state to a final state and where the event's temporal attribute is qualified with regards to a temporal reference.
- An event which triggers a change from an initial state to a final state and where the event's spatial attribute is qualified with regards to a spatial object from the environment.
- An event which triggers a change from an initial state to a final state and where the event's spatial attribute is qualified with regards to a spatial object and the event's temporal attribute is qualified with regards to a temporal reference.

From a representation perspective, a simple pattern is a knowledge structure which has the following attributes:

- A pair (*simple pattern type, simple pattern referent*) where simple pattern referent is used to identify a simple pattern instance.
- A Simple Pattern Propositional Content (*SPPC*) which is a knowledge structure specifying the simple pattern and describing the qualified spatiotemporal situation.

4.3   Complex Pattern

A software agent can be programmed to monitor/observe dynamic phenomena. During its observation activity the agent may detect interesting configurations that usually involve several STSs. A single STS can be interpreted and qualified to form a simple pattern. However, this interpretation can change when one STS is related to one or even several other STSs. When several STSs are related using spatial or temporal relations, they form what we call a *Complex Pattern*. Hence, a complex pattern is a set of simple patterns related by spatial and/or temporal relations. Two possible configurations can be present in complex patterns:

- Case 1: When a spatial relation links two simple patterns, the relation is established between the events' spatial attribute of each simple pattern.
- Case 2: When a temporal relation links two simple patterns, the relation is established between the events temporal attributes of each simple pattern.

From a representation perspective, a complex pattern is a knowledge structure which has the following attributes:

- A pair (complex pattern type, complex pattern referent) where complex pattern referent used to identify a complex pattern instance.
- A Complex Pattern Propositional Content (CPPC) which is a knowledge structure describing the complex pattern. The content of this knowledge structure is basically a set of simple patterns related by temporal and/or spatial relations. It is worth to note that both relation types are binary.

4.4   Contextual Information

In this section, we discuss how contextual information can be added to spatiotemporal patterns to enhance an agent's reasoning capabilities. After giving an overview of the notion of context and its usage, we propose an extension of our pattern formalism using contextual information.

According to Bazire and Brezillon [26], it is difficult to find a general definition of context that can be applied to any discipline. Brezillon, a well-known researcher in the context domain, suggested that *"instead of modeling context, it is better to design a knowledge model that takes into account the context and helps in making decision"*[28]. Freksa reached the same conclusion and proposed an approach for context modeling from a cognitive perspective, rather than providing an exact definition of it [15]. Considering context aware systems in the ubiquitous computing domain, researchers relate contextual information to spatial information. Context helps software agents when tracking objects' positions and adapting their goals, taking into account spatial location [27]. In the Complex Event Processing (CEP) domain, context is defined with respect to events: a context "specifies the relevance of the events participating in pattern detection" [22]. Three types of contexts are considered: temporal context, space-based context, semantic-based context. In CEP systems, contexts are used to partition the event cloud in multiple streams and are used for filtering purposes [21]. For this reason, they do not play the role of complementary information.

In our model, we adopt Brezillon's and Freksa's approaches which consider the context from a cognitive perspective. Contexts are associated with the agent's knowledge structures and provide additional information that helps agents when reasoning about spatiotemporal patterns. We use Freksa's approach to define the relationship between contexts and spatiotemporal patterns. Hence, contextual information may change the meaning of a spatiotemporal pattern. In our formalism, we include contextual information in the knowledge structure that specifies a spatiotemporal pattern.

Formally, a context is a conceptual graph with the following attributes:

- A pair *(context type, context referent)* where *context type* is a concept type whose values depend on the application domain. The most used context types in the ubiquitous computing and event processing [3] are *semantic context, temporal context* and *spatial context*. Semantic context is a non-temporal and non-spatial information that can be added to characterize the related pattern. "National Holiday" or "Vacations" are examples of a semantic context that can be used to characterize patterns. Temporal context is a temporal information structure that can be a time interval or a single time reference. Spatial context is a spatial information structure. *Context referent* is the identifier of an instance of the context type.

• Context Propositional Content (*CPC*) which is a knowledge structure describing the contextual information according to the context type.

5   Case Study: Outage Management Systems

In this section, we apply our approach to the domain of Electric Distribution and in particular to Outage Management Systems (OMS) introduced in Section 2. We define qualitative spatiotemporal patterns to help agents make decisions about crew assignment and the determination of outage status in a distribution network. We only provide some pattern examples and some simplifications are proposed. For example, we assume that pattern instances are already populated in a given knowledge base and are accessible by agents. Issues related to data collection from the event cloud and pattern matching aspects are not addressed here.

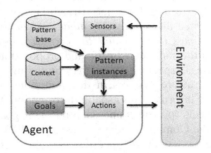

Fig. 4. Agent's reasoning model adapted from [14]

The main function of an OMS is to help managing crews and to update customers with the real time outage status in their respective regions. To achieve these goals, a software agent should correlate information collected from several data sources and recognize relationships between them in order to detect interesting situations. Data sources which provide information about the environment may be static (such as sensor states and geographic data) or dynamic. Dynamic data sources (mostly sensors) provide information about outage events. Each software agent has its own contextual information and may correlate it with data provided by other sources in order to detect interesting situations from its point of view. Based on the result of this interpretation, an agent can suggest solutions for crew assignment and/or for the characterization of the outage status. A goal-based agent is used in the proposed approach. Fig. 4 illustrates our agent's reasoning model adapted from [14].

The system architecture is described in Fig. 5. Outage sensors are deployed throughout the distribution network. They are mounted on power cables and they report an outage event by sending a message, using an industrial communication protocol such as DNP3 (Distributed Network Protocol) or Modbus, to a centralized SCADA using a wireless cell communication (GSM or CDMA). The retrieved messages are located using a GIS which also models the distribution network. The Amine Knowledge base [7] (KB) is used to store the pattern specifications and the detected pattern instances. The Amine Prolog +CG language is used by the agent to find relevant patterns and to update the environment using actions such as updating the crew assignment status and updating an outage state.

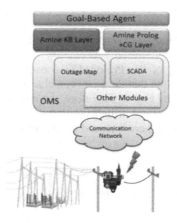

Fig. 5. System architecture for outage management system and the interaction with goal-based agent

5.1   Pattern Specifications

We propose two types of patterns: simple patterns and complex patterns. Simple patterns are used to represent situations where only one event occurs. Complex patterns are used for more complex situations where there are temporal and spatial relations between situations.

The simple pattern which is expressed using Amine's linear CG notation and illustrated in Fig. 6 represents the following situation:

"An event type OutageEvent occurs near an area (here defined as Outage Area), the resulting state of the cable is Broken and the crew status is free".

Thanks to temporal relations between events and states (Fig. 6), it becomes possible to represent the change that occurred on a specific cable after the occurrence of the outage event. Such a feature is very important in this kind of systems to enhance the users' decision making capabilities. In some cases, an outage can be reported but it may not necessarily be related to a cable break. Hence, a user may need to investigate other elements in the network. When an agent detects an instance of such a pattern, it performs two actions: First, it assigns a crew to the corresponding area to fix the problem. Second, it updates the outage map using the outage information. Both actions can be achieved using Prolog+CG programs:

1. Assign_Free_Crew:-SimplePattern_With_ContextInfo
   (_Simple_Pattern,Environment) which assigns the crew to the outage event,
2. UpdateOutageMap(_Simple_Pattern,Environment) which updates the outage map
   with the information on the related event information.

[Normal_Outage_Pattern]-
-Pattern_Description->[Proposition : [[OutageEvent]-
                                -Before->[[CableState]-status->[Broken]]]
                    -near->[Outage_Area]]
-Context_Description->[Proposition : [ElectricCrew]-attr->[CrewStatus="Free"]]

Fig. 6. An example of simple pattern with linear notation

Fig. 8 shows an example of the Prolog+CG console output when the action As-signFreeCrew is performed. Note that the agent can change the crew status and retrieve it using CG rules as illustrated by Fig. 7. Therefore, the next time an outage event occurs, the agent's contextual information will indicate that the crew belonging to this specific area is already assigned and that there is a need to find another alternative.

If Normal_Break Pattern and [Crew ="Team2"]-attr->[CrewStatus = "Free"]
Then
[Crew ="Team2"]-attr->[CrewStatus = "Assigned"]

Fig. 7. An example of rule used for crew assignment

In some cases a user may be interested in situations which involve more than one event. For example, the complex pattern illustrated in Fig. 9 represents the following situation: *"An event type OutageEvent occurs near an area (here defined as _Area). Before that, a communication error event occurred on a sensor in the same area"*.

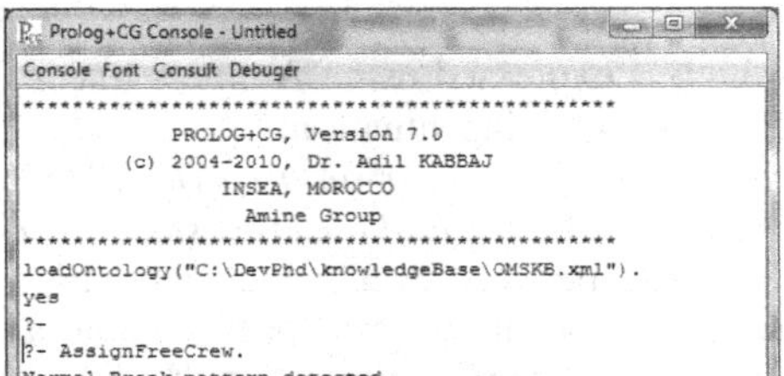

Fig. 8. Output of the action AssignFreeCrew in Prolog+CG console

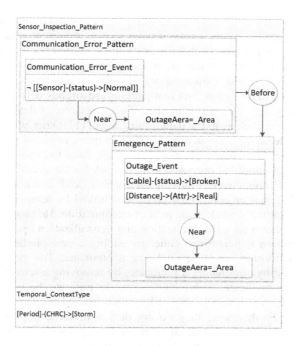

Fig. 9. An example of complex pattern representation

This is a typical case for monitoring systems. A sensor can be out of service and the system detects an outage event in the same area, but reported by another sensor. Such a situation can occur during a storm where sensors are still working but communication links are failed. Hence, the storm can be used as contextual information in the pattern definition. When such a pattern is detected, the crew should repair the outage and inspect the sensor and fix the issue. Note the presence of the negation operator in the structure of the communication error event which means that the non-occurrence of an event can be considered as an event by a user.

6   Discussions and Conclusion

Modeling and reasoning about situations of interest is a multidisciplinary task since it requires various expertises. In addition to the spatial and temporal aspects, the situations should reflect the reality of the environment in which agents evolve and interact with, as well as the characteristics of the application domain. Spatiotemporal patterns have been used to model situations of interest. They have been studied in the context of three main approaches: the data mining approach, the logical approach and complex event processing. The data mining approach aims at discovering patterns in databases containing large amounts of data. Once found, these patterns (or structures) are stored in specific tables called "Field View" existing in spatiotemporal databases [2]. The main shortcoming of this approach is that most of the obtained patterns are represented in a query-based form, which is not suitable to support the qualitative aspects of human reasoning. Moreover, using Conceptual Graphs and qualitative techniques, our approach offers an expressive tool to represent patterns in human readable form. The logical approach focuses on the spatiotemporal pattern definition and proposes models for pattern detection and querying. Several works in this area proposed languages to build spatiotemporal queries, assuming that spatiotemporal data is already stored in databases; and that patterns are already defined by users. The proposed query languages do not seem to be efficient because of their lack of generality and expressiveness such as in the data mining area [11]. Other works concentrated on reasoning and detecting spatiotemporal patterns using artificial intelligence techniques. For example, first order logic has been widely used to express spatiotemporal patterns [5] [6] [4]. Lattner [8] proposed a framework for the prediction of situations of interest related to mobile robots moving on a soccer field. The patterns are defined by a sequence of spatial or conceptual predicates linked by temporal relations. The original aspect of Lattner's work is the pattern manipulation. He proposes a hierarchy of patterns and defines a set of specialization and generalization operations. The specialization of a pattern is performed either by adding a new predicate, by adding a new temporal relationship, or by specializing a predicate. The generalization of a pattern is performed by eliminating a predicate, by removing a temporal relationship or by generalizing a predicate. Although these works mitigate the lack of expressiveness of classical query languages, they still remain limited because they are usually application-driven. Furthermore, they do not deal with some important aspects for experts such as the use of contextual information. Our approach attempts to overcome these limitations by integrating contextual information in the agent's cognitive model to be used when reasoning about patterns. The Complex event processing community (CEP) developed during the past 10 years new models and applications to solve database performance problems [9]. Originally, CEP has been used in the financial industry to predict market developments and exchange rate trends. CEP patterns emerge from relationships between event attributes such as time, cause (causal relation between events) and aggregation (significance of an event's activity in relation to other events). Although works on patterns with CEP addressed the issue of processing spatiotemporal events and pattern matching, they are limited in terms of expressiveness and spatial representation. Most of the patterns are SQL-based and a user needs to be knowledgeable about complex query languages [9]. Furthermore, current CEP formalisms support only events in their pattern definitions. Hence, dynamic phenomena are not easy to represent. Thanks to our extension of Haddad's formalism, our pattern model allows for using and reasoning about dynamic spatiotemporal phenomena and temporal relations can be specified between events and states. All these aspects enable our model to reduce the semantic gap between human's perception of situations of interest and spatiotemporal definitions. From a computational perspective, our model allows for enhancing agents with reasoning capabilities and makes it applicable in a various number of industries thanks to its generality. The current step of this research is to implement the proposed formalism in a complex scenario provided by the industry. In this scenario a Complex Event Processing engine is used to manage the stream of events from different data sources and to detect spatiotemporal patterns instances.

References

1. Political Psychology: Special issue on Neuroscientific contributions to political psychology. Political Psychology 24(4) (2003)
2. Erwig, M.: Toward Spatiotemporal patterns. In: Spatiotemporal Databases, pp. 1–26. Springer (2003)
3. Etzion, O., Zolotorvesky, N.: Spatial Perspectives in Event Processing. In: Sachs, K., Petrov, I., Guerrero, P. (eds.) Buchmann Festschrift. LNCS, vol. 6462, pp. 85–107. Springer, Heidelberg (2010)
4. Gehrke, J., Lattner, A., Herzog, O.: Qualitative Mapping of Sensory Data for Intelligent Vehicles. In: Visser, U., Lakemeyer, G., Vachtesevanos, G., Veloso, M. (eds.) Workshop on Agents in Real-Time and Dynamic Environments at the 19th International Joint Conference on Artificial Intelligence (IJCAI 2005), Edinburgh, UK, pp. 51–60 (2005)
5. Holzmann, C.: Rule based reasoning about qualitative spatiotemporal relations. In: Proceedings of the 5th International Workshop on Middleware for Pervasive and Ad-hoc Computing: Held at the ACM/IFIP/USENIX 8th International Middleware Conference (MPAC 2007), pp. 49–54. ACM, New York (2007)
6. Holzmann, C., Ferscha, A.: A framework for utilizing qualitative spatial relations between networked embedded systems. Journal Pervasive and Mobile Computing 6(3), 362–381 (2010)
7. Kabbaj, A.: Development of intelligent systems and multi-agents systems with Amine platform. In: Schärfe, H., Hitzler, P., Øhrstrøm, P. (eds.) ICCS 2006. LNCS (LNAI), vol. 4068, pp. 286–299. Springer, Heidelberg (2006)
8. Lattner, A.D., Miene, A., Visser, U., Herzog, O.: Sequential Pattern Mining for Situation and Behavior Prediction in simulated Robotic Soccer. In: Bredenfeld, A., Jacoff, A., Noda, I., Takahashi, Y. (eds.) RoboCup 2005. LNCS (LNAI), vol. 4020, pp. 118–129. Springer, Heidelberg (2006)
9. Luckham, D.C.: The Power of Events: An Introduction to Complex Event Processing in Distributed Enterprise Systems. Addison-Wesley (2002)
10. Open Geospatial Consortium, http://www.opengeospatial.org
11. Gorawski, M., Jureczek, P.: A proposal of spatio-temporal pattern queries. In: Proceedings of the International Conference on Complex, Intelligent and Software Intensive Systems (CISIS 2010), pp. 587–593. IEEE Computer Society, Washington, DC (2010)
12. Cole, S., Hornsby, K.: Modeling Noteworthy events in a geospatial domain. In: Rodríguez, M.A., Cruz, I., Levashkin, S., Egenhofer, M. J. (eds.) GeoS 2005. LNCS, vol. 3799, pp. 77–89. Springer, Heidelberg (2005)
13. Moulin, B.: Temporal contexts for discourse representation: An extension of the conceptual graph approach. Applied Intelligence 7, 225–227 (1997)
14. Russell, S., Norvig, P.: Artificial Intelligence: A Modern Approach. Prentice Hall (1995) ISBN 0-13-604259-7
15. Freksa, C., Klippel, A., Winter, S.: A cognitive perspective on spatial context. In: Cohn, A.G., Freksa, C., Nebel, B. (eds.) Spatial Cognition: Specialization and Integration, vol. 05491 (2007) ISSN 1862-4405
16. Lind, J.: Patterns in agent-oriented software engineering. In: Giunchiglia, F., Odell, J., Weiss, G. (eds.) AOSE 2002. LNCS, vol. 2585, pp. 47–58. Springer, Heidelberg (2003)
17. Grenon, P., Smith, B.: SNAP and SPAN: Towards Dynamic Spatial Ontology. Spatial Cognition and Computation 4(1), 69–103 (2004)

18. Desclés, J.-P.: Représentation des connaissances: Archétypes cognitifs, schèmes conceptuels et schémas grammaticaux. In: Actes Sémiotiques - Documents VII, 69/70. EHESS, Paris (1985)
19. Haddad, H., Moulin, B.: A Framework to Support Qualitative Reasoning about COAs in a Dynamic Spatial Environment. Journal of Experimental & Theoretical Artificial Intelligence, 1–40 (First published on: July 27, 2010 (iFirst)), vol. 22(4), 341–380, Taylor & Francis (2010)
20. Wikipedia, http://www.wikipedia.org
21. Adi, A., Biger, A., Botzer, D., Etzion, O., Sommer, Z.: Context Awareness in Amit. Active Middleware Services 2003, 160–167 (2003)
22. Sharon, G., Etzion, O.: Event Processing Networks – model and implementation. IBM System Journal 47(2), 321–334 (2008)
23. Galton, A.: Fields and Objects in Space, Time, and Space-time. Spatial Cognition and Computation 4, 39–68 (2004)
24. Cohn, A., Renz, J.: Qualitative Spatial Reasoning. In: Van Harmelen, F., Lifschitz, F.V., Porter, B. (eds.) Handbook of Knowledge Representation. Elsevier (2007)
25. Toussaint, G.T.: The use of context in pattern recognition. Pattern Recognition 10, 189–204 (1978)
26. Bazire, M., Brézillon, P.: Understanding Context Before Using It. In: Dey, A., Kokinov, B., Leake, D., Turner, R. (eds.) CONTEXT 2005. LNCS (LNAI), vol. 3554, pp. 29–40. Springer, Heidelberg (2005)
27. Chen, G., Kotz, D.: A survey of context-aware mobile computing research. Technical report. Dartmouth College Hanover, NH, USA (2000)
28. Brézillon, P.: Context in Artificial Intelligence: II, Key elements of context. Computing and Informatics 18(5) (1999)
29. Sowa, J.F.: Conceptual Structures: Information Processing in Mind and Machine. Addison-Wesley, Massachusetts (1984)

Combining Restricted Boltzmann Machine and One Side Perceptron for Malware Detection

Răzvan Benchea[1,2] and Dragoş Teodor Gavriluţ[1,2]

[1] "Alexandru Ioan Cuza" University, Faculty of Computer Science, Iaşi, România
[2] Bitdefender Laboratories, Iaşi, România

Abstract. Due to the large increase of malware samples in the last 10 years, the demand of the antimalware industry for an automated classifier has increased. However, this classifier has to satisfy two restrictions in order to be used in real life situations: high detection rate and very low number of false positives. By modifying the perceptron algorithm and combining existing features, we were able to provide a good solution to the problem, called the one side perceptron. Since the power of the perceptron lies in its features, we will focus our study on improving the feature creation algorithm. This paper presents different methods, including simple mathematical operations and the usage of a restricted Boltzmann machine, for creating features designed for an increased detection rate of the one side perceptron. The analysis is carried out using a large dataset of approximately 3 million files.

1 Introduction

During the last years 10 years, malware landscape has changed dramatically. With more than 200.000 [1] samples appearing daily it is impossible for anyone to analyze them by hand. That is why an automated malware classification algorithm is needed. However, in order to be practical, this algorithm needs to satisfy the market demands: it needs to perform fast during testing and it also needs to have an extremely low false positive rate. The later restriction is based on the fact that a clean file (for example an operating system file) that is classified as infected and later deleted can make the system impracticable.

Based on these restrictions we have previously developed an algorithm, called One Side Perceptron (OSP) [2], that can be trained with zero false positive. In order to increase its speed we have only used a small number of features, selected with F2 [3] score from all available. Even though this method achieves a good detection rate we believe it can be improved by modifying the way it uses the features. Since the algorithm is based on linear classifier it doesn't try to find any relations between the features.

In this paper we will use a neural network, more exactly a restricted Boltzmann machine [4], to create a new set of features from existing ones. We will use these to train an OSP algorithm and will compare the results with those obtained from training the algorithm with the best selected features according to F2 score. Since the restricted Boltzmann machine is a probabilistic network,

N. Hernandez et al. (Eds.): ICCS 2014, LNAI 8577, pp. 93–103, 2014.
DOI: 10.1007/978-3-319-08389-6_9, © Springer International Publishing Switzerland 2014

the features can be outputted as a probability value or as a sample generated based on this probability. During testing we have observed that using a sample instead of a probability does not yield good results, so in this paper we will refer to the second method.

This paper is organized in 3 sections. Section 2 describes other methods used to solve the problem as well as their advantages and disadvantages. Section 3 starts with the theory behind the algorithm and continues with describing our method of training, testing and results. Finally, in section 4 we present our conclusions and future work.

2   Related Work

A good categorization of malware detection techniques was carried out by the authors in [5]. Apart from setting two main categories, anomaly-based and signature-based, the authors also split the detection algorithms by the method they use to collect information: static, dynamic and hybrid. The anomaly-based technique has the advantage of detecting types of malware that were not present in the training set, while the static method of extracting features can be used in very large data sets. The authors in [6] were the first to use Machine Learning techniques for the detection of different malwares based on their binary codes. Using this method they were able to create a proactive malware detector that was twice as good as signature based one in detecting new samples.

Most of the research regarding malware detection was carried out using features obtained from extracting n-grams patterns from binary files or from executed API calls. Authors in [7] have used common n-gram in order to create a profile for each class of malware then assigned a new sample to the class using k-nearest neighbor. Using n-grams authors in [8] have tested a variety of inductive methods, including naive Bayes, decision trees, support vector machines(SVM), and boosting and concluded that boosted decision trees outperform other methods (in regard with the true positive detection). On the other hand, by comparing different feature selection and classifiers, authors in [9] have concluded that support vector machines are superior in terms of prediction accuracy. Using different classifiers on variable length n-grams, authors in [10] managed to obtain a 98.4% detection rate with 1.9% false positive. Even though one might consider that 1.9% is an acceptable false positive rate, it is still too big to be used in industry. The authors in [11] used features based on n-grams of opcodes obtained after disassembling the file. They achieved a 95% detection rate and a 0.1% false positive. However, the dataset size is quite small (26093) and was obtained after removing packed or compressed files from the database.

Another method frequently found in malware detection research is to extract API calls from the binary files, apply some filtering algorithm, and use them as features in a classifier. However, most of the time, these features are extracted by executing the file in a controlled environment, like a virtual machine. Even though this method extracts some features that could not be extracted by other methods it is impractical when used in large datasets due to the amount of time required to run each sample. The authors in [12] observed this problem and applied an algorithm, called fuzzy pattern recognition, that was initially developed by [13] to work on dynamic extracted features. Even so, the test was still carried out on a very small dataset (120 benign and 100 malicious). By assuming that api calls and string provide important semantics and can show the executable intent, the authors in [14] have used this information to create features. After using a Max-Relevance algorithm to keep the most 500 important features a SVM ensemble was trained. This method achieved a 92% detection rate and very few false positives. Using api call as features, selecting only the relevant ones and using a SVM as a classifier was also addressed by [15]. A very high detection rate (99%) was achieved by statically extracting api calls, selecting important ones based on their information gain, and using them to train a decision tree. By statically collecting information from file headers as well as the dlls and api calls that a sample might use, the authors in [16] created 25592 features from which only 2450 were selected. Even though the detection rate was 99.6%, the false positive rate was 2.7% and the data set used was not relevant for a real life situation since it contained 20 times more malicious samples than clean.

A similar work to ours was presented in a recent paper [17]. The authors first reduced the number of features using principal component analysis or random projection and then trained a system made up of a neural network and a logistic regression algorithm. The authors tested a neural net with several hidden layers and stated that the one with a single hidden layer provides the best two-class error rate as well as the lowest false positive rate (0.83%). The main difference from their work is how false positives are treated. Instead of developing an algorithm that best classifies a collection of samples and then applies several others (9 in the above paper) to reduce the number of false positives, we use a classifier that obtains zero false positives during training. Even though the detection rate is smaller, we consider usage of (OSP)[2] for practical purposes because of its low false positives.

3   Framework Architecture

In anti-malware industry, having a false positive is considered to be worse than a lower detection rate. The main argument is that a false positive may affect some system or personal files. Since the default action for most anti-malware is to disinfect the infected component, if a false positive appears there is a risk of deleting personal files or removing system files that may lead to a full system shut down. Having this limitation means that some machine learning algorithms (such as support vector machine, neural networks and so on) cannot be used because of the high risk of false positives. There are several solution for this problem; the one that we have adopted is to have a 2 layer filter; the first one is a type of a neural network, called restricted Boltzmann machine, while the second layer consists in usage of (OSP) [2] algorithm (Algorithm 1).

The purpose of the neural network is to increase detection rate and lower the number of false positives by finding correlations between features and transforming them from discrete values to continuous. The OSP has a similar role: to find the best hyper plane that gives no false positives. The reason for choosing a restricted Boltzmann machine is because it has proved successful in other fields like image classification, video action recognition and speech recognition [18],[19],[20] and because it has a structure that permits learning to be done in parallel, making suitable for newer hardware that can perform distributed calculations, like a graphical processing unit(GPU).

The entire framework is constructed in such a way that the hidden layer of the neural network will make up the new set of features for the perceptron. The neurons from the visible layer are initialized with the original features (Fig1) . By doing so, we add a small performance overhead , but we are able to train the classifier with new features that are created in a non-linear way from the original ones. The usage of this method brings another advantage. Even though original perceptron features are boolean values, they can be transformed by the neural network in continuous ones by using outputted probability of the restricted Boltzmann machine as feature for the perceptron.

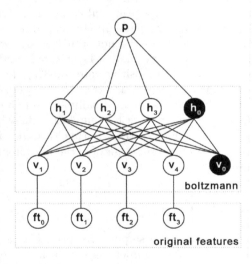

Fig. 1. Neural network architecture

To better understand how the algorithms that make our framework work, we will describe them in the following two subsection.

3.1   One Side Perceptron (OSP)

The OSP algorithm was design to ensure that the number of false positives from the training step is zero, while increasing the detection rate. Having a 0 false positive rate in the training step ensures a very low false positive on a different database. The algorithm adds a secondary step in the training function by retesting the model only on the clean files until every clean file is correctly classified.

To facilitate writing of the algorithm, we consider these abbreviations:

1. $Record \rightarrow R$
2. $Features \rightarrow F$
3. $Label \rightarrow L$
4. $maxIterations - maxI$
5. $SingleSetClass - S - all\ records\ belonging\ to\ a\ specific\ class$

Algorithm 1. One Side Class perceptron - OSC1

1: $w \leftarrow [w_1, w_2, \ldots, w_m], w_i \leftarrow 0, i \leftarrow \overline{1, m}$;
2: $b \leftarrow 0; \lambda \leftarrow a\ learning\ rate$
3: $iteration \leftarrow 0$
4: $S \leftarrow S \bigcup\limits_{i=1}^{|R|} R_i, where\ R_i.L\ =\ benign\ files$
5: **repeat**
6:     $TrainFunction(R, w, b)$
7:     **repeat**
8:         $TrainFunction(S, w, b)$
9:         $errors \leftarrow 0$
10:         **for** $i\ =\ 1 \rightarrow |S|$ **do**
11:             **if** $(not\ IsCorrectClassified(S_i, w, b))$ **then**
12:                 $errors \leftarrow errors + 1$
13:             **end if**
14:         **end for**
15:     **until** ($errors = 0$)
16:     $iteration \leftarrow iteration + 1$
17: **until** $(iteration \geq maxI)or(every\ record\ is\ correctly\ classified)$

The algorithm ends when a specific number of iterations is reached or when every records is correctly calcified. In practice, even after removing most of the noise from a database, the second criteria (every record is correctly clarified) never happens. The maximum number of iterations is usually something determined dynamically (for example, if no growth in detection rate was observed during the last iterations). It is better to stop the algorithm in such situation in order to avoid overfitting. However the tests in this paper were made with a constant number of iterations (derive from the previous observation) so that we can properly compare the results.

While this approach provides a small number of false positives, it also reduces the detection rate significantly. Even though algorithms are not designed to be used alone in an anti-malware products but as a complementary solution to a more deterministic set of algorithms (signature based detection, automata,

---

Algorithm 2. Train Function

1: **function** $Train(Records, w, b)$
2:     **for** $i = 0 \rightarrow |Records|$ **do**
3:         **if** $Records_i.L \times ((\sum_{j=1}^{m} Records_i.F_j \times w_j) + b) \leq 0$ **then**
4:             **for** $j = 1 \rightarrow m$ **do**
5:                 $w_j \leftarrow w_j + Records_i.L \times \lambda \times Records_i.F_j$
6:             **end for**
7:             $b \leftarrow b + Records_i.L \times \lambda$
8:         **end if**
9:     **end for**
10: **end function**

---

etc.), the increase of detection for such an algorithm usually translates in a better proactivity [21] for the product.

There are several methods that can be used to increase the detection rate of such an algorithm. The simplest one is to perform a better feature selection for the attributes that will be used. Another one is to combine different features to create a new set that would produce a better linear separability for the test data set. The secondary approach is called mapping and consists in creating a new set of features by combining the initial features one with each other(Alg 3:

1. sort the entire set of existing feature based on a specific score (usually F2)
2. choose the first n
3. combine every selected feature with the other ones using the AND operator

---

Algorithm 3. MapAllFeatures algorithm

1: **function** $MapAllFeatures(R)$
2:     $newFeatures \leftarrow \phi$
3:     **for** $i = 1 \rightarrow |R.F|$ **do**
4:         **for** $j = 1 \rightarrow |R.F|$ **do**
5:             $newFeatures \leftarrow newFeatures \cup R.F_i \otimes R.F_j$
6:         **end for**
7:     **end for**
8: **end function**
        where $\otimes$ can be any operator. using AND we obtained the best results.

When choosing the value of n in the previous algorithm, some practical restriction should be taken under consideration. The first is that if n is too large, than the weight vector may be too big for a product update (usually an anti-malware product tries to have small updates less than 15-20Kb). If n is too small, than in some cases clean and malware files will have the same set of features. Obviously, for that algorithm to work, these samples should be excluded; to be able to preserve the low false positive rate, usually the excluded samples are the malware ones so that we can be sure we would not detect any clean file.

This means that a small n translates in a lower detection rate. Finally, because the mapped algorithm produces $n \times (n+1)/2$ features, the n should be chosen carefully so that the total number of newly created features is small enough for a product update. Even though using a mapped features does help to increase the detection rate, not being able to use too many of them is a strong limitation. The next logical step was to search for another way of obtaining a new set of features.

3.2 Restricted Boltzmann Machine

A restricted Boltzmann machine is a stochastic neural network made up of two layers: one of visible and one of hidden units. It is called restricted because of its bipartite structure such that each visible neuron is connected to each hidden neuron and each hidden neuron is connected to each visible one. There are no connections between elements on the same layer. Figure 2 shows a restricted Boltzmann with 4 visible units, 3 hidden units and 2 bias units. By using these restrictions, each neuron can be learnt in parallel making the model suitable for training on distributed platforms, like GPUs.

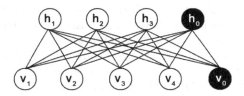

Fig. 2. Restricted Boltzmann Machine

The visible and hidden units have binary states $\{0, 1\}$ and interconnected by a symmetric weight matrix $\{w\}$. Let v_i and h_j represent the states of visible and hidden unit i and j. The state probabilities of the units are:

$$p(h_j = 1|v) = \frac{1}{1 + e^{-\sum\limits_{v_i \in v} v_i * w_{ij}}} \tag{1}$$

$$p(v_i = 1|h) = \frac{1}{1 + e^{-\sum\limits_{h_j \in h} h_j * w_{ij}}} \tag{2}$$

The main objective of this network is to learn binary features that capture high-order structure in the data (visible units). The most often used algorithm for training a restricted Boltzmann machine is contrastive divergence [22]. In contrastive divergence learning, the training is done in two steps. The first step consists in obtaining an initial pair of visible and hidden units based on data. It works by initializing the visible units with a datum (v_i) from the training set and obtaining the set of initial hidden units (h_j) by sampling each neuron according to the probabilities obtained with equation (1).The next step requires to obtain another pair of visible and hidden units ($\hat{v}_i$ and $\hat{h}_j$) but this time the visible units are generated using the previous generated hidden units using equation 2. The contrastive divergence update equation is:

$$\Delta w_{ij} = \lambda(< v_i h_j > - < \hat{v}_i \hat{h}_j)) \tag{3}$$

where $< \cdot >$ refers to the mean over the training data.

4   Configuration and Results

We have collected clean and malicious files for a period of two month (January and February 2014). For each file collected we have ve extracted a set of features. Most of the features ar boolean values and indicates a specific behavior for that particular file (for example: there are features that indicate network activity, Windows registry alteration, file deletion and other). We have also extracted some features that define a value that is specific to a file (for example: file size, file entropy, number of sections, file entry point position, etc). These files were also used to create some boolean features (for example, if a file has more than 10 sections).

Our initial database consists in 3087200 files (clean and malicious) with 3299 features extracted for each record. One problem that we have to deal with is the presence of noise in the database. There are several reason for why this happens. The most common one are the grayware file. The most known grayware files are considered to be adware, spyware, keygens, craks and so on. While their malicious behavior is debatable, they lack the standard components that a malware has (such as a packer, an obfuscated code, etc). Without at least some of this components and no specific dynamic behavior it is really difficult to distinguish such a file from a clean one. Another similar case are file infectors. In this case, the malware infects a clean file. However, when the system extracts the features it will also extract a lot of features from the host file as well, leading to a set of features that is very close (and sometimes identical for small file infectors) to the original clean file.

All of this cases had to be filtered out. After this step we computed the F2 score for each feature and we only kept the first 300 features. This was done to create a model that can be used in practice. Having a large number of features means that one needs to compute all of them to be able to evaluate a proposed model since scanning a file should not have a big impact on system performance we have to reduce the number of features that we compute. If we are to keep only 300 features, some of the records that in the original database had a different set of features, will now be identical (from the features set point of view). In this cases, only one record was kept in the database. If we found cases where records that have the same set of features have a different label we decided to keep only one that is labeled as a clean file (this ensures that our system will not produce false positives). While this measure will reduce the detection rate and there is a chance that some records that are consider to be malicious to be labeled as clean, the practical purpose of this algorithm is first to provide a very low false positive rate and only then to increase the detection rate. If we were to remove the clean records, there is a high probability to get higher false positive. Removing both records (the clean and the malicious one) has a smaller chance of false positives but still a chance. The only way to be sure that the clean file with the same set of features as the malware one is correctly classified is to change the label of the malware one to clean. In theory, both of the records can be manually analyzed and the labels to be properly decided. However, in practice and in particular when dealing with large databases, analyzing hundreds of this pairs is not feasible. At the end of this iteration, the final database had 1260543 files (31507 malicious files and 1229036 clean files) with 300 features for each record.

In order to see the improvement brought by the use of the neural network we used as baseline the results obtained by the OSP and the ones obtained by OSP-mapped algorithm using a 3-fold validation method. All of the algorithms were trained using 10000 iterations.

The training and testing was carried out using a system with 24 Gb of Ram, Intel Xeon CPU E5-2440 (2.4 GHz) and Windows Server 2008 R2. Training of the Boltzmann machine was done on a Ubuntu 12.0 system, using the GPU of an ATI Radeon 7970. Development and training on the GPU was done using the OpenCL [23] framework. Since OpenCL doesn't come with a random number generator that can be used on a distributed framework, we have used the one provided by David Thomas [24]. The learning rate used in training was 0.1 for the BoltzMann machine and 0.01 for the OSP.

Table 1 shows the results obtained using these 3 algorithms. We were in particular interested in the low false positive rate. It should be mentioned that none of these algorithm were to be used by themselves for malware detection but as a complementary method to a more deterministic one (signature detections). OSP is the one side perceptron algorithm tested on the database with 300 features, OSP-MAP is the same algorithm but with all of the 300 features mapped and OSP-RBM is the one side perceptron algorithm tested on the same database where the 300 features were built using a restricted Boltzmann machine from the original ones.

Even though the OSP-MAP algorithm obtained a slightly better detection rate than OSP-RBM, the number of false positives is very high. This is why we consider the OSP-RBM to be better when it comes to be used in industry. The difference of 1.35% in detection rate can be supported by other deterministic algorithms.

Table 1. The results of the 3-fold cross-validation for OSP, OSP-MAP, OSP-RBM algorithms

Algorithm	Detection	False Positive	Accuracy
OSP	73.11%	20 (0.00162%)	99.32 %
OSP-RBM	88.83%	3 (0.00024%)	99.72 %
OSP-MAP	90.18%	87(0.00707%)	99.72 %

5 Conclusions and Future Work

By using a restricted Boltzmann machine to bring relation between features we have managed to get very close to our goal in obtaining a zero false positive classifier that can have a good detection rate. However we still believe there are some things left to test.

There are different configurations of the training network that we didn't test and could improve the classification score. First, we have only used contrastive divergence in one step, but this could be changed to an arbitrary number. Second, different algorithms could be tested to train the restricted Boltzmann machine, one of them being persistent contrastive divergence. We also believe that by using more layers of restricted Boltzmann machine could also improve detection. However, this method might come with the cost of more memory usage and an increased testing time.

One of the most important factor that restricts the detection rate is the fact that the algorithm is trained for zero false positive. By relaxing this, and allowing for a few false positives that could be filtered by other methods, a better detection rate could be achieved.

References

1. www.av-test.org: http://www.av-test.org/en/statistics/malware/
2. Gavrilut, D., Benchea, R., Vatamanu, C.: Optimized zero false positives perceptron training for malware detection. In: SYNASC, pp. 247–253. IEEE Computer Society (2012)
3. Chen, Y.W., Lin, C.-J.: Combining SVMs with various feature selection strategies. In: Guyon, I., Nikravesh, M., Gunn, S., Zadeh, L.A. (eds.) Feature Extraction. STUDFUZZ, vol. 207, pp. 315–324. Springer, Heidelberg (2006)
4. Paul, S.: Information processing in dynamical systems: Foundations of harmony theory. Parallel Distributed Processing: Explorations in the Microstructure of Cognition 1, 194–281 (1986)
5. Idika, N., Mathur, A.P.: A survey on malware detection techniques. PhD thesis. Purdue University (February 2007)
6. Schultz, M.G., Eskin, E., Zadok, E., Stolfo, S.J.: Data mining methods for detection of new malicious executables. In: IEEE Symposium on Security and Privacy, pp. 38–49. IEEE Computer Society (2001)
7. Abou-Assaleh, T., Cercone, N., Keselj, V., Sweidan, R.: N-gram-based detection of new malicious code. In: COMPSAC Workshops, pp. 41–42. IEEE Computer Society (2004)
8. Kolter, J.Z., Maloof, M.A.: Learning to detect and classify malicious executables in the wild. Journal of Machine Learning Research 6, 2721–2744 (2006)
9. Cai, D.M., Gokhale, M., Theiler, J.: Comparison of feature selection and classification algorithms in identifying malicious executables. Computational Statistics & Data Analysis 51(6), 3156–3172 (2007)
10. Siddiqui, M.A.: Data mining methods for malware detection (2008)
11. Shabtai, A., Moskovitch, R., Feher, C., Dolev, S., Elovici, Y.: Detecting unknown malicious code by applying classification techniques on opcode patterns. Security Informatics 1(1), 1–22 (2012)

12. Hung, T.C., Lam, D.X.: A feature extraction method and recognition algorithm for detection unknown worm and variations based on static features (2011)
13. Zhang, B., Yin, J., Hao, J.: Using fuzzy pattern recognition to detect unknown malicious executables code. In: Wang, L., Jin, Y. (eds.) FSKD 2005, Part I. LNCS (LNAI), vol. 3613, pp. 629–634. Springer, Heidelberg (2005)
14. Ye, Y., Chen, L., Wang, D., Li, T., Jiang, Q., Zhao, M.: Sbmds: an interpretable string based malware detection system using svm ensemble with bagging. Journal in Computer Virology 5(4), 283–293 (2009)
15. Dai, J., Guha, R.K., Lee, J.: Efficient virus detection using dynamic instruction sequences. JCP 4(5), 405–414 (2009)
16. Baldangombo, U., Jambaljav, N., Horng, S.J.: A static malware detection system using data mining methods. CoRR abs/1308.2831 (2013)
17. Dahl, G.E., Stokes, J.W., Deng, L., Yu, D.: Large-scale malware classification using random projections and neural networks. In: ICASSP, pp. 3422–3426. IEEE (2013)
18. Lee, H., Grosse, R.B., Ranganath, R., Ng, A.Y.: Convolutional deep belief networks for scalable unsupervised learning of hierarchical representations. In: Danyluk, A.P., Bottou, L., Littman, M.L. (eds.) ICML. ACM International Conference Proceeding Series, vol. 382, p. 77. ACM (2009)
19. Taylor, G.W., Fergus, R., LeCun, Y., Bregler, C.: Convolutional learning of spatio-temporal features. In: Daniilidis, K., Maragos, P., Paragios, N. (eds.) ECCV 2010, Part VI. LNCS, vol. 6316, pp. 140–153. Springer, Heidelberg (2010)
20. Rahman Mohamed, A., Dahl, G.E., Hinton, G.E.: Acoustic modeling using deep belief networks. IEEE Transactions on Audio, Speech & Language Processing 20(1), 14–22 (2012)
21. Cimpoesu, M., Gavrilut, D., Popescu, A.: The proactivity of perceptron derived algorithms in malware detection. Journal in Computer Virology 8(4), 133–140 (2012)
22. Hinton, G.E.: Training products of experts by minimizing contrastive divergence. Neural Computation 14(8), 1771–1800 (2002)
23. Khronos group, http://www.khronos.org/opencl/
24. Thomas, D.: http://cas.ee.ic.ac.uk/people/dt10/research/rngs-gpu-mwc64x.html

Knowledge Management and Human Trafficking: Using Conceptual Knowledge Representation, Text Analytics and Open-Source Data to Combat Organized Crime

Ben Brewster[1], Simon Polovina[1], Glynn Rankin[2], and Simon Andrews[1]

[1] CENTRIC, Sheffield Hallam University, UK
{B.Brewster,S.Polovina,S.Andrews}@SHU.ac.uk
[2] Rankin Kinsella Associates, UK
GlynnRankin@RKA.org.uk

Abstract. Globalization, the ubiquity of mobile communications and the rise of the web have all expanded the environment in which organized criminal entities are conducting their illicit activities, and as a result the environment that law enforcement agencies have to police. This paper triangulates the capability of open-source data analytics, ontological knowledge representation and the wider notion of knowledge management (KM) in order to provide an effective, inter-disciplinary means to combat such threats, thus providing law enforcement agencies (LEA's) with a foundation of competitive advantage over human trafficking and other organized crime.

1 Introduction

Globalization, the ubiquity of mobile communications and the emergence of the web have all aided in the stimulation of trade links, international communications and increased global mobility [1]. Despite the positive potential of this new global, digital environment, it also provides organized criminal entities with new avenues to exploit when conducting illicit activities, and as a result provides law enforcement agencies (LEA's) with an ever-widening environment to police. The new environment defined by mobile communications and the web provides LEAs with a valuable open-source platform that can be applied to enhance existing intelligence based investigations. The use of analytical approaches such as natural language processing (NLP), sentiment analysis and other text mining techniques enables LEAs to follow the breadcrumb trail of evidence leading towards illegal activity.

In 2012, social media service Twitter reported that on average more than 500 million posts were made to its service per day, and its popularity is continuing to grow year over year [2]. In addition to the sustained ubiquity of established social media, new services continue to emerge, targeting differing market segments and offering varying forms of functionality. The popularity of these services, along with other web sources presents a potentially unrivalled intelligence platform (in terms of size) that LEAs can access in order to combat and prevent a variety of organized criminal threats. Ontological approaches to conceptual knowledge representation and the wider philosophy of knowledge management (KM) can provide the infrastructure needed for the intelligence-led operations of LEAs to utilize open source-data.

The EU, FP7 funded ePOOLICE project aims to make use of open-source repositories such as the web to identify, prevent and even predict emergent organized crime threats to better inform LEAs and the actions and decisions they make in relation to these threats. Through the application and development of an ontological knowledgebase, the system will make use of semantically managed domain expertise, provided by LEA partners in order assess the validity of any data mined from open-source repositories [3]. To facilitate the mining of open-source data, ePOOLICE will use NLP text mining techniques such as concept extraction and content categorization in order to manage the acquisition, identification and filtration of relevant data crawled online, identifying poignant indicators and information. 'Privacy by Design' is also implemented; most notably by anonymizing or removing potentially sensitive data that could be used to explicitly identify individuals. To enable knowledge to be effectively managed, domain expertise such as indicators and their relationships will be codified using conceptual graphs and embedded within a larger knowledgebase.

The triangulation between open-source data analytics, ontological knowledge representation and the wider notion of KM potentially provides LEAs with a foundation of competitive advantage over a range organized criminal threats such as drug trafficking, cyber-crime, terrorism and human trafficking; An advantage not easily reciprocated by criminal groups [4]. In this paper we outline the growing problem of human trafficking, and present an overview of the potential capability of ePOOLICE in utilizing open-source data through the use of KM, conceptual graphs and text analytics in order to aid law enforcement in strategically combatting the human trafficking problem.

2   Human Trafficking

Human Trafficking is a complex, global issue that affects millions of people worldwide each year, and as such requires a similarly global and comprehensive response in order to combat its continual spread [5]. Article 3 of the United Nations Protocol to Prevent, Suppress and Punish Trafficking in Persons provides the first agreed definition of human trafficking:

"Trafficking in persons" shall mean the recruitment, transportation, transfer, harboring or receipt of persons, by means of the threat or use of force or other forms of coercion, of abduction, of fraud, of deception, of the abuse of power or of a position of vulnerability or of the giving or receiving of payments or benefits to achieve the consent of a person having control over another person, for the purpose of exploitation. Exploitation shall include, at a minimum, the exploitation of the prostitution of others or other forms of sexual exploitation, forced labor or services, slavery or practices similar to slavery, servitude or the removal of organs' [6].

Human trafficking is a global criminal business that impacts on every country in the world. It is estimated to have a global worth of $32 billion and is recognized as a high profit, low risk crime. In 2012, The United Kingdom's National Crime Agency reported a 9% in the number of trafficking victims it had identified, with the same report indicating there had been a 130% growth of individuals being illegally trafficked to the UK for Cannabis Cultivation purposes [7]. Frequently in these cases the entire trafficking process is controlled by criminal organizations aiming to further exploit trafficked victims through inducting them into a variety of other illicit activities, such as forced labor and prostitution [7], [8]. EUROPOL and UNODC have also published statistics indicating that human trafficking is one of the most prominent and complex security issues across the globe today. [8], [9].

In order to improve the human trafficking defense architecture; data acquisition, application, and trans-border co-operation must be significantly improved, with a view to fostering an environment where knowledge is a key aspect of LEA's arsenal in combating perpetrators of trafficking [5]. To be truly effective, human trafficking identification must take a multidisciplinary approach, triangulating the requirements to maximize indicator identification, validation and perpetrator reprehension [10]. However, human trafficking indicators are often well concealed, with traffickers going to great lengths to obscure their activities, hiding the true size and nature of the trafficking problem [11], [12], [13]. The circumstances under which it takes place are also often complex and significantly varied from case to case [5]. Human trafficking, and particularly trafficking for the purposes of forced labor often involves victims that are considered vulnerable due to a variety of factors that include poverty, unemployment, inequality and political conflicts within their own states [7]. These indicators can be applied to strategically anticipate new destinations and transit routes taken by victims.

Human trafficking is traditionally considered to be under-reported, meaning that efforts to combat the problem cannot rely on observer reports alone [14]. The diverse range of potential criminal activity to which human trafficking is associated (i.e. money laundering, prostitution, forced labor and drug cultivation [15]) means that it provides an ideal use case upon which the application of KM, and specifically ontological knowledge representation and textual extraction through open-source scanning for combatting emergent organized criminal threats can be demonstrated.

In the UK, a number of recent LEA operations have uncovered instances where certain 'nail bars' have been used as fronts for trafficking activities [16]. The use of nail bars, and massage parlors as fronts for trafficking and the crimes to which it is associated is not solely isolated to the UK, with US law enforcement officials reporting similar issues [17], [18]. In the UK alone, ninety salons have been raided and prosecuted in the last five years for employing illegal migrants, concealing cannabis cultivation farms, prostitution rings and money laundering operations [19].

In response, open-source data mining has the ability to abstract data from official reports and news articles to provide a strategic outlook of the trafficking environment and combine it with operational level indicators derived from open social media to map potential hotspots of activity, and validate potential weak indicators of operational level activity. Further, the aggregation of operational data itself presents a strategic outlook of the current environment, enabling current indicators patterns ('primary' indicators) to be temporally extrapolated to indicate future or emergent threats [20].

The short narrative presented previously around the use of nail-bars as fronts for trafficking activity provides sufficient scope to demonstrate the potential capability of conceptual graphs, text analytics and the wider notion of KM in aiding the intelligence-led fight against human trafficking. As an emergent situation in the UK, the scenario described provides sufficient resource in terms of media reports to identify potential trends at a strategic level - such as patterns in migration, or hotspots of drug seizures at specific locations that may indicate trafficking. In addition, pages and posts indicating references of nail bars, and unusual activity and behaviors at the locations derived from the media information can also be applied to identify operational indicators of trafficking, and organized crime.

3   Knowledge Management

To ensure that the knowledge derived from open-source is utilized effectively we propose the use of KM, as a mechanism for LEA's to acquire, apply and store knowledge in a more efficient manner. Knowledge has increasingly become a key strategic resource for LEA's. The September 11th 2001 attacks in New York and Virginia simultaneously identified inadequacies in the current operations of public safety organizations, and the future requirement that knowledge be considered a key determinant of organizational success [21], [22]. Taking this rationale into consideration it is clear that knowledge has become a key strategic resource that has a tangible and quantifiable value in the intelligence-led fight against organized criminal and terrorist acts [23]. More generally, KM aims to gainfully exploit the intellectual capital of the individuals and organizations where it is being employed through the "systematic and explicit management" of all knowledge related activities including the practices, programs and policies that are set and followed by the organization [24]. The various disciplines that are holistically referred to as KM effectively exist across three domains; the organization (its structure, processes and controlling mechanisms), its culture (the informal, tacit practices of its individuals and occupational groups), and its technology (IT based systems) [25, pp.7-8]. Only through the effective management of all three of these domains can knowledge be leveraged effectively by decision makers.

Within the public sector, LEA's are increasingly acknowledging that due to the trans-national nature of organized crime, and in particular those of a trafficking related nature, there is a requirement for a coordinated, multi-agency approach to tackling said offences. UNODC have developed an intelligence sharing system in collaboration with global LEA's in order to facilitate intelligence sharing across borders and jurisdiction in order to aid in the fight against the growing threat of human trafficking. The resulting system, called the VRS-MSRC (Voluntary Reporting System on Migrant, Smuggling and Related Conduct), aims to provide an online outlet for LEA's to create and manage strategic knowledge that can be applied in the formulation of regional and national trafficking policies, informed by data from across global agencies through the analysis of routes used, transportation methods, financial transactions, victim and offender profiles, and other potentially relevant information [26].

Within the applied context of what they describe as 'Intelligence Management', Akhgar & Yates [23] define KM as a process of creating value added learning processes so that knowledge becomes a key strategic resource of LEAs with 'measurable and quantifiable value' in successfully combatting organized crime or an act of terrorism. In this view, in order to enable the understanding of dynamic, ever evolving real world problems a KM approach based upon a number of value added learning processes is required to enable a comprehensive understanding of the problem situation at hand. This approach is firmly grounded in epistemology, and places value on both human and machine learning processes, acknowledging the potential added value of human, tacit inquisition in strategic intelligence.

There is some debate among KM scholars however as to whether it is possible to abstract the tacit, un-codified aspects of human knowledge at all, with some (see [27]) arguing that in order to retain the contextual value of tacit knowledge, it cannot be codified or 'converted' into an explicit, expressible format, such as what is exemplified in Nonaka, Toyama & Konno's [28] SECI knowledge spiral. Boisot's model [29], named the i-Space suggests that in order to be converted, and disseminated, tacit knowledge must be codified and stripped of some of its contextual value in order to be recorded, and further disseminated explicitly. This poses a potential point of issue for ontology driven knowledge representation and other technological systems that make claim to being focused on, or influenced by KM.

Conceptual approaches to KM, such as the knowledge cycles proposed by Dalkir [25], Meyer & Zack [30] and Wiig [31], capture and model the lifecycle of knowledge assets within an organization from their acquisition right through to their application in aiding decision makers and solving problems. Within this, the technologically oriented aspect of KM most commonly targets the capture, creation and codification of knowledge [25, pp.77-97]. This 'explicit' knowledge must be formally expressible through the use of formal, systematic language so that it can be tangibly stored and shared through written mediums such as conceptual graphs and more commonly through items such as training manuals, codes of practice or mathematical formulae [28], [32]. Ontologies and the use of conceptual graphs provide one such approach to knowledge codification.

4   Ontology Driven Knowledge Representation

Ontologically driven knowledge frameworks are static by nature, and thus can only provide solutions to known problems, based on the domain knowledge and relationships that have been modelled within them. This limits any potential impact of any ontologically driven solutions, as although the frameworks themselves are static, the environments upon which they are based are dynamic and ever evolving [23]. However, despite this and the earlier discussion around knowledge conversion, the role of ontologically underpinned knowledge cannot be ignored due to the requirement for technological systems to capture, model, organize and retrieve the constituent, abstracted parts of knowledge - i.e. act as a repository for codified organizational domain knowledge. Sowa [33] addresses this point of issue to some extent through emphasizing the dynamic and changing nature of information, and the complexities that these attributes cause for computer systems.

As a result, Technology and IT systems have to be constructed in a way that facilitates their evolution along with the environment. Increasingly, modern initiatives towards intelligence-led policing practice have also sought to integrate information systems within intelligence strategies. In the past these systems have comprised of what we may now consider to be simplistic document management systems and databases of existing policing records. Approaches such as that offered by IBM's 'COPLINK' [34], [35] have enabled LEA's to share and analyze vast quantities of existing records and information from across the intelligence community to aid national and border security. More recently, contemporary intelligence-led approaches have begun to integrate data analytics in intelligence-led activity for the detection, and predication of potential criminal activity based not only on historical crime records but also on emergent indicators in the environment. This is the mandate of the ePOOLICE project.

Conceptual graphs provide one such environment for ontology driven knowledge representation, providing a means to represent knowledge that is both understandable by software and people [36]. Ontology driven knowledge representation, and specifically conceptual graphs deal with data acquisition and KM though the use of an ontological vocabulary, and the subsequent relationships that are defined between the concepts in that vocabulary, providing a means to explore and model the semantics of domain knowledge [37].

4.1   Applying CGs Using CoGui

Conceptual graphs (CGs) provide a valuable research tool for identifying and expressing the relationships between organized crime indicators, such as those defined by the International Labor Office Delphi survey [38], and the United Nations anti-human trafficking manual [10], and the behaviors and activity that can be identified through crawling other open sources and the text analytics that underpin it. One such tool for representing domain knowledge through the use of conceptual graphs (CGs) is CoGui. Aimed at enabling the visual expression of knowledge in the form of CG's, CoGui is heavily based upon Sowa's [37] work on conceptual structures and knowledge representation. CoGui enables the composition of a set of ontology powered CG's capable of representing assertions [39]. In this section CoGui will be used to demonstrate the utility and capability of conceptual graphs in enhancing the value of the ePOOLICE project. In representing the ontological relationships between entities, CG's enable the definition of 'rules', as shown in figure 1. Rules enable the visualization of implied knowledge using an if-then (hypothesis-conclusion) scenario. The following section revisits the scenario presented in section 2 around the use of nail bars to demonstrate the applicability of conceptual graphs. The rule presented in figure 1 hypothesizes a generalized migration pattern, potentially facilitated by organized crime groups.

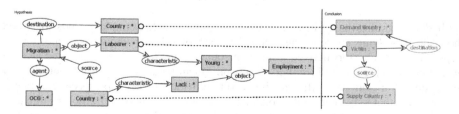

Fig. 1. Victim Route Rule

Figure 2 shows a fact CG that is derived from the textual extraction described in section 5. The CG identifies how the relationships between locations and nail-bars (the concepts extracted from open-source) are modeled to reveal their linkages with domain knowledge that exists behind the indicators. The fact CG represents a specific scenario, in this case migration movements between two specific locations, an event that is known to coincide with increases in trafficking.

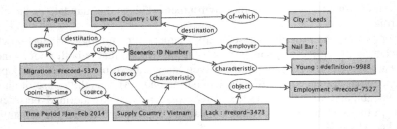

Fig. 2. Fact CG

The rule identified in figure 2 can then be applied to the fact in figure 1 to show the reasoning behind it. Figure 3 identifies through projection that the rule is indeed a potential instance of migration from Vietnam, with Leeds, UK the likely destination. The resulting fact graph is then used to update, and further enrich the wider knowledgebase. It is worth noting that in this particular example, although the activities highlighted, and relationships made between them potentially provide valuable information on events that may indicate trafficking activity, there is no evidence that suggests illegal activity is taking place in the case provided.

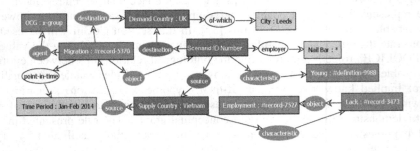

Fig. 3. Output

In demonstrating the application of CG's using this simple example it is possible to see how the underlying concepts can applied as part of a larger knowledge repository to represent how indicators derived from open-source, using textual extraction and analysis techniques (such as those identified in section 5) can be abstracted to not only give an indication of operational trafficking, but also to provide a strategic perspective of migration patterns that may indicate the potential for a future, or emergent trafficking threat. In order to extract the indicators of such illicit activity from the open-sources, we introduce the use of textual analytics into the information retrieval process.

5   Textual Extraction and Categorization

Text analytics encompasses a wide variety or semantic and linguistic disciplines such as sentiment analysis, data mining, concept and contextual extraction, and content categorization [40] and is used to identify trends and previously unknown insights about their content [41]. Existing work in the area has focused on the extraction of named entities and concepts from unstructured closed-sources such as police case reports, enabling improved access to and co-ordination of such information [42], [43].One such aspect of text analytics; content categorization, is designed to create, maintain and apply semantic infrastructure to previously unstructured content [44]. In applying text analytics to open-source data it is possible to discern insight into groups and networks that was not previously possible due to the scale and level of access that provided by the web [45]. As human trafficking encompasses such a wide variety of potential issues, from forced labor, drug cultivation and distribution, the sexual exploitation of women and children, and numerous other forms of exploitation a taxonomy driven approach to categorizing disparate content enables decision makers to separate potentially relevant information from the noise of irrelevant open-source data.

Two approaches have been considered for categorizing content in this way; statistical and rule based. Statistical categorization relies on the acquisition of a robust corpus of training documents that accurately represent each content category, with documents matched to that particular category based upon their closeness to the statistical signature of the training corpus. The statistical signature is based upon measures such as the frequency at which particular words or phrases appear within documents. Although considered easy to develop, statistical models are entirely reliant on external content mined from the open-source environment being represented accurately by the training corpus. Despite having the benefit of being relatively simple to develop, the user has less direct control over the output, and thus the accuracy of a statistical approach tends to be lower [46, pp.201-202]. Due to the disparate nature of open-source data and the broad scope of the ePOOLICE project in categorizing data based on a wide variety of criteria and criminal activity, a rule-based approach consisting of both Boolean rules and linguistic terms is pursued. This approach enables the use of regular expressions, and linguistic and Boolean rules to define rules for each category in the defined taxonomy, thus enabling a higher degree of precision. Boolean rules although considered the most difficult and time intensive to develop, are also the easiest to refine and maintain, and potentially the most powerful and therefore accurate categorization method [46, pp.269-270]. In order to categorize content effectively, a representative taxonomy is first required based upon the variety and subject matter of the source material to be categorized as well as what the intended outcome of the categorization is [41].

5.1   Content Categorization

In order to establish a comprehensive understanding of the environment, it is first necessary to design a taxonomy that encompasses all domain relevant characteristics, in order to facilitate the categorization of content against these factors. For this purpose, a taxonomy based upon a number of existing publications including the International Labor

Organization's 'Delphi' survey Indicators [38], UNODC's Anti-Human Trafficking manual for criminal justice practitioners [10] has been tailored to represent the specifics of identifying illicit nail bars in UK towns and cities. In this example (shown in Figure 4), documents related to mentions of young female workers will be categorized using Boolean syntax that uses logic stating that if a mention of an individual being 'young' or expressions that indicate youth appears in the same sentence, and within three words of syntax referencing a child or woman then the content is likely to be in reference to underage workers. Alone, this information does not necessarily provide any meaningful insights, however when content is similarly matched as being related to a location, or business such as a nail-bar then it provides a potential weak indicator of illicit activity.

```
(OR,
  (AND,
    (SENT,
      (DIST_3,
        (OR,"young","youthful","under-age","under age","under 16","under 18"
        ),
        (OR,
          (OR,"girl","girls"
          ),
          (OR,"child","children"
          ),
          (OR,"women","woman"
          ),
          (OR,"ladies","Ladies","Lady","lady"
          ),
          (OR,"she"
          )
        )
      )|
    )
  )
)
```

Fig. 4. Young Workers Categorizer

5.2   Concept Extraction

A further application of text analytics is referred to as named-entity or concept extraction. In this instance, a list of location 'classifiers' are used in combination with Boolean rules in order extract explicit mentions of 'locations' in sentences that infer that something, or someone is transferred to or from it, the outcome of which could be applied to geographically map trafficking activity from reports and news articles to give a strategic view of the existing trafficking environment, and known routes.

Classifiers consist of a list of raw text concepts. In this instance these are made up of a list of countries and major towns and cities from across the globe extracted from MaxMind - a free database of major world cities [47].

Additional Concepts are then added to the taxonomy to identify the relationship between the 'Location' concepts and a particular use of language; in this case travelling to or from a destination. The syntax defined using Boolean rules states that if a term that indicates that someone/ or something is 'arriving' or 'destined' appears in a sentence with a concept from the 'LOCATION' rule then a positive match is returned to indicate a possible destination location, with the same logic applied to the OriginLocation rule. The '_c' signifies the value to be returned, in this instance referring to the 'LOCATION' concept. The Boolean category rule qualifier '@' is then used as a suffix in order to expand the search to contain all versions of the word forms it follows. For example, 'arriving' is expanded to include 'arrive', 'arrived' and 'arrives'.

DestinationLocation Concept:

```
CONCEPT_RULE:(AND,(SENT,(DIST_4,(OR,"destined@","destinat
ion@","going@","arriving@","going@"),"_c{LOCATION}")))
```

OriginLocation Concept:

```
CONCEPT_RULE:(AND,(SENT,(DIST_4,(OR,"origin@","originates
@","source@","sourced@"),"_c{LOCATION}")))
```

Figure 5 shows these concept rules applied to a dummy 'news report' describing an artificial report of traffickers being apprehended in the UK city of Leeds. The blue-tagged data references the DestinationLocation concept rule, whereas the red-tagged data represents data matching the OriginLocation rule.

Upon arriving in <DESTINATION_LOC><LOCATION>Leeds</LOCATION></DESTINATION_LOC>
traffickers were apprehended by police. The nail salon workers were identified as
being trafficking victims originating from
<ORIGIN_LOC><LOCATION>Vietnam</LOCATION></ORIGIN_LOC>.

Best Matches	
Concept	Matches
Top/LOCATION	2
Top/DESTINATION_LOC	1
Top/ORIGIN_LOC	1

Fig. 5. Concept Rule Matches

Concept extraction, as demonstrated above can be applied in order extract known trafficking routes from textual data sources such as news articles and official reports and mapped geographically to show current, and historical trafficking hotspots, information which could be of potential benefit in validating weak indicators of emergent activity.

5.3   Weak Signals in the Environment

Further, similar syntax is applied to detect weak signals in the environment. The example in figure 6 demonstrates how tweets that are collated and analyzed from a micro-blog may provide indicators of labor exploitation, with nail-bars used as a front for such illicit activity. In this instance, after the content is filtered to remove information that may be used to identify individuals, the concepts and categorizer's defined in the taxonomy are used to analyze incoming data during its acquisition, with any rule-matches resulting the addition of the <Concepts> and <Categories> tags. The terms identified may aid in the detection of migration patterns, and as a result trafficking activity, due to appearance of new nail bars in one location. Further rules would then be defined to pick upon specific indicators of underage work, forced labor and sexual exploitation such as the mention of 'young looking girls' in figure 6.

```
<?xml version="1.0" encoding="UTF-8"?>
<article>
<query>"Nail Salon" "nail bar"</query>
<authortimezone>London</authortimezone>
<doclang>en</doclang>
<body>The girls working at that new nail-bar in Leeds sure look young. I must ask them what their
secrets are for young looking skin!</body>
<LOCATION>Leeds</LOCATION>
<Categories>top\NailBar</Categories>
</article>

<?xml version="1.0" encoding="UTF-8"?>
<article>
<query>"Nail Salon" "nail bar"</query>
<authortimezone>London</authortimezone>
<doclang>en</doclang>
<body>Must have walked past 2 or 3 nail bars on the way to Elland Road this afternoon, they are
springing up all over the place!</body>
<LOCATION>Elland Road;Leeds</LOCATION>
<Categories>top\NailBar</Categories>
</article>
```

Fig. 6. XML Document Markup

Although it is necessary to add further complexity, and comprehensiveness to the categorization and extraction rules presented to offer actual utility in the detection of organized crime signals in open-source data, it is clear that a developed version of such concepts can have real potential in the identification of tangible crime indicators in open-source data. Further, the underlying relationship between indicators and entities enables them to be generalized and extrapolated (i.e. codified and stripped of context) and applied to wider organized crime scenarios. In addition, the aggregation of operational 'primary' (i.e. current threat) indicators such as those identified previously can be used to map strategic trends and patterns in activity to identify emergent and future threats through the idea of temporal proximity [20].

6   Concluding Remarks

In this paper we have presented human trafficking as a potential use case for semantic knowledge representation in aiding to re-enforce both the strategic and operational capability of law enforcement agencies in combatting organized crime threats as part of a wider knowledge management strategy. KM has been identified and rationalized as an enabler of more effective intelligence practice, enhancing law enforcement agencies capacity to integrate new intelligence streams such as ePOOLICE into existing practice through improving communication and information sharing channels, knowledge acquisition and retrieval, and knowledge co-ordination and availability. Conceptual graphs are used to demonstrate one way in which state-of-the-art technologies can stimulate more effective KM practice and as a result improve LEAs response to organized crime threats. In addition, the utility of applying text analytics in extracting valuable information from disparate information sources has been demonstrated, and the requirement for further research as part of the ePOOLICE project identified in order to expand the basic concepts outlined here in order to have real, tangible impact in extracting organized crime indicators from 'primary' open-source data, and ultimately in informing domain decision makers.

Disclaimer
In this document, terms indicating origin or ethnicity are not being used to imply anything general or stereotypical of that origin or ethnic group. Such terms are only used as instances of factual reporting and are not be taken as a reference to any race or ethnic group as a whole. Nevertheless, for a system to monitor organized crime to operate effectively, the identification of certain elements such as gender, nationality and ethnicity, in addition to the explicit identification of crime gangs, victim groups and modus operandi are often important. For instance, Vietnamese victims are trafficked into Europe (supported by several reliable sources) and, therefore, the reference to the Vietnamese origin of criminal groups is crucial to investigate such cases as it forms a key characteristic of the phenomena described. Furthermore, when sensitive or personal data is being handled, it will be done so in accordance with laws protecting the privacy and human rights of individuals, including data protection laws.

Acknowledgement. This research has received funding from European Union Seventh Framework Programme FP7/2007 - 2013 under grant agreement n° FP7-SEC-2012-312651.

References

1. Elliott, A., Urry, J.: Mobile lives. Routledge (2010)
2. Krikorian, R.: Twitter Engineering Blog: New Tweets per Second Records, and How!
   `https://blog.twitter.com/2013/new-tweets-per-second-record-and-how`
3. ePOOLICE. ePOOLICE - About,
   `https://www.epoolice.eu/EPOOLICE/about.jsp`
4. Gottschalk, P.: Knowledge management technology for organized crime risk assessment. Inf. Syst. Front. 12(3), 267–275 (2010),
   `http://dx.doi.org/10.1007/s10796-009-9178-8`
5. Rankin, G., Kinsella, N.: Human trafficking–The importance of knowledge information exchange. In: Intelligence Management, pp. 159–180. Springer (2011)
6. United Nations Office on Drugs and Crime (UNODC), United Nations Convention Against Transnational Organised Crime and the Protocols Thereto (2004)
7. Serious Organised Crime Agency (SOCA), UKHTC: A Strategic Assessment on the Nature and Scale of Human Trafficking in 2012 (2013)
8. Europol. Organised Crime Threat Assessment,
   `https://www.europol.europa.eu/sites/default/files/publications/octa2011.pdf`
9. United Nations Office on Drugs and Crime (UNODC), Global Report on Trafficking in Persons (2012)
10. United Nations Office on Drugs and Crime (UNODC), Anti-Human Trafficking Manual for Criminal Justice Practitioners (2009)
11. Logan, T.K., Walker, R., Hunt, G.: Understanding human trafficking in the united states. Trauma. Violence Abuse 10(1), 3–30 (2009)
12. Laczko, F.: Enhancing Data Collection and Research on Trafficking in Persons. In: Measuring Human Trafficking, pp. 37–44. Springer (2007)

13. Tyldum, G., Brunovskis, A.: Describing the Unobserved: Methodological Challenges in Empirical Studies on Human Trafficking. In: Laczko, F., Godziak, E. (eds.) Data and Research on Human Trafficking: A Global Survey. IOM International Organisation for Migration, Geneva (2005)
14. United Nations Office on Drugs and Crime (UNODC), Global Report on Trafficking in Persons (2009)
15. EUROPOL. EU Serious and Organised Crime Threat Assessment, https://www.europol.europa.eu/sites/default/files/publications/socta2013.pdf
16. Human Trafficking Foundation. Modern Day Slavery in British Nail Bars, http://www.humantraffickingfoundation.org/news/2013/modern-day-slavery-british-nail-bars
17. Martin, P.: Nail Salons and Human Trafficking, http://www.huffingtonpost.com/phillip-martin/nail-salons-and-human-tra_b_669076.html
18. Siskin, A., Wyler, L.S.: Trafficking in persons: US policy and issues for congress. Springer (2012)
19. Arbuthnott, D.: Beauty and the beasts: Slaves in Britain. The Sunday Times Magazine (2013) (August 18, 2010)
20. Criminal Intelligence Service Canada (CISC). Strategic Early Warning for Criminal Intelligence, http://www.cisc.gc.ca/products_services/sentinel/document/early_warning_methodology_e.pdf
21. Chaves, T.D., Pendleton, M.R., Bueerman, C.J.: Knowledge management in policing. Community Oriented Policing Services (COPS) US Department of Justice (2005)
22. Hughes, R.G., Stoddart, K.: Hope and fear: Intelligence and the future of global security a decade after 9/11. Intelligence and National Security 27(5), 625–652 (2012), http://dx.doi.org.lcproxy.shu.ac.uk/10.1080/02684527.2012.708518
23. Akhgar, B., Yates, S.: Intelligence Management: Knowledge Driven Frameworks for Combating Terrorism and Organized Crime. Springer (2011)
24. Wiig, K.M.: Knowledge management: An emerging discipline rooted in a long history. In: Knowledge Horizons: The Present and the Promise of Knowledge Management, pp. 3–26 (2000)
25. Dalkir, K.: Knowledge management in theory and practice. Routledge (2013)
26. United Nations Office on Drugs and Crime (UNODC). Voluntary Reporting System on Migrant Smuggling and Related Conduct (VRS-MSRC), https://www.unodc.org/southeastasiaandpacific/en/vrs-msrc.html
27. Cook, S.D., Brown, J.S.: Bridging epistemologies: The generative dance between organizational knowledge and organizational knowing. Organization Science 10(4), 381–400 (1999)
28. Nonaka, I., Toyama, R., Konno, N.: SECI, ba and leadership: A unified model of dynamic knowledge creation. Long Range Plann. 33(1), 5–34 (2000)
29. Boisot, M.: Knowledge assets: Securing competitive advantage in the knowledge economy (1998)
30. Meyer, M., Zack, M.: The design and implementation of information products. Sloan Management Review 37(3), 43–59 (1996)
31. Wiig, K.: Knowledge management foundations. In: Thinking about Thinking. How People and Organizations Create, Represent, and Use Knowledge, Arlington, TX, USA (1993)

32. Nonaka, I., Takeuchi, H.: The knowledge-creating company: How Japanese companies create the dynamics of innovation. Oxford University Press, New York (1995)
33. Sowa, J.F.: The challenge of knowledge soup. In: Research Trends in Science, Technology and Mathematics Education, pp. 55–90 (2006)
34. Chen, H., Zeng, D., Atabakhsh, H., Wyzga, W., Schroeder, J.: COPLINK: Managing law enforcement data and knowledge. Communications of the ACM 46(1), 28–34 (2003)
35. IBM. IBM i2 COPLINK: Accelerating Law Enforcement, http://www-01.ibm.com/common/ssi/cgi-bin/ssialias?subtype=BR&infotype=PM&appname=SWGE_ZZ_ZZ_USEN&htmlfid=ZZB03016USEN&attachment=ZZB03016USEN.PDF
36. Polovina, S.: An introduction to conceptual graphs. In: Priss, U., Polovina, S., Hill, R. (eds.) ICCS 2007. LNCS (LNAI), vol. 4604, pp. 1–14. Springer, Heidelberg (2007)
37. Sowa, J.F.: Conceptual structures: information processing in mind and machine. Addison-Wesley Longman Publishing Co., Inc. (1984)
38. ILO (International Labour Office), Operational Indicators of Traffcking in Human Beings (2009)
39. CoGui. CoGui in a Nutshell, http://www.lirmm.fr/cogui/nutshell.php
40. Feldman, R., Sanger, J.: The text mining handbook: advanced approaches in analyzing unstructured data. Cambridge University Press (2007)
41. Reamy, T.: Enterprise content categorization - how to sucessfully choose, develop and implement a semantic strategy. Knowledge Architecture Professional Services, KAPS Group (2010)
42. Chau, M., Xu, J.J., Chen, H.: Extracting meaningful entities from police narrative reports. In: Proceedings of the 2002 Annual National Conference on Digital Government Research, pp. 1–5 (2002)
43. Chen, H., Chung, W., Xu, J.J., Wang, G., Qin, Y., Chau, M.: Crime data mining: A general framework and some examples. Computer 37(4), 50–56 (2004)
44. Reamy, T.: Enterprise content categorization - the business strategy for a semantic infrastructure. Knowledge Architecture Professional Services, KAPS Group (2010)
45. Hu, X., Liu, H.: Text analytics in social media. In: Aggarwal, C.C., Zhai, C. (eds.) Mining Text Data, pp. 385–414. Springer (2012)
46. SAS Institute Inc. SAS Content Categorization Studio 12.1: User's Guide, http://support.sas.com/documentation/onlinedoc/ccs/12.1/ccsug.pdf
47. MaxMind. Free World Cities Database, http://www.maxmind.com/en/worldcities

Default Reasoning Implementation in CoGui

Patrice Buche[1], Jérôme Fortin[2], and Alain Gutierrez[3]

[1] INRA IATE, LIRMM GraphIK, France
Patrice.Buche@supagro.inra.fr
[2] Université Montpellier II
IATE/LIRMM GraphIK, France
fortin@polytech.univ-montp2.fr
[3] CNRS LIRMM, France
alain.gutierrez@lirmm.fr

Abstract. This is an application paper in which we propose to present the actual implementation of default reasoning under conceptual graph formalism using CoGui. CoGui is a free graph-based visual tool, developed in Java, for building Conceptual Graph knowledge bases. We present the extension of this application to define and represent default CG rules (a CG-oriented subset of Reiter's default logics) and how to use these rules in skeptical or credulous reasoning.

1 Introduction

Default conceptual graph rules have been introduced in [1] in order to model expert knowledge, especially in agronomy applications. Default CG rules encode a subset of Reiter's default logic [2] and deal with knowledge of the form "if an hypothesis is proved true, a conclusion is generally true unless something that we know prevent us to infer this conclusion. Dealing with default is a kind of non monotonic reasoning because adding some new information to a knowledge base may prevent to apply some default.

The contribution of this paper is the presentation of the actual implementation of default reasoning under conceptual graph formalism using CoGui. CoGui is a free graph-based visual tool, developed in Java, for building Conceptual Graph knowledge bases.

The paper is organized as follows : Section 2 introduces classical notions of CG and CG rule. It gives some clues about their implementation in CoGui. Section 3 recalls basic definition of Reiter's defaults and introduces conceptual default rules. Section 4 is devoted to the presentation of the deduction algorithm using default CG rules.

2 Conceptuals Graphs in Cogui

In this section, we recall main notations and results required for the default CG rules used in this paper. In Section 2.2, we present the simple CGs of [3]. In Section 2.3, the CG rules of [4].

For each of these subsections we present how CoGui is designed to model these different notions. Figure 1 shows the structure of Cogui packages. CoGui is composed of four different layers associated with four different eclipse projects :

N. Hernandez et al. (Eds.): ICCS 2014, LNAI 8577, pp. 118–129, 2014.
DOI: 10.1007/978-3-319-08389-6_11, © Springer International Publishing Switzerland 2014

- the project *fr.lirmm.graphik.cogui.core* contains the model of knowledge representation, and all algorithms that permit to explore this model. It also contains tools to serialize data objects in xml, and to transpose a model in Datalog+ [5,6];
- the project *fr.lirmm.graphik.cogui.rdf* permits to make import and export from the CoGui internal model to RDF(S) and an OWL fragment [7];
- the project fr.lirmm.graphik.cogui.edit contains the user interface;
- the project *fr.lirmm.graphik.cogui.appli* is the upper layer of CoGui containing its entry point in the *CoGuiApplication* Class.

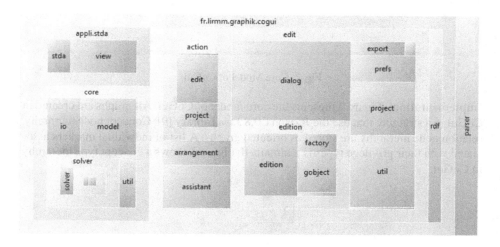

Fig. 1. Packages Structure of CoGui

2.1 Support

Syntax. With the simple CGs of [3], a knowledge base is structured into two objects: the vocabulary (also called support) encodes hierarchies of types, and the conceptual graphs (CGs) themselves represent entities and relations between them. Simple CGs are extended to handle *conjunctive types*, as done in [8].

Definition 1 (Vocabulary). *We call* vocabulary *a tuple* $\mathcal{V} = (\mathcal{C}, \mathcal{R} = (\mathcal{R}_1, \ldots, \mathcal{R}_k)$, $\mathcal{M}_I, \mathcal{M}_G)$ *where* $\mathcal{C}$ *is a partially ordered set of* concept types *that contains a greatest element* $\top$, *each* $\mathcal{R}_i$ *is a partially ordered set of* relation types *of arity* i, $\mathcal{M}_I$ *is a set of* individual markers, *and* $\mathcal{M}_G$ *is a set of* generic markers. *Note that all these sets are pairwise disjoint, and that we denote all the partial orders by* $\leq$.

Definition 2 (Conjunctive types). *A* conjunctive concept type *over a vocabulary* $\mathcal{V}$ *is a set* $T = \{t_1, \ldots, t_p\}$ *(that we can note* $T = t_1 \sqcap \ldots \sqcap t_p$ *of arity* p. *If* $T = \{t_1, \ldots, t_p\}$ *and* $T' = \{t'_1, \ldots, t'_q\}$ *are two conjunctive concept types, then we also note* $T \leq T' \Leftrightarrow \forall t'_i \in T', \exists t_j \in T$ *such that* $t_j \leq t'_i$.

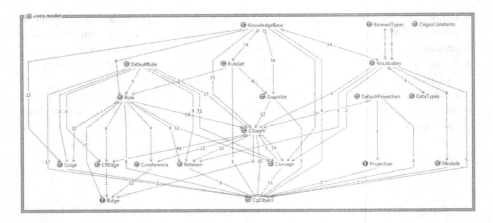

Fig. 2. Core Model of CoGui

Implementation. Figure 2 presents the core model of CoGui. All graphs are created in CoGui in a structure that use the JGraphT 0.8 Java Library [9]. Concept node hierarchy, relation node hierarchy are stored as oriented graphs. A list of individual markers associated with their privilege types is maintained. Figure 3 shows a concept type hierarchy in CoGui.

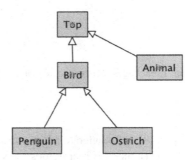

Fig. 3. A concept node hierarchy represented in CoGui

2.2 Simple Conceptual Graphs

A simple conceptual graph, also often called fact is defined as follows :

Definition 3 (Simple CGs). *A simple CG is a tuple* $G = (C, R, \gamma, \lambda)$ *where* C *and* R *are two finite disjoint sets (*concept nodes *and* relations*) and* γ *and* λ *two mappings:*

- $\gamma : R \to C^+$ *associates to each relation a tuple of concept nodes* $\gamma(r) = (c_1, \ldots, c_k)$ *called the* arguments *of* r, $\gamma_i(r) = c_i$ *is its* ith *argument and* $degree(r) = k$.
- λ *maps each concept node and each relation to its* label. *If* $c \in C$ *is a concept node, then* $\lambda(c) = (type(c), marker(c))$ *where* $type(c)$ *is a conjunctive concept type and*

marker(c) *is either an individual marker of* $\mathcal{M}_I$ *or a generic marker of* $\mathcal{M}_G$. *If* $r \in R$ *is a relation and* degree$(r) = k$, *then* $\lambda(r)$ *is a relation type of arity* k.

A simple CG is said to be normal *if all its concept nodes have different markers. Any simple CG G can be put into its* normal form nf(G) *in linear time.*

Semantics. We associate a first order logics (FOL) formula $\Phi(\mathcal{V})$ to a vocabulary $\mathcal{V}$, and a FOL formula $\Phi(G)$ to a simple CG G. These formulae are obtained as follows:

Interpretation of a Vocabulary. Let $\mathcal{V} = (\mathcal{C}, \mathcal{R} = (\mathcal{R}_1, \ldots, \mathcal{R}_k), \mathcal{M}_I, \mathcal{M}_G)$ be a vocabulary. We can consider each concept type of $\mathcal{C}$ as a unary predicate name, each relation type of $\mathcal{R}_i$ as a predicate name of arity i, each individual marker of $\mathcal{M}_I$ as a constant, and each generic marker of $\mathcal{M}_G$ as a variable. For each pair (t, t') of concept types of $\mathcal{C}$ such that $t \leq t'$, we have a formula $\phi((t, t')) = \forall x(t(x) \rightarrow t'(x))$. For each pair (t, t') of relation types of arity i such that $t \leq t'$, we have a formula $\phi((t, t')) = \forall x_1 \ldots \forall x_i(t(x_1, \ldots, x_i) \rightarrow t'(x_1, \ldots, x_i))$. Then the FOL interpretation $\Phi(\mathcal{V})$ of the vocabulary $\mathcal{V}$ is the conjunction of the formulae $\phi((t, t'))$, for every pair of types (t, t') such that $t' < t'$.

Interpretation of a Simple CG. Let $G = (C, R, \gamma, \lambda)$ be a simple CG. We can associate a formula to each concept node and relation of G: if $c \in C$ and $type(c) = t_1 \sqcap \ldots \sqcap t_k$, then $\phi(c) = t_1(marker(c)) \wedge \ldots \wedge t_k(marker(c))$; and if $r \in R$, with $\gamma(r) = (c_1, \ldots, c_q)$ and $\lambda(r) = t$, then $\phi(r) = t(marker(c_1), \ldots, marker(c_q))$. We note $\phi(G) = \bigwedge_{c \in C} \phi(c) \wedge \bigwedge_{r \in R} \phi(r)$. The FOL formula $\Phi(G)$ associated with a simple CG is the existential closure of the formula $\phi(G)$.

Implementation. Simple CG are of course stored as graphs in CoGui. Each node of a Fact graph has a given label. Depending on the type of the considered node, two different types of label are used. For relation nodes, it is enough to label the node by its relation name. For concept nodes, type and marker are needed. A marker can be an individual marker or a generic marker. In a given fact Graph, each generic marker must have a unique name. In order to perform reasoning, one extra attribute is given for each node (concept and relation nodes). This attribute permit to identify uniquely any node of a CoGui project, stored in any kind of graphs (Support, Simple CG, Rules...).

2.3 Conceptual Graph Rules

CG rules form an extension of CGs with knowledge of form "if *hypothesis* then *conclusion*". Introduced in [10], they have been further formalized and studied in [4].

Syntax. A usual way to define CG rules is to establish *co-reference relations* between the hypothesis and the conclusion. We have used here instead *named generic markers*: generic nodes with same marker representing the same entity.

Definition 4 (CG rule). *A conceptual graph rule, defined on a vocabulary* $\mathcal{V}$, *is a pair* $R = (H, C)$ *where H and C are two simple CGs, respectively called the* hypothesis *and the* conclusion *of the rule.*

Semantics. We present here the usual Φ semantics of a CG rule, and recall the equivalent semantics Φ^f using function symbols introduced in [11]. This equivalent semantics makes for an easier definition of default rules semantics: since default rules are composed of different formulas, we cannot rely upon the quantifier's scope to link variables, and thus have to link them through functional terms.

Let $R = (H, C)$ be a CG rule. Then the FOL interpretation of R is the formula $\Phi(R) = \forall x_1 \ldots \forall x_k(\phi(H) \rightarrow (\exists y_1 \ldots \exists y_q \phi(C)))$, where $x_1, \ldots, x_k$ are all the variables appearing in $\phi(H)$ and $y_1, \ldots, y_q$ are all the variables appearing in $\phi(C)$ but not in $\phi(H)$. If $\mathcal{R}$ is a set of CG rules, then $\Phi(R) = \bigwedge_{R \in \mathcal{R}} \Phi(R)$.

As an alternate semantics, let G be a simple CG and X be a set of nodes. We denote by $F = \{f_1, \ldots, f_p\}$ the set of variables associated with generic markers that appear both in G and in X. The formula $\phi_X^f(G)$ is obtained from the formula $\phi(G)$ by replacing each variable y appearing in $\phi(G)$ but not in F by a functional term $f_G^y(f_1, \ldots, f_p)$. Then the FOL interpretation (with function symbols) of a rule $R = (H, C)$ is the formula $\Phi^f(R) = \forall x_1 \ldots \forall x_k(\phi(H) \rightarrow \phi_X^f(C))$ where X is the set of nodes appearing in H. If $\mathcal{R}$ is a set of CG rules, then $\Phi^f(\mathcal{R}) = \bigwedge_{R \in \mathcal{R}} \Phi^f(R)$. Note that this semantics translates in a straightforward way the skolemisation of existentially quantified variables.

Implementation. A rule is stored in CoGui as a single bi-colored graph with two connected components :

- one component stands for the hypothesis of the rule and is tagged by one color,
- the second component models the conclusion of the rule, and is tagged with the other color.

As some nodes of the hypothesis need to correspond to some nodes of the conclusion, a list of coreference links is needed for each rule. The justification and the conclusion must be two connected component. This choice have been made in CoGui in order to optimize some graph operations.

Remark that it wouldn't be sufficient to model a rule by a single bi-colored connected graph, because this representation would not permit to specialize (according to the concept type hierarchy) in the conclusion any node of the Hypothesis. For example, one can express that each Animal that flies is a Bird. This rule is shown in Figure 4.

Computing Deduction. We present here the forward chaining mechanism used to compute deduction with CG rules. In general, this is an undecidable problem. The reader can refer to [12] for an up-to-date cartography of decidable subclasses of the problem.

Definition 5 (Application of a rule). *Let G be a simple CG, $R = (H, C)$ be a rule, and π be a homomorphism from H to G. The application of R on G according to π produces a normal simple CG $\alpha(G, R, \pi) = \text{nf}(G \oplus C_\pi)$ where:*

- *C_π is a simple CG obtained as follows from a copy of C: (i) associate to each generic marker x that appears in C but not in H a new distinct generic marker $\sigma(x)$; (ii) for every generic concept node c of C whose marker x does not appear*

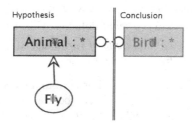

Fig. 4. A rule that specialize a node of the hypothesis

in H, replace marker(c) *with* $\sigma(x)$*; and (iii) for every generic concept node c of C, if* marker(c) *also appears in H, then replace* marker(c) *with* marker($\pi(c)$).
- *the operator* $\oplus$ *generates the* disjoint union *of two simple CGs G and G′: it is the simple CG whose drawing is the juxtaposition of the drawings of G and G′.*

Theorem 1. *Let G and Q be two simple CGs, and $\mathcal{R}$ be a set of CG rules, all defined on a vocabulary $\mathcal{V}$. Then the following assertions are equivalent:*

- $\Phi(\mathcal{V}), \Phi(G), \Phi(\mathcal{R}) \vdash \Phi(Q)$
- $\Phi(\mathcal{V}), \Phi(G), \Phi^f(\mathcal{R}) \vdash \Phi(Q)$
- *there exists a sequence $G_0 = \mathrm{nf}(G), G_1, \ldots, G_n$ of simple CGs such that:*
  (i) $\forall 1 \leq i \leq n$, there is a rule $R = (H, C) \in \mathcal{R}$ and a homomorphism π of H to G_{i-1} such that $G_i = \alpha(G_{i-1}, R, \pi)$;
  and (ii) there is a homomorphism from Q to G_n.

Note that the forward chaining algorithm that relies upon the above characterization is ensured to stop when the set of rules involved is *range restricted, i.e.* their logical semantics Φ does not contain any existentially quantified variable in the conclusion (see [13] for more details about tractable cases).

The "functional semantics" can provide us with an alternate rule application mechanism α^f. Let us begin by *"freezing"* the graph G, *e.g.* by replacing each occurrence of a generic marker by a distinct individual marker that not appears yet in G. Then, when applying a rule R on G (or a graph derived from G) according to a projection π, we consider the formula $\Phi^f(R)$ associated with R. Should the application of $R = (H, C)$ produce a new generic node c from the copy of a generic node having marker y, we consider the functional term $f_C^y(x_1, \ldots, x_k)$ associated to the variable y. Then the marker of c becomes $f_C^y(\pi(x_1), \ldots, \pi(x_k))$. Thanks to the previous theorem, α^f makes for an equivalent forward chaining mechanism, that has an added interest. It allows to have a functional constant identifying every concept node generated in the derivation. This feature will be used to explain default rules reasonings in an easier way than in [1].

3 Default Rules

3.1 Default CG Rules

A Brief Introduction. Let us recall some basic definitions of Reiter's default logics. For a more precise description and examples, the reader should refer to [14,2].

Definition 6 (Reiter's default logic). *A Reiter's default theory is a pair* (Δ, W) *where* W *is a set of FOL formulae and* Δ *is a set of* defaults *of form* $\delta = \frac{\alpha(\overrightarrow{x}):\beta_1(\overrightarrow{x}),\cdots,\beta_n(\overrightarrow{x})}{\gamma(\overrightarrow{x})}$, $n \geq 0$, *where* $\overrightarrow{x} = (x_1, \cdots, x_k)$ *is a tuple of variables,* $\alpha(\overrightarrow{x})$, $\beta_i(\overrightarrow{x})$ *and* $\gamma(\overrightarrow{x})$ *are FOL formulae for which each free variable is in* $\overrightarrow{x}$.

The intuitive meaning of a default δ is "For all individuals $(x_1, \cdots, x_k)$, if $\alpha(\overrightarrow{x})$ is believed and each of $\beta_1(\overrightarrow{x}), \cdots, \beta_n(\overrightarrow{x})$ can be consistently believed, then one is allowed to believe $\gamma(\overrightarrow{x})$". $\alpha(\overrightarrow{x})$ is called the *prerequisite,* $\beta_i(\overrightarrow{x})$ are called the *justifications* and $\gamma(\overrightarrow{x})$ is called the *consequent.* A default is said *closed* if $\alpha(\overrightarrow{x})$, $\beta_i(\overrightarrow{x})$ and $\gamma(\overrightarrow{x})$ are all closed FOL formulae.

Intuitively, an *extension* of a default theory (Δ, W) is a set of formulae that can be obtained from (Δ, W) while being consistently believed. More formally, an extension E of (Δ, W) is a minimal deductively closed set of formulae containing W such that for any $\frac{\alpha:\beta}{\gamma} \in \Delta$, if $\alpha \in E$ and $\neg\beta \notin E$, then $\gamma \in E$. The following theorem provides an equivalent characterization of extensions that we use here as a formal definition.

Theorem 2 (Extension). *Let* (Δ, W) *be a closed default theory and* E *be a set of closed FOL formulae. We inductively define* $E_0 = W$ *and for all* $i \geq 0$, $E_{i+1} = Th(E_i) \cup \{\gamma \mid \frac{\alpha:\beta_1\cdots,\beta_n}{\gamma} \in \Delta, \alpha \in E_i$ *and* $\neg\beta_1, \cdots, \neg\beta_n \notin E\}$, *where* $Th(E_i)$ *is the deductive closure of* E_i. *Then* E *is an extension of* (Δ, W) *iff* $E = \cup_{i=0}^{\infty} E_i$.

Note that this characterization is not effective for computational purposes since both E_i and $E = \cup_{i=0}^{\infty} E_i$ are required for computing E_{i+1}.

The following reasoning tasks come with Reiter's Default Logic:

- EXTENSION: Given a default theory (Δ, W), does it have an extension?
- SKEPTICAL DEDUCTION: Given a default theory (Δ, W) and a formula Q, does Q belong to all extensions of (Δ, W)? In this case we note $(\Delta, W) \vdash_S Q$.
- CREDULOUS DEDUCTION: Given a default theory (Δ, W) and a formula Q, does Q belong to an extension of (Δ, W)? In this case we note $(\Delta, W) \vdash_C Q$?

Syntax of Default CGs

Definition 7 (Default CGs). *A default CG, defined on a vocabulary* $\mathcal{V}$, *is a tuple* $D = (H, C, J_1, \ldots, J_k)$ *where* $H, C, J_1, \ldots,$ *and* J_k *are simple CGs respectively called the hypothesis, conclusion, and justifications of the default.*

Semantics. The semantics of a default CG $D = (H, C, J_1, \ldots, J_k)$ is expressed by a closed default $\Delta(D)$ in Reiter's default logics.

$$\Delta(D) = \frac{\phi(H) \ : \ \neg\phi_X^f(C), \phi_{X \cup Y}^f(J_1), \ldots, \phi_{X \cup Y}^f(J_k)}{\phi_X^f(C)}$$

where X is the set of nodes of the hypothesis H and Y is the set of nodes of the conclusion C. If $D = (H, C, J_1, \ldots, J_k)$ is a default, we note $std(D) = (H, C)$ its *standard part*, which is a CG rule.

Implementation. A default rule is stored as a single multi–colored graph. As for classical rules, one color is needed for the hypothesis, one for the conclusion, and n other colors are used to tag each of the n justifications. Figure 4 shows a default rule that means "unless a birds is known as a penguin or an ostrich, we can suppose that it flies".

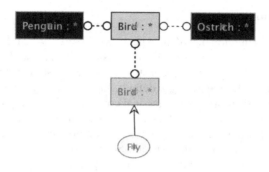

Fig. 5. Representation of a default rule

4 Default Reasoning with Default CG Rules in CoGui

In this section, we explain the algorithm that permits to compute deduction in the alternate derivation mechanism α^f introduced in [11]. This mechanism is introduced for an easier description of the sound and complete reasoning mechanism of [1]. Moreover, this alternate derivation mechanism permits to get a strait–forward way to implement default reasoning in CoGui. Let G and Q be two simple CGs, $\mathcal{R}$ be a set of CG rules, and $\mathcal{D}$ be a set of default CG rules, all defined over a vocabulary V. A node of the default derivation tree DDT($\mathcal{K}$) of the knowledge base $\mathcal{K} = ((V, G, \mathcal{R}), \mathcal{D})$ is always labelled by a simple CG called *fact* and a set of simple CGs called *constraints*. A node of DDT($\mathcal{K}$) is said *valid* if there is no homomorphism of one of its constraints or the constraints labelling one of its ancestors into its fact. Let us now inductively define the tree DDT($\mathcal{K}$):

– its root is a node whose fact is G and whose constraint set is empty;
– if x is a valid node of DDT($\mathcal{K}$) labelled by a fact F and constraints $\mathcal{C}$, then
　　for every rule D in $\mathcal{D}$, for every homomorphism π of the hypothesis of D into a simple CG F' $\mathcal{R}$-derived from F,
　　　　x admits a successor whose fact is the fact $\alpha^f(F', std(D), \pi)$, and whose constraints are the $\pi(J_i)$ iff that successor is valid.

Theorem 3. *Let G and Q be two simple CGs, $\mathcal{R}$ be a set of CG rules, and $\mathcal{D}$ be a set of default CG rules, all defined over a vocabulary V. Then $\Phi(Q)$ belongs to an extension of the Reiter's default theory $(\{\Phi(V), \Phi(G), \Phi(\mathcal{R})\}, \Delta(\mathcal{D}))$ iff there exists a node x of DDT$((V, G, \mathcal{R}), \mathcal{D})$ labelled by a fact F such that $\Phi(V), \Phi(F), \Phi(\mathcal{R}) \vdash \Phi(Q)$.*

Intuitively, this result [1] states that the leaves of $DDT(\mathcal{K})$ encode extensions of the default. What is interesting in this characterization is that: 1) though our default CGs are not normal defaults in Reiter's sense, they share the same important property: every default theory admits an extension; and 2) if an answer to a query is found in any node of the default derivation tree, the same answer will still be found in any of its successors.

4.1   Implementation

In order to compute deduction in CoGui, a new class `CoguiDefaultRuleApplyer` has been created. This class contains a list of Classical rules, a list of default rules and a stack of j–constraints. This stack of j–constraints is used to store each justification that has been "supposed" before applying a default. In practice, a j–constraint is stored as any conceptual graph. Contrary to the classical constraints in conceptual graphs [13], j–constraints need to be attached to the fact on a precise place. The semantic of a classical constraint is the following : if a constraint C can be projected in a graph G, then the knowledge base $\mathcal{K} = (G, C)$ is inconsistent. So if a constraint contains some generic marker, then the constraint can be projected in any place of the graph G.

When dealing with default rules, j–constraints have a different semantics. This is due to the fact that a given justification is precisely linked to an hypothesis of a default.

Let us consider the default $D = (Bird(x), fly(x), Penguin(x))$. This default means that if we find a bird in a database, then we can deduce that this bird flies unless we know that it is a penguin. So if in a fact graph G, we find a concept node $Bird(y)$, associated with the generic marker y, we can try to apply the default D. After adding the conclusion of the default, we have to keep in mind for future deduction (with classical rules or default rules) that the bird y (and only this one) can not be considered in any way as a penguin. It is not possible to use the classical constraint $Penguin(x)$ because it would mean that any individual can be a penguin, which is obviously not what we want to model. So the difference between a classical constraint and a j–constraint is just in its semantics : When adding the j–constraint $Penguin(y)$, it just says that any node that is labeled by the generic marker y (and only this particular generic marker) can represent a penguin.

The main algorithm implemented in CoGui to find extensions of a given knowledge base is based on a backtrack algorithm. We begin with an initial fact graph G, we choose one default and one projection, we apply this default on the projection and run a saturation round by classical rules. If no more default can be applied, no constraint is violated and no default rule is active, the current state is an extension. Otherwise, we continue our exploration of the default derivation tree. Once an extension is found, the current state is saved as a new extension, and a backtrack is operated. This backtrack is possible with the help of the reverse actions implemented in CoGui.

$createJConstaint(J_i, \pi)$ is a procedure that creates a j–constraint by renaming each generic marker of J_i by the corresponding label of the projection π if the considered node is linked with the hypothesis in D, and with a new generic marker otherwise. The procedure $createAddConclusionAction$ permits to create a single defeasible action that permits to add a graph (conclusion of the default) to another graph according to a certain projection.

Algorithm 1. Main deduction algorithm

Data: $\mathcal{K} = ((V, G, \mathcal{R}), \mathcal{D})$
Result: The set of extensions of $\mathcal{K}$
begin

> /* Initialisation of global variables   */
> $currentGraph \longleftarrow G$;
> $constraintStack \longleftarrow \emptyset$;
> $currentStackReverseActions \longleftarrow \emptyset$;
> $stackReverseActions \longleftarrow \emptyset$;
> $extensionsSet \longleftarrow \emptyset$;
> $applyMaster()$;

Algorithm 2. applyMaster

Data: $\mathcal{K}$ (in), currentGraph (in/out); constraintStack (in/out); currentStackReverseActions
   (in/out); stackReverseActions (in/out)
begin

> $ruleApplied \longleftarrow false$;
> **for** $D = (H, C, J_1, \cdots, J_n) \in \mathcal{D}$ **do**
>> **for** π *projection of H in currentGraph* **do**
>>> $applied \longleftarrow applyDefault(D, \pi)$;
>>> **if** *applied* **then**
>>>> $ruleApplied \longleftarrow true$;
>>>> $A \longleftarrow createSaturateAction(currentGraph, \mathcal{R})$;
>>>> applyAction(A,currentGraph);
>>>> stackReverseAction.put(reverse(A));
>>>> $applyMaster()$;
>>>> $A \longleftarrow stackReverseAction.pop(A)$;
>>>> applyAction(A);
>
> **if** $not(ruleApplied)$ **then**
>> extensionSet.add(New copy of current graph);

Algorithm 3. applyDefault

Data: Default D, projection π (and currentGraph; constraintStack;
        currentStackReverseActions ; stackReverseActions as (global variables))
Result: Boolean : true iff D has been applied
begin

   /* Checking justifications   */
   applicable $\longleftarrow$ *true*;
   for J_i *justification of* D **do**
      if π *can be extended to project* J_i **then**
         applicable $\longleftarrow$ *false*;

   if *applicable* **then**
      for J_i *justification of* D **do**
         $C \longleftarrow createJConstraint(D, J_i, \pi)$;
         $constraintStack.put(C)$;
      $A \longleftarrow createAddConclusionAction(currentGraph, C, \pi)$;
      applyAction(A,currentGraph);
      stackReverseActions.put(reverse(A));
      return *true*;
   else
      return *false*;

---

5 Conclusion

It is now possible to model default CG rule with CoGui and to make default reasoning. Default CG rules model a subset of classical Reiter's default rule, in which the conclusion is always seen as an implicit justification. The reasoning implementation permit to compute all extensions of a given knowledge base. The construction of the extension is done by exploring the default deviation tree by a deep first search, so there is no need to store all the derivation trees to compute deduction.

From this list of extensions it is simple to check if a fact is skeptically or credulously entailed by the knowledge base : CoGui permits to manipulate any CoGui object with a devoted script language. Scripts are object stored in CoGui that permit to creates easily automatic tasks that manipulates any CoGui objects. Scripts are develloped using a Java-like langage. This functionality has not been described in this paper in order to focus on the default reasoning, but scripts can be easily used to perform skeptical and credulous deduction. One further work is to directly integrate this functionality to the CoGui native possibilities.

Default rules in CoGui are now used for some applications projects. We use it in particular to model expert knowledge in some agronomy applications [15]. In this domain, an application called Capex (for expert capitalization) has been developed over CoGui and its default reasoning possibilities.

References

1. Baget, J.-F., Croitoru, M., Fortin, J., Thomopoulos, R.: Default conceptual graph rules: Preliminary results for an agronomy application. In: Rudolph, S., Dau, F., Kuznetsov, S.O. (eds.) ICCS 2009. LNCS (LNAI), vol. 5662, pp. 86–99. Springer, Heidelberg (2009)
2. Reiter, R.: A logic for default reasoning. Artificial Intelligence 13, 81–132 (1980)
3. Sowa, J.F.: Conceptual graphs for a database interface. IBM Journal of Research and Development 20(4), 336–357 (1976)
4. Salvat, E., Mugnier, M.L.: Sound and complete forward and backward chaining of graph rules. In: Eklund, P., Ellis, G., Mann, G. (eds.) ICCS 1996. LNCS, vol. 1115, pp. 248–262. Springer, Heidelberg (1996)
5. Calì, A., Gottlob, G., Lukasiewicz, T., Pieris, A.: Datalog+/-: A family of languages for ontology querying. In: de Moor, O., Gottlob, G., Furche, T., Sellers, A. (eds.) Datalog 2010. LNCS, vol. 6702, pp. 351–368. Springer, Heidelberg (2011)
6. Mugnier, M.-L.: Ontological Query Answering with Existential Rules. In: Rudolph, S., Gutierrez, C. (eds.) RR 2011. LNCS, vol. 6902, pp. 2–23. Springer, Heidelberg (2011)
7. Baget, J.-F., Croitoru, M., Gutierrez, A., Leclère, M., Mugnier, M.-L.: Translations between RDF(S) and conceptual graphs. In: Croitoru, M., Ferré, S., Lukose, D. (eds.) ICCS 2010. LNCS, vol. 6208, pp. 28–41. Springer, Heidelberg (2010)
8. Baget, J.-F.: Simple Conceptual Graphs Revisited: Hypergraphs and Conjunctive Types for Efficient Projection Algorithms. In: Ganter, B., de Moor, A., Lex, W. (eds.) ICCS 2003. LNCS (LNAI), vol. 2746, pp. 229–242. Springer, Heidelberg (2003)
9. Naveh, B., Contributors: Jgraph (March 2014), http://jgrapht.org/
10. Sowa, J.F.: Conceptual Structures: Information Processing in Mind and Machine. Addison–Wesley (1984)
11. Baget, J.-F., Fortin, J.: Default conceptual graph rules, atomic negation and tic-tac-toe. In: Croitoru, M., Ferré, S., Lukose, D. (eds.) ICCS 2010. LNCS, vol. 6208, pp. 42–55. Springer, Heidelberg (2010)
12. Baget, J.F., Leclère, M., Mugnier, M.L., Salvat, E.: Extending decidable cases for rules with existential variables. In: Proceedings of the 21st International Joint Conference on Artificial Intelligence, Pasadena, California, USA, July 11-17, pp. 677–682 (2009)
13. Chein, M., Mugnier, M.L.: Graph-based Knowledge Representation: Computational Foundations of Conceptual Graphs, 1st edn. Springer Publishing Company, Incorporated (2008)
14. Brewka, G., Eiter, T.: Prioritizing default logic: Abridged report. In: Festschrift on the Occasion of Prof. Dr. W. Bibel's 60th Birthday. Kluwer (1999)
15. Buche, P., Cucheval, V., Diattara, A., Fortin, J., Gutierrez, A.: Implementation of a knowledge representation and reasoning tool using default rules for a decision support system in agronomy applications. In: Croitoru, M., Rudolph, S., Woltran, S., Gonzales, C. (eds.) GKR 2013. LNCS (LNAI), vol. 8323, pp. 1–12. Springer, Heidelberg (2014)

Extracting Threshold Conceptual Structures from Web Documents

Gabriel Ciobanu[1], Ross Horne[1], and Cristian Văideanu[2]

[1] Romanian Academy, Institute of Computer Science, Iaşi, Romania
[2] A.I.Cuza University of Iaşi, Faculty of Mathematics, Romania
gabriel@info.uaic.ro, {ross.horne,cvaideanu}@gmail.com

Abstract. In this paper we describe an iterative approach based on formal concept analysis to refine the information retrieval process. Based on weights for ranking documents we define a weighted formal context. We use a Galois connection to introduce a new type of formal concept that allows us to work with specific thresholds for searching words in Web documents. By increasing the threshold, we obtain smaller lattices with more relevant concepts, thus improving the retrieval of more specific items. We use techniques for processing large data sets in parallel, to generate sequences of Galois lattices, overcoming the time complexity of building a lattice for an entire large context.

1 Introduction

Formal concept analysis (FCA) is a data analysis technique based on lattice theory which provides effective methods for conceptual clustering and knowledge representation [16]. In the last decade, FCA has been used for various applications like cluster analysis, semantic Web, image processing and knowledge discovering. It has been proved useful particularly for information retrieval (IR) applications, providing a support structure which has improved search strategies like query refinement, ranking, document classification and combinations of various views for semistructured data.

In FCA-based information retrieval applications, the documents usually serve as formal objects and the index terms as formal attributes. Thus, the formal context coincides in fact with the so-called document-term matrix. The concept lattice (Galois lattice) associated can be interpreted as a search space that can be explored using different retrieval strategies. A query submitted by the user is assimilated with a concept intent, while the documents retrieved by the system are its extent. There are many approaches which have been developed methods to explore the concept lattice in order to find the relevant documents for a user's query. Thus in [1] the retrieved documents are found using the distance from the concepts of the lattice to the "query concept", i.e. the concept whose intent coincides with the query. In [11] the lattice-based IR techniques for classifying and searching relevant bioinformatic data sources is used to find the needed information exploring only the superconcepts of the query concept in the conceptual structure. In [4] it is proposed a FCA-based approach for semantic indexing and

N. Hernandez et al. (Eds.): ICCS 2014, LNAI 8577, pp. 130–144, 2014.
DOI: 10.1007/978-3-319-08389-6_12, © Springer International Publishing Switzerland 2014

retrieval using the Galois lattice as a semantic index and as a search space to model terms.

Organising the data as a lattice structure has many advantages from the perspective of information retrieval: an enhanced browsing retrieval, better possibilities for extracting knowledge from the conceptual hierarchy and for identifying the conceptual associations for query reformulation. However, these approaches have one important limitation: the size of the concept lattice can be very large with respect to their underlying context. Some researchers have proposed different solutions to overcome this problem. Thus, the system CREDO generates a small portion of the lattice, typically consisting of the query concept and its neighbours [2]. In [14] several formal methods of combining a document collection with a multi-faceted thesaurus are presented. The lattice-based system FaIR improve the retrieval by partitioning the set of documents into smaller sets.

Here we address the issue of extracting conceptual structures from large collections of Web documents at a lower complexity. Thus, we propose a model which add to the lattice-based IR approach some relevance conditions based on terms weight. First of all, we use a formal context having an extended incidence relation that includes some weights associated with each term and document. Based on this context type, we generate a Galois lattice which connects terms and relevant documents. Using a relevance condition for documents which depends on a threshold value, we get a new type of formal concepts, namely threshold formal concepts (t-concepts). Our dynamical structure offers a more flexible way to extract, organise and represent the information contained in Web pages. Every time the threshold value t is increased, a new threshold Galois connection between the document set and the term set is created. Thus, we build a sequence of t-concept lattices which can iteratively refine the set of retrieved documents, as a result of the user's query. By modifying the terms weight threshold, the density of the formal context can be rapidly adjusted such that the complexity of the corresponding Galois lattice to be decreased. We also use the t-concept lattices to define a new method for ranking the documents returned by the system. We develop here a technique combining the navigation through the system of t-concept lattices built and a relevance condition based on terms weight. Firstly, the rank of a document is computed as the length of the shortest path from the query concept to the concepts whose extent contains the document. When many documents are equally relevant, we refine the ranking by using the sequence of t-Galois lattices.

As we already noticed, the complexity of the concept lattice is one of the major problems of any lattice-based IR applications. In this paper we overcome this shortcoming by using the power of parallel distributed calculus applied to our model of threshold concept lattices. The MapReduce software framework allows us to generate in parallel the concepts of every fixed t-lattice from the sequence, and then all the t-lattices in the sequence. The power of the parallel calculus across a distributed cluster of processors help us to overcome the complexity of the generating process.

The rest of the paper is organised as follows. In Section 2 we briefly overview the FCA results. Then we introduce the notion of threshold concept lattice in Section 3. In Section 4 we describe the iterative retrieval process, including some notes about the MapReduce algorithm, and the ranking method for the documents. Finally, our conclusions are presented in Section 5.

2   Basics of Formal Concept Analysis

Formal concept analysis was proposed as a mathematical method of data analysis [16]. This approach takes a binary relation between a set of objects and a set of attributes, which are properties of the objects, and then creates a space of "concepts".

Definition 1. *A formal context is a triple* (X, Y, I) *with* X, Y *being abstract sets and* $I \subseteq X \times Y$ *a relation between* X *and* Y. *We call the elements of* X *objects, those of* Y *attributes and* I *the incidence relation of the context* (X, Y, I). *If* xIy, *we say that "object* x *has the attribute* y".

Definition 2. *Let* (X, Y, I) *be a formal context. The applications* $\alpha : (\mathcal{P}(X), \subseteq) \to (\mathcal{P}(Y), \subseteq)$ *and* $\beta : (\mathcal{P}(Y), \subseteq) \to (\mathcal{P}(X), \subseteq)$ *defined by*

$$\alpha(A) = \{y \in Y \mid \forall x \in A, \; xIy\}, \; A \neq \varnothing, \; \alpha(\varnothing) = Y$$
$$\beta(B) = \{x \in X \mid \forall y \in B, \; xIy\}, \; B \neq \varnothing, \; \beta(\varnothing) = X$$

are called the derivation operators.

It is known that the derivation operators α, β form an antitone Galois connection between the ordered sets $(\mathcal{P}(X), \subseteq)$ and $(\mathcal{P}(Y), \subseteq)$ i.e. α, β are decreasing and the operators $\beta \circ \alpha$, $\alpha \circ \beta$ are extensive.

Definition 3. *Let* (X, Y, I) *be a formal context. A pair* $(A, B) \in \mathcal{P}(X) \times \mathcal{P}(Y)$ *is said to be a formal concept of* (X, Y, I) *if* $\alpha(A) = B$ *and* $\beta(B) = A$. *The sets* A, B *are called the extent and the intent of the formal concept* (A, B), *respectively. The set of all formal concepts is denoted by* $\mathcal{C}(X, Y, I)$.

If (A, B) is a formal concept, then B can be interpreted as a set of terms appearing in a query, while A represents the documents retrieved. These sets are maximal according to the properties stated in the definition. On the set of formal concepts we can define a hierarchical order, denoted as $\leq$; thus, we say that the concept (A_1, B_1) is less (more specific) than the concept (A_2, B_2) or that (A_2, B_2) is greater (more general) than (A_1, B_1)) if $A_1 \subseteq A_2$ (equivalently $B_1 \supseteq B_2$). The basic theorem of FCA states that the set of formal concepts ordered by the $\leq$ relation forms a complete lattice [7].

Theorem 1. *The set* $\mathcal{C}(X, Y, I)$ *endowed with the relation* $\leq$ *is a complete lattice called concept lattice or Galois lattice, and is denoted by* $\underline{\mathcal{C}}(X, Y, I)$.

Example 1. An example of formal context and its concept lattice is presented in Figure 1. We find the concepts which belong to the concept lattice following the method presented in [7] :

$c_1^1 = (\{d_1, d_2, d_3, d_4, d_5, d_6\}, \varnothing)$ (the top concept)

$c_2^1 = (\{d_2, d_3, d_4, d_5, d_6\}, \{ring\})$, $c_3^1 = (\{d_1, d_4, d_5, d_6\}, \{gold\})$

$c_4^1 = (\{d_2, d_3, d_4\}, \{ring, algebra\})$, $c_5^1 = (\{d_2, d_3, d_5, d_6\}, \{ring, planet\})$

$c_6^1 = (\{d_4, d_5, d_6\}, \{ring, gold\})$, $c_7^1 = (\{d_4\}, \{ring, gold, algebra\})$

$c_8^1 = (\{d_5, d_6\}, \{ring, gold, planet\})$, $c_9^1 = (\{d_2, d_3\}, \{ring, algebra, planet\})$

$c_{10}^1 = (\varnothing, \{ring, gold, algebra, planet\})$ (the bottom concept)

	ring	gold	algebra	planet
d_1	0	1	0	0
d_2	1	0	1	1
d_3	1	0	1	1
d_4	1	1	1	0
d_5	1	1	0	1
d_6	1	1	0	1

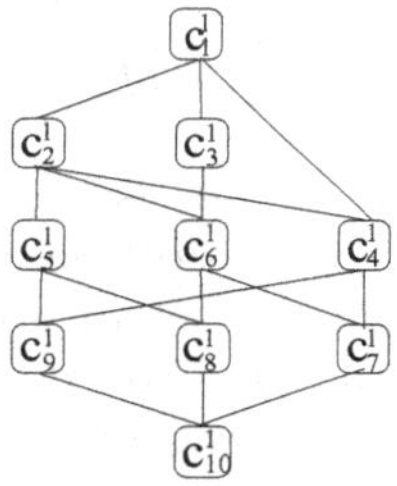

Fig. 1. A Formal Context and its Concept Lattice

In [1], based on the subconcept-superconcept relation $\leq$, a so-called neighbour relation is defined. Thus, if c_1, c_2 are two concepts in the context (X, Y, I), we say that c_1 is the nearest neighbour of c_2 if either $c_1 < c_2$ and $c_1 \leq c < c_2$ implies $c = c_1$ or $c_2 < c_1$ and $c_2 < c \leq c_1$ implies $c = c_1$. It is clear that the nearest neighbour represents the minimal refinement or enlargement of a concept in a given formal context.

3   Threshold Formal Concepts

In what follows, we consider X to be a set of documents (Web pages) and Y a set of terms (descriptors). For example, for each document we can take as attributes some selected indexed words obtained by eliminating all stopwords and very common terms, stemming words to their roots or limiting them to nouns and few descriptive adjectives and verbs [9].

We define now our model. Firstly, a new type of derivation operators, called threshold-derivation operators, are described. Let $y \in Y$ be a term. A function $w_y : X \to [0, \infty)$ is called *weight* of the term y. The weight $w_y(x)$ for $x \in X$ can be, for example, the normalised frequency of the term y in the document x, but we consider $w_y(x)$ to be the term frequency-inverse document frequency of the term y (see [10]). We consider the weighted incidence relation $\tilde{I} : X \times Y \longrightarrow [0, \infty)$, $\tilde{I}(x, y) = w_y(x)$, for all $x \in X$, $y \in Y$ and the weighted formal context $(X, Y, \tilde{I})$.

Definition 4. *Let $t \in [0, \infty)$. We define the function $\psi_t : X \to \mathcal{P}(Y)$ by*

$$\psi_t(x) = \{y \in Y | w_y(x) \geq t\}, \forall x \in X.$$

The set $\psi_t(x)$ represents all terms in Y such that their weight in the document x is greater than the threshold t. We now define a new type of derivation operators, namely the threshold-derivation operators.

Definition 5. *Let $t \in [0, \infty)$. $\alpha_t : \mathcal{P}(X) \to \mathcal{P}(Y)$ and $\beta_t : \mathcal{P}(Y) \to \mathcal{P}(X)$*

$$\alpha_t(A) = \bigcap_{x \in A} \psi_t(x), \ \forall A \subseteq X, A \neq \varnothing, \ \alpha_t(\varnothing) = Y \ and$$
$$\beta_t(B) = \{x \in X \mid B \subseteq \psi_t(x)\}, \forall B \subseteq Y$$

are called threshold-derivation operators (t-derivation operators).

In fact, $\alpha_t(A)$ is the set of terms which belong to all documents in A with weights in each of these documents greater than the threshold t, while $\beta_t(B)$ represents the set of documents such that the weight of any term from B is greater than t. The t-derivation operators generalise the operators defined in the previous paragraph. If we take $0 < t \leq \min\{w_y(x) \mid (x, y) \in X \times Y\}$, then we obtain $\alpha_t(A) = \alpha(A)$, for all $A \subseteq X$ and $\beta_t(B) = \beta(B)$ for all $B \subseteq Y$.

Example 2. Let $X = \{d_1, d_2, .., d_6\}$ be a set of documents, $Y = \{ring, gold, algebra, planet\}$ a set of attributes, and $\tilde{I}$ the incidence relation given in Fig.2:

	ring	gold	algebra	planet
d_1	0	3	0	0
d_2	4	0	3	1
d_3	3	0	2	4
d_4	1	1	2	0
d_5	4	2	0	3
d_6	2	4	0	2

Fig. 2. A Weighted Formal Context

We have
$$\alpha_{1.5}(\{d_2, d_3\}) = \psi_{1.5}(d_2) \cap \psi_{1.5}(d_3) =$$
$$\{ring, algebra\} \cap \{ring, algebra, planet\} = \{ring, algebra\},$$

which means that the attributes "$ring, algebra$" belong to documents d_2, d_3 and their weights are greater than the threshold $t_2 = 1.5$. We have $\alpha_{1.5}(\{d_2, d_3\}) \neq \alpha(\{d_2, d_3\}) = \{ring, algebra, planet\}$, which proves that α_t generalises α, the derivation operator defined in Example 1. If $t_3 = 2.5$, then we obtain

$$\alpha_{2.5}(\{d_2, d_3\}) = \psi_{2.5}(d_2) \cap \psi_{2.5}(d_3) =$$
$$\{ring, algebra\} \cap \{ring, planet\} = \{ring\}.$$

Because $\beta_{1.5}(\{ring, algebra\}) = \{d_2, d_3\}$, it results that d_2, d_3 represent all documents which have "$ring, algebra$" as attributes with weights greater than 1.5. We also notice that

$$\beta_{1.5}(\{ring, algebra\}) \neq \beta(\{ring, algebra\}) = \{d_2, d_3, d_4\}.$$

To define our model of formal concepts, the following result is fundamental.

Proposition 1. *The pair (α_t, β_t) is an antitone Galois connection between the ordered sets $(\mathcal{P}(X), \subseteq), (\mathcal{P}(Y), \subseteq)$.*

Proof. (i) We first prove that α_t is decreasing (i.e., $A_1 \subseteq A_2$ implies $\alpha_t(A_2) \subseteq \alpha_t(A_1)$). We have $\alpha_t(A_2) = \bigcap_{x \in A_2} \psi_t(x) \subseteq \bigcap_{x \in A_1} \psi_t(x) = \alpha_t(A_1)$.

(ii) Next, we show that the operator $\beta_t \circ \alpha_t$ is extensive (i.e., $A \subseteq \beta_t(\alpha_t(A))$).
We have $\beta_t(\alpha_t(A)) = \beta_t\left(\bigcap_{x \in A} \psi_t(x)\right) = \left\{x' \in X \mid \bigcap_{x \in A} \psi_t(x) \subseteq \psi_t(x')\right\}$.
If $x \in A$, it follows that $\bigcap_{x \in A} \psi_t(x) \subseteq \psi_t(x)$, hence $x \in \beta_t(\alpha_t(A))$.

(iii) We prove that the t-operator β_t is decreasing (i.e. $B_1 \subseteq B_2$ implies $\beta_t(B_2) \subseteq \beta_t(B_1)$). Let $B_1 \subseteq B_2$. If $x \in \beta_t(B_2)$, then $B_2 \subseteq \psi_t(x)$, hence $B_1 \subseteq \psi_t(x)$.

(iv) Finally, we show that the operator $\alpha_t \circ \beta_t$ is extensive (i.e. $B \subseteq \alpha_t(\beta_t(B))$). The inclusion $B \subseteq \psi_t(x)$ is true for all $x \in \beta_t(B)$, and consequently,
$$B \subseteq \bigcap_{x \in \beta_t(B)} \psi_t(x) = \alpha_t(\beta_t(B)).$$

We study how the operators α_t, β_t depend on the threshold t.

Proposition 2. *Let $t \in [0, \infty)$, $A \subseteq X$ be a subset of documents and $B \subseteq Y$ a subset of terms. Then α_t, β_t are decreasing with respect to t, i.e.*
$$\alpha_{t+1}(A) \subseteq \alpha_t(A) \quad and \quad \beta_{t+1}(B) \subseteq \beta_t(B), \forall t \in [0, \infty).$$

Proof. In the case $A \neq \varnothing$ we have:
$$\psi_{t+1}(x) = \{y \in Y \mid w_y(x) \geq t + 1\} \subseteq \{y \in Y \mid w_y(x) \geq t\} = \psi_t(x),$$
for all $x \in X$, which implies that $\bigcap_{x \in A} \psi_{t+1}(x) \subseteq \bigcap_{x \in A} \psi_t(x)$. If $A = \varnothing$ the relation is obvious. The second inclusion results in a similar way.

We define the notion of formal concept which allows to develop more efficient methods for information retrieval.

Definition 6. *Let $t \in [0, \infty)$. A pair $(A, B) \in \mathcal{P}(X) \times \mathcal{P}(Y)$ is said to be a threshold-formal concept (t-formal concept) if*
$$\alpha_t(A) = B \quad and \quad \beta_t(B) = A.$$

The set A is called threshold-extent, and the set B threshold-intent of the t-formal concept (A, B); we denote $A = ext((A, B))$, and $B = int((A, B))$. The set of t-formal concepts is denoted by $\mathcal{C}_t(X, Y, \tilde{I})$. For each document $x \in X$, we define the associate document concept $(\beta_t(\alpha_t(x)), \alpha_t(x))$, and for each word $y \in Y$, the associate term concept $(\beta_t(y), \alpha_t(\beta_t(y)))$.

Example 3. We consider the context from Example 2. We find that

$$\alpha_{1.5}(\{d_2, d_3\}) = \{ring, algebra\} \text{ and } \beta_{1.5}(\{ring, algebra\}) = \{d_2, d_3\},$$

hence the pair $(\{d_2, d_3\}, \{ring, algebra\})$ is a t-formal concept corresponding to $t_2 = 1.5$. This concept is different from the "classical" formal concept $(\{d_2, d_3\}, \{ring, algebra, planet\})$ (see Figure 1).

We now study how the t-derivation operators act on t-formal concepts.

Proposition 3. *Let $t \in [0, \infty)$, J an index set and $(A_j, B_j) \in \mathcal{C}_t(X, Y, \tilde{I})$, for all $j \in J$. Then, we have:*

(i) $\alpha_t(\bigcup\limits_{j \in J} A_j) = \bigcap\limits_{j \in J} \alpha_t(A_j)$
   (every intersection of intents is an intent);
(ii) $\beta_t(\bigcup\limits_{j \in J} B_j) = \bigcap\limits_{j \in J} \beta_t(B_j)$
   (every intersection of extents is an extent).

Proof. (i)

$$\alpha_t(\bigcup\limits_{j \in J} A_j) = \bigcap\limits_{x \in \cup A_j} \psi_t(x) = \bigcap\limits_{j \in J} \bigcap\limits_{x \in A_j} \psi_t(x) = \bigcap\limits_{j \in J} \alpha_t(A_j).$$

(ii) We have

$$\beta_t(\bigcup\limits_{j \in J} B_j) = \{x \in X \mid \bigcup\limits_{j \in J} B_j \subseteq \psi_t(x)\} = \bigcap\limits_{j \in J} \{x \in X \mid B_j \subseteq \psi_t(x)\}$$

$$= \bigcap\limits_{j \in J} \beta_t(B_j).$$

Corollary 1. *(i) Every extent of a t-formal concept is the intersection of some terms extents.*
(ii) Every intent of a t-formal concept is the intersection of some documents intents.

The hierarchical order in the t-Galois lattices can be introduced in a similar way as for the Galois lattices.

Definition 7. *Let $t \in [0, \infty)$ and (A_1, B_1), $(A_2, B_2) \in \mathcal{C}_t(X, Y, \tilde{I})$. Then (A_1, B_1) is a subconcept of (A_2, B_2) whenever $A_1 \subseteq A_2$, and we denote this by $(A_1, B_1) \leq (A_2, B_2)$. The set of all formal concepts ordered by $\leq$ is denoted by $\underline{\mathcal{C}}_t(X, Y, \tilde{I})$.*

We prove the basic theorem on threshold-concept lattices:

Theorem 2. *Let $t \in [0, \infty)$. Then the poset $\underline{\mathcal{C}}_t(X, Y, \tilde{I})$ of t-formal concepts is a complete lattice.*

Proof. The proof is similar to the classical one. If J is an index set and $\{(A_j, B_j)|j \in J\} \subseteq C_t(X, Y, \tilde{I})$ is a subset of t-formal concepts, we define the t-infimum and, respectively, the t-supremum of this set by:

$$\bigwedge_{j \in J} (A_j, B_j) = (\bigcap_{j \in J} A_j, \alpha_t(\beta_t(\bigcup_{j \in J} B_j)))$$

$$\bigvee_{j \in J} (A_j, B_j) = (\beta_t(\alpha_t(\bigcup_{j \in J} A_j)), \bigcap_{j \in J} B_j).$$

We first prove the formula for the infimum. Since $A_j = \beta_t (B_j)$ for all $j \in J$, and according to Proposition 3, we have

$$(\bigcap_{j \in J} A_j, \alpha_t(\beta_t(\bigcup_{j \in J} B_j))) = (\beta_t(\bigcup_{j \in J} B_j), \alpha_t(\beta_t(\bigcup_{j \in J} B_j))),$$

which means that this pair is a t-formal concept. Since $\inf\{A_j \mid j \in J\} = \bigcap_{j \in J} A_j$, and taking into account the order relation for t-concepts, it follows that the infimum of the subset $\{(A_j, B_j)|j \in J\}$ is $\bigwedge_{j \in J} (A_j, B_j)$.

The proof for the supremum formula is similar.

It is worth noting that, though the t-concepts set is a complete lattice (as in the classical case), it is qualitatively different. The t-concepts are built using a relevance condition over the terms; this dependence can be exploited to find better ways to search through a collection of Web documents.

4   Iterative Retrieval Process

Using a sequence of t-concept lattices and techniques based on querying and navigating through the corresponding lattices, we present an iterative procedure which can improve the the process of extracting information from data sets. At each step, using a relevance condition, we build a new Galois lattice through the MapReduce programming model. Thus, we obtain a decreasing sequence of concepts (c^i) such that the intents of c^i are the successive queries Q_i submitted to the system, while the extents represent the retrieved documents.

Let $t_1 = 0$ be the initial threshold, $K_1 = (X, Y, \tilde{I})$ a given weighted formal context, and Q_1 the initial query. Let us describe the step i of the procedure ($i \in \mathbb{N}^*$). Let Q_{i-1} be the subset of terms which represents the user query, obtained in the step $(i-1)$ and $t_i > t_{i-1}$, the new threshold value. Next, using the MapReduce model we build the concept lattice $C_i(X, Y, \tilde{I})$, $i \in \mathbb{N}^*$ associated to the new context K_i. There are a lot of algorithms to find the Galois lattice, for example Ganter's "Next Closure Algorithm" [7], or the algorithm proposed in [8]. As inserting or deleting documents or terms in the formal context is an usual operation, it is clear that we need an incremental algorithm to build the

t-conceptual structure. Therefore, we think that the AddIntent algorithm [12] is a good choice, in order to generate the t-Galois lattice.

In what follows we describe how we used the MapReduce software framework to implement our model. First, using SPARQL, a Python script harvests strings describing resources on the Web. The Python script also tokenizes the strings, removes stop words and reduces words to stems using the platform NLTK for language processing. The data is then fed into a MapReduce pipeline. The pipeline first calculates the term frequency-inverse document frequency for the collection of documents. It filters out the top n terms for each document. Finally, it calculates the intents of all t-formal concepts. To achieve this task, it just takes two passes: the first pass gathers all the intents and extents of individual objects and attributes, and the second pass calculates the t-formal concepts. 1. First pass:

```
map phase (object, attribute) : (uri, string)
   emit (object, attribute) and (attribute, object)
reduce phase (x, ys) : (a, [b])
   pipe into second pass
```

2. Second pass:

```
map phase (x, ys) : (a, [b])
   for each y in ys
      emit (y, ys)
reduce phase (y, yss) : (a, [[a]])
   store (y, intersection yss).
```

Let us consider an example. Let $\{(a,t),(b,t),(b,u),(c,u)\}$ be the incidence relation. In one very efficient pass, we can obtain the extents and intents: $\{(a,t),(b,t,u),(c,u)\}, \{(t,a,b),(u,b,c)\}$. From this we obtain the document concepts and the term concepts: $\{(t,t),(u,u)\}, \{(a,a,b),(b,b),(c,b,c)\}$. The massively parallel nature of MapReduce with the grouping of keys between the map and reduce phase overcomes the complexity problem.

The sequence of Galois lattices obtained allows gradual refinement or enlargement of a query. Let us describe how we can construct the new query Q_{i+1}. When the query Q_i is submitted by the user, there are two possibilities: either the Galois lattice $C_i(X, Y, \tilde{I})$ contains a concept $c^i = (ext(c^i), int(c^i))$ such that the query can be interpreted as the intent of c^i ($int(c^i) = Q_i$), or this property is not satisfied. In the later case, following the ideas from [1], we augment the Galois lattice $C_i(X, Y, \tilde{I})$ with a pseudo-concept denoted pc^i such that the intent of pc^i equals the query Q_i and as extent, we consider a set which contains one pseudodocument pd^i. We can navigate through $C_i(X, Y, \tilde{I})$ starting from the concept c^i or from the pseudoconcept pc^i, and choose another concept of the lattice, say $\bar{c}^i$. Then we take $Q_{i+1} = int(\bar{c}^i)$. The concept $\bar{c}^i$ is usually chosen such that to give a minimal refinement or enlargement of c^i so that, in fact, $\bar{c}^i$ is one of the neighbours nodes (parents or children) of c^i or pc^i. One can also take $Q_{i+1} = Q_i$. It is worth noting that if c^i is a concept in $C_i(X, Y, \tilde{I})$, it generally does not result that c^i is a concept in $C_{i+1}(X, Y, \tilde{I})$; neither the converse property is generally true. We emphasise that, in order to traverse the conceptual hierarchy starting from the query concept, we use the breadth-first search algorithm implemented again on a parallel architecture.

We argue now that the query Q_i can be reformulated at each step such that the sequence of the retrieved extents $\left(ext\left(c^i\right)\right)_{i \in \{1,2,..,p\}}$, $p \in \mathbb{N}^*$, which depends on the threshold t_i, to decrease. At step i we must take into account three situations. In the first one we consider $Q_i = Q_{i+1}$. In the second one, we choose $\bar{c}^i$ to be a child of the concept c^i or of the pseudoconcept pc^i (we enlarge the query) and we have $ext\left(c^i\right) \supseteq ext\left(c^{i+1}\right)$, which means that we refine the information retrieved.

In the third situation, $\bar{c}^i$ is a parent of c^i or pc^i (the query is refined), hence we have $ext\left(c^i\right) \subseteq ext\left(c^{i+1}\right)$ which is the minimal enlargement of the document set.

For example, if at each step i we are in the first case, the sequence of sets $\left(ext\left(c^i\right)\right)_{i \in \{1,2,..,p\}}$ (the extents of concepts c^i) is decreasing
$$ext\left(c^{i-1}\right) \supseteq ext\left(c^i\right), \forall i \in \{1, 2, .., p\},$$
and the procedure stops when $ext\left(c^i\right)$ become $\varnothing$.

In the general case, the sequence of retrieved documents can be split into a finite number of decreasing "subsequences":

$$ext(c^1) \supseteq ext\left(c^2\right) \supseteq ... \supseteq ext\left(c^{i_1}\right),$$
$$ext(c^{i_1+1}) \supseteq ext\left(c^{i_1+2}\right) \supseteq ... \supseteq ext\left(c^{i_2}\right),$$
$$ext(c^{i_2+1}) \supseteq ext\left(c^{i_2+2}\right) \supseteq ... \supseteq ext\left(c^{i_3}\right)...$$

such that at steps i_1, i_2,.. the user chooses the parent node associated to the query concept.

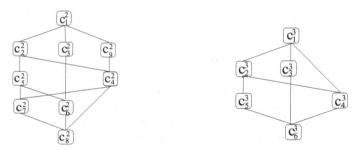

Fig. 3. Threshold $t_2 = 1.5$

Example 4. We illustrate our procedure on a small example. Let $K_1 = (X, Y, \tilde{I})$ be the formal context given in the table presented in Example 2, and the initial threshold $t_1 = 0$. The t-concept lattice $\underline{\mathcal{C}}_1\left(X, Y, \tilde{I}\right)$ is represented in Figure 1. At step 1, we choose the threshold t_2 to be 1.5.

In Figure 3, the diagram of the concept lattice $\underline{\mathcal{C}}_2(X, Y, \tilde{I})$ is sketched; the concepts are: $c_1^2 = (\{d_1, d_2, d_3, d_4, d_5, d_6\}, \varnothing)$, $c_2^2 = (\{d_2, d_3, d_5, d_6\}, \{ring\})$, $c_3^2 = (\{d_1, d_5, d_6\}, \{gold\})$, $c_4^2 = (\{d_2, d_3\}, \{ring, algebra\})$, $c_5^2 = (\{d_3, d_5, d_6\}, \{ring, planet\})$, $c_6^2 = (\{d_5, d_6\}, \{ring, gold, planet\})$, $c_7^2 = (\{d_3\},$

$\{ring, algebra, planet\})$, $c_8^2 = (\varnothing, \{ring, gold, algebra, planet\})$ and $c_9^2 = (\{d_2, d_3, d_4\}, \{algebra\})$. Next, we consider $t_3 = 2.5$ and $t_4 = 3.5$, respectively. The Hasse diagrams of the suitable Galois lattices $\underline{\mathcal{C}}_i(X, Y, \tilde{I})$, $i \in \{3, 4\}$ can be found in a similar way, and they are depicted in Figure 3 and Figure 4, respectively.

The concepts for $t_3 = 2.5$ and $t_4 = 3.5$ are $c_1^3 = (\{d_1, d_2, d_3, d_4, d_5, d_6\}$, $\varnothing)$, $c_2^3 = (\{d_2, d_3, d_5\}, \{ring\})$, $c_3^3 = (\{d_1, d_6\}, \{gold\})$, $c_4^3 = (\{d_2\}, \{ring, algebra\})$, $c_5^3 = (\{d_3, d_5\}, \{ring, planet\})$, $c_6^3 = (\varnothing, \{ring, gold, algebra, planet\})$, respectively $c_1^4 = (\{d_1, d_2, d_3, d_4, d_5, d_6\}, \varnothing)$, $c_2^4 = (\{d_2, d_5\}, \{ring\})$, $c_3^4 = (\{d_6\}, \{gold\})$, $c_4^4 = (\{d_3\}, \{planet\})$ and $c_5^4 = (\varnothing, \{ring, gold, algebra, planet\})$.

Let us see the procedure at work. One of the simplest sequences of queries is obtained, for example, by choosing $Q_1 = \{gold\}$ and take $Q_i = Q_{i+1}$, $i \in \{1, 2, 3\}$; in this case we must mention that at every step there exists a concept c^i such that $int(c^i) = Q_i, i \in \{1, 2, 3, 4\}$. The set of documents retrieved at each search, represented by $ext(c^i)$, $i \in \{1, 2, 3, 4\}$, is decreasing: $c_3^1 = \{d_1, d_4, d_5, d_6\}$, $c_3^2 = \{d_1, d_5, d_6\}$, $c_3^3 = \{d_1, d_6\}$, $c_3^4 = \{d_6\}$, $\varnothing$. We notice again that as t_i increases the documents corresponding to the query become more relevant. For example, the more relevant documents for the query $\{gold\}$ are d_6, d_1, d_5, d_4 (in this order).

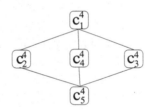

Fig. 4. Threshold $t_4 = 3.5$

Let us analyse a more complicated example. If we start with the query $Q_1 = \{ring, algebra\}$, we obtain in $\underline{\mathcal{C}}_1$ the formal concept $c_4^1 = (\{d_2, d_3, d_4\}$, $\{ring, algebra\})$. Since in the lattice $\underline{\mathcal{C}}_2$, there is no concept such that its intent equals Q_1, we must add to $\underline{\mathcal{C}}_2$ a pseudoconcept $pc^2 = (pd^2, \{ring, algebra\})$ and choose Q_2 to be the intent of the nearest neighbour concept of pc^2. Thus, if we choose $Q_2 = \{ring, algebra, planet\}$ the extent $ext(c_7^2) = \{d_3\}$ of the obtained concept decreases. If $Q_2 = \{ring\}$ we get the concept $c_2^2 = (\{d_2, d_3, d_5, d_6\}, \{ring\})$, and so the set of retrieved documents $ext(c_2^2) = \{d_2, d_3, d_5, d_6\}$ does not decrease. Thus, the user either can find new relevant documents (in our case d_5, d_6), or can deduce documents d_2, d_4 to be less important than d_3. For the next step of the procedure we take $Q_3 = \{ring\}$, and find in the lattice $\underline{\mathcal{C}}_3$ the extent $ext(c_2^3) = \{d_2, d_3, d_5\}$, which means that the sequence of retrieved documents decreases again. Finally, in step 4, we can take $Q_4 = \{ring\}$ and we obtain $ext(c_3^4) = \{d_2, d_5\})$. Hence, at each step, the user has the possibility to choose the query such that the sequence $\left(ext(c^i)\right)_{i \in \{1, 2, 3, 4\}}$ decreases.

4.1   Documents Ranking

It is obvious that for large collections, the set of documents which is associated with the query could be very big. Using the hierarchical structure of the t-Galois lattices, we propose a method to rank the documents retrieved by the system.

The FCA-based IR model defined in [15] implements an explicit relevance feedback. When the query concept has too many parent or child concepts, it is difficult for the user to inspect all the intents of these concepts to reformulate the query. To fix this problem, the authors define an order relation on the set of these children (or parents) nodes, a so-called *preference* relation. By means of this relation, the user can select the most relevant documents. This model is different from that developed in [13], where the choice of the relevant documents which are used to the query expansion is made directly by the user. In [1] the query is merged into the concepts document space and the similarity between the query and a document is computed as the length of the shortest path linking the concept query with the concept whose extent equals the set of attributes of the document.

To compute the rank of a document, we approach a slightly different method combined with the use of the t-concept lattices. Let $p \in \mathbb{N}^*$ and $F = \{t_1, t_2, ..., t_p\} \subset [0, \infty)$, $t_1 < t_2 < ... < t_p$ be the set of values of the threshold (the optimal choice for the set F depends on the document collection and is established through experiments). For each $t_i \in F$, $i \in \{1, 2, .., p\}$, we denote with $\mathcal{C}_i$, C_i the Galois lattice and the set of concepts that belong to $\mathcal{C}_i$, respectively. We remind that the distance between two concepts C_i^1, $C_i^2 \in C_i$ is defined by:

$$d_i : C_i \times C_i \to [0, \infty), \ d_i \left(C_i^1, C_i^2 \right) = \text{length of the shortest path from } C_i^1 \text{ to } C_i^2.$$

Now, let $t_i \in F$ be a fixed threshold, $Q \subset Y$ the query, and $d \in X$ a document. We define the similarity between d and Q in the threshold lattice $\mathcal{C}_i$ as being the least of the distances from the query concept C_Q to the nodes C_i^d which contain d and are superconcepts of C_Q:

$$sim_i : X \to [0, \infty), \ sim_i (d, Q) = \min\{d_i \left(C_i^d, C_Q \right) \mid C_i^d \in \mathcal{C}_i^d, \ C_i^d \supseteq C_Q\}.$$

In fact this similarity is given by the number of minimal refinements to modify the query such that to equal the intent of a concept C_i^d. It is natural to consider a document that better matches the query, to have a greater rank.

Definition 8. *A document d_1 is ranked ahead of d_2 in the lattice $\mathcal{C}_i$, related to a query Q, if $sim_i (d_1, Q) < sim_i (d_2, Q)$.*

Example 5. In example 4, if we set $Q = \{ring, algebra\}$ the document d_3 is ranked better than d_5 in the lattice $\mathcal{C}_1$ (see Figure 1) because $sim_1 (d_3, Q) = 0$ and $sim_i (d_5, Q) = 1$.

Taking into account the sequence of t-Galois lattices and using the method previously described, we order the documents in the concept lattice $\mathcal{C}_1$. The set of documents which have the same rank still could be very large. We now augment the defined rank method with a criterion based on terms' weight in documents. Let d_1 and d_2 be two documents of equal rank, related to Q. We use the iterative process described in previous section. Let $(c^i)_{i \in \{1,2,..,p\}}$ be a sequence of concepts obtained by applying the above mentioned procedure, $c^i = (ext(c^i), int(c^i))$, $ext(c^i) \supseteq ext(c^{i+1})$. Since when i increases the concepts c^i become more relevant, it is natural to define:

Definition 9. *Let d_1, $d_2 \in X$ with $sim_1(d_1, Q) = sim_1(d_2, Q)$. The document d_1 is ranked ahead d_2 related to the query Q and we denote by $d_1 \succ d_2$, if there is $i \in \{1, 2, .., p\}$ such that $d_1 \in ext(c^i)$ and $d_2 \notin ext(c^i)$.*

Example 6. Using Example 4, we take $Q = \{ring, algebra\}$. The documents d_2, d_3, d_4 have, in lattice $\underline{C}_1$, the rank 0, while d_1, d_5, d_6 the rank 1. We choose the concepts c_4^1 and c_4^3, which intents equal the query, $ext(c_4^1) = \{d_2, d_3, d_4\} \supseteq ext(c_4^3) = \{d_2\}$. Because c_4^3 is more relevant than c_4^1 and $d_2 \in ext(c_4^3)$, $d_3, d_4 \notin ext(c_4^3)$ it results that d_2 is more relevant than d_3 and d_4. Now, we consider the sequence $(c_2^i)_{i \in \{1,2,3,4\}}$, $c_2^i \in C_i$, for all $i \in \{1, 2, 3, 4\}$ of parent concepts of c_Q. The documents that belong to the most relevant concepts have the bigger rank. We have $\{d_2, .., d_6\} \supseteq \{d_2, d_3, d_5, d_6\} \supseteq \{d_2, d_3, d_5\} \supseteq \{d_2, d_5\}$, hence we find that $d_3 \succ d_4$ and $d_5 \succ d_6 \succ d_1$.

Once the top ranked documents are returned in response to the user's request Q, the system can add to the query the most relevant terms selected from these documents, thus improving the retrieval process.

5   Conclusion

In this paper we developed a theoretical model which integrates some relevance conditions to an IR model based on concept lattices. Using the weights of the terms, we described a new type of concepts, namely threshold formal concepts, which offers a more efficient way to extract the information from a collection of Web documents. The sequence of hierarchical structures we built provides a dynamical IR model. During the search process, as the threshold increases the t-formal concepts become more relevant, thus obtaining a more rapid access to the documents needed. Depending on the feedback, the user can navigate, not only in the same t-concept lattice, but in any lattice in the sequence and dynamically refine or enlarge the query. Lattices generated with FCA techniques can be complex, thus difficult to use in practical applications. Due to the weighted form of the context, even if we work with big data tables, by increasing the threshold t we obtain a sequence of t-formal contexts with a lower density which can decrease the complexity of the method.

The novelty of our approach comes also from the use of a parallel and distributed system to implement the model. We used the MapReduce software framework which allows to achieve the NLP filtering, and to generate the sequence of the t-concept lattices. The preliminary results are very promising. Using parallel computation, we overcome the complexity of the iterative process, even in the case of large and dense t-formal contexts.

We have also introduced a new ranking method which combines hierarchical navigation in a t-Galois lattice with relevance conditions. Thus, output documents are ranked according to their similarity with the query based on the structure of the t-lattices, and using the fact that documents which occur in more relevant concepts are more relevant. As future work, we intend to improve our experimental studies over large, dense contexts. Thus, we would like to study what values of the threshold t should the user choose in order to optimise the described iterative process. By experimenting with a wide range of values, we hope to observe which features of the weighted formal context yield a fast retrieval behaviour.

Acknowledgements. We thank the anonymous reviewers for their comments. The work was supported by a grant of the Romanian Authority for Scientific Research, project number PN-II-ID-PCE-2011-3-0919.

References

1. Carpineto, C., Romano, G.: Order-Theoretical Ranking. Journal of the American Society for Information Science 51, 587–601 (2000)
2. Carpineto, C., Romano, G.: Exploiting the Potential of Concept Lattices for Information Retrieval with CREDO. J. Univ. Comput. Sci. 10, 985–1013 (2004)
3. Cigarrán, J.M., Gonzalo, J., Peñas, A., Verdejo, F.: Browsing Search Results via Formal Concept Analysis: Automatic Selection of Attributes. In: Eklund, P. (ed.) ICFCA 2004. LNCS (LNAI), vol. 2961, pp. 74–87. Springer, Heidelberg (2004)
4. Codocedo, V., Lykourentzou, I., Napoli, A.: A Contribution to Semantic Indexing and Retrieval Based on FCA; An Application to Song Datasets. In: Proceedings CLA. CEUR Workshop, vol. 972, pp. 257–268 (2012)
5. Dau, F., Ducrou, J., Eklund, P.: Concept Similarity and Related Categories in SearchSleuth. In: Eklund, P., Haemmerlé, O. (eds.) ICCS 2008. LNCS (LNAI), vol. 5113, pp. 255–268. Springer, Heidelberg (2008)
6. El Qadi, A., Aboutajdin, D., Ennouary, Y.: Formal Concept Analysis for Information Retrieval. Int'l Journal of Computer Science and Information Security 7, 119–125 (2010)
7. Ganter, B., Wille, R.: Formal Concept Analysis. Mathematical Foundations. Springer (1999)
8. Godin, R., Missaoui, R., Alaoui, H.: Incremental Concept Formation Algorithms Based on Galois Lattices. Computational Intelligence 11, 246–267 (1995)
9. Karp, D., Schabes, Y., Zaidel, M., Egedi, D.: A Freely Available Wide Coverage Morphological Analyzer for English. In: Proceedings 14th COLING, pp. 950–955 (1992)
10. Manning, C.D., Raghavan, P., Schütze, H.: An Introduction to Information Retrieval. Cambridge University Press (2009)
11. Messai, N., Devignes, M.-D., Napoli, A., Smaïl-Tabbone, M.: Querying a Bioinformatic Data Sources Registry with Concept Lattices. In: Dau, F., Mugnier, M.-L., Stumme, G. (eds.) ICCS 2005. LNCS (LNAI), vol. 3596, pp. 323–336. Springer, Heidelberg (2005)

144   G. Ciobanu, R. Horne, and C. Văideanu

12. van der Merwe, D., Obiedkov, S., Kourie, D.: AddIntent: A New Incremental Algorithm for Constructing Concept Lattices. In: Eklund, P. (ed.) ICFCA 2004. LNCS (LNAI), vol. 2961, pp. 372–385. Springer, Heidelberg (2004)
13. Nauer, E., Toussaint, Y.: Dynamical Modification of Context for an Iterative and Interactive Information Retrieval Process on the Web. In: Proceedings CLA. CEUR Workshop, vol. 331, 12 p. (2007)
14. Priss, U.: Lattice-Based Information Retrieval. Knowledge Organization 27, 132–142 (2000)
15. Spyratos, N., Meghini, C.: Preference-Based Query Tuning Through Refinement/Enlargement in a Formal Context. In: Dix, J., Hegner, S.J. (eds.) FoIKS 2006. LNCS, vol. 3861, pp. 278–293. Springer, Heidelberg (2006)
16. Wille, R.: Restructuring Lattice Theory: An Approach Based on Hierarchies of Concepts. In: Rival, I. (ed.) Ordered Sets, pp. 445–470. Reidel (1982)

A Formal Topology of Web Classification

Gabriel Ciobanu[1] and Dănuţ Rusu[2]

[1] Romanian Academy, Institute of Computer Science, Iaşi, Romania
[2] "A.I.Cuza" University of Iaşi, Faculty of Mathematics, Romania
gabriel@iit.tuiasi.ro, drusu@uaic.ro

Abstract. The World Wide Web is a graph in which the nodes are the pages and the edges are web links. A classification associates to each web page a set of documents. This paper presents a topological approach of the web classification, aiming to describe classifications and search processes over the web. An original feature is provided by the distinctness operators which are able to detect when a document is not in a certain classification class. We prove that there is a bijection between regular distinctness operators and regular topologies. Adding some properties to a regular distinctness operator, we associate it to a regular Alexandrov topology.

1   Introduction

A web system can be represented as a directed graph in which the nodes are the web pages, and the edges represent the HTML links between the web pages. Surfing the web in order to find certain information is done by direct navigation, namely by following the links in the web documents. Web search engines are based on indexing. In both navigation and web searching, we can use the notion of classification as given in Definition 1. Having a classification, it is necessary to decide if a web document is part of a class in that classification. If a document x belongs to the class of a document y, we consider that the class of x is a subset of the class of y. For instance, if x is a document containing information regarding differential geometry, and y is a document containing information regarding geometry (in general), then the class of x (differential geometry) is included in the class of y (general geometry). In order to provide a formal approach, we describe web classifications in terms of graphs, domain theory and topology.

It is important to detect that a document is an element of a class. This paper deals with differentiating between web documents, namely detecting when a document is not in a certain class. For this we introduce and study a distinctness operator. A web page can contain several types of information, and so a web document can belong to more than one class. Together with the possibility of finding these classes, it would be useful to be able to differentiate, to distinguish between the documents, and exclude a document from certain classes. This process of excluding a document from a class can be done by comparing sets of keywords, and by using an inequality relation. Classical inequality (defined as the logical negation of equality) is not appropriate because we can have two different documents which can belong to the same class. This approach leads to a

N. Hernandez et al. (Eds.): ICCS 2014, LNAI 8577, pp. 145–158, 2014.
DOI: 10.1007/978-3-319-08389-6_13, © Springer International Publishing Switzerland 2014

notion of "distinctness" which is related to that of "apartness". Apartness spaces were introduced in [3] as a foundation for constructive topology, and a set-set apartness relation is closely related to the classical notion of proximity, modelling the concept of nearness of sets in a metric or topological environment. In this paper we define and study a distinctness relation. We associate an Alexandrov topology to each classification in a one-to-one manner. Then we investigate regular distinctness operators, and prove a bijection between regular distinctness operators and regular topologies. By adding some properties to a regular distinctness operator, we can associate it to a regular Alexandrov topology.

2   Formal Classification via Domain Theory

A web classification can be seen as a binary relation between web documents, which groups the documents into several classes. As expected, this relation should be both reflexive (since each document must belong to its own class), and transitive (since if a document belongs to a particular class C, than the class associated with that document should also belong to C). A classification can also be interpreted as a preorder over the set of web documents. Alternatively, it can simply be treated as a function which associates each web document with its own class.

Definition 1. *Let $\mathcal{W}$ be a set, interpreted as the collection of web documents, and $f : \mathcal{W} \to \mathcal{P}(\mathcal{W})$ a function. f is called reflective if $x \in f(x)$, for all $x \in \mathcal{W}$; f is called hereditary if $y \in f(x)$ implies $f(y) \subseteq f(x)$; f is called a classification of $\mathcal{W}$ if it is reflective and hereditary.*

If f is a web classification, then $f(x)$ is called the class of the document x, and x is a generic keyword of the class $f(x)$. A web classification f is called *unambiguous* if every class has a unique generic keyword. Therefore, a web classification f is unambiguous iff f is injective.

Proposition 1. *A function $f : \mathcal{W} \to \mathcal{P}(\mathcal{W})$ is a classification iff $y \in f(x)$ is equivalent to $f(y) \subseteq f(x)$.*

We denote by $\mathcal{C}(\mathcal{W})$ the set of all web classifications, and by $\mathcal{C}_0(\mathcal{W})$ the set of all unambiguous web classifications. We define a relation over $\mathcal{C}(\mathcal{W})$ by $f \leq g$ iff $g(x) \subseteq f(x)$ for all $x \in \mathcal{W}$. This is a partial order on $\mathcal{C}(\mathcal{W})$; moreover, $(\mathcal{C}(\mathcal{W}), \leq)$ is a complete lattice.

Proposition 2.   *1. Let $\bot, \top : \mathcal{W} \to \mathcal{P}(\mathcal{W})$ defined by $\bot(x) = \mathcal{W}$ and $\top(x) = \{x\}$ for all $x \in \mathcal{W}$. $\bot$ is the least element (trivial classification) of $(\mathcal{C}(\mathcal{W}), \leq)$. $\top$ is the largest element (discrete classification) of $(\mathcal{C}(\mathcal{W}), \leq)$.*
   2. For all $A \subseteq \mathcal{C}(\mathcal{W})$, there exists the supremum defined by $\sqcup A(x) = \cap_{f \in A} f(x)$.
   3. If $A \subseteq \mathcal{C}(\mathcal{W})$ is a chain, then there exists the infimum $\sqcap A$ defined by $\sqcap A(x) = \cup_{f \in A} f(x)$.

Example 1. In order to explain how we can employ these classifications, we use a dictionary D of terms/words which appear in the classification process. Such a D could be related to a specific language, and could be developed through a learning process. To each document $x \in \mathcal{W}$ we associate a list $L(x)$ of its terms which are meaningful for the content of the document. Such a list could be provided to the indexing robots by the use of a specific label `meta`; alternatively, if these labels are missing, the list could be derived by considering the links contained in each document (this is a weaker version of the list indicated by the user with the help of `meta` labels). The terms of the list are used to build expressions according to the semantic rules of the language; we denote by $E_n(x)$ the set of expressions of length $n \in N^* = N \setminus \{0\}$ obtained by the terms of $L(x)$ which are present in the content of x. It is clear that $E_1(x) = L(x)$. We denote by $E(x)$ the set of expressions which are present in the content of x; since starting from a certain length n we have $E_n(x) = \emptyset$, we can write $E(x) = \cup_{n \in N^*} E_n(x)$. Using a specific separator "," between terms, each expression of length n can be viewed as an n-tuple of the Cartesian product $L(x)^n$, where the order of the terms is given by the semantic rules of the language. Thus $E_n(x) \subseteq L(x)^n$, and $E(x) \subseteq \cup_{n \in N^*} L(x)^n$.

In order to clarify the order of the terms imposed by the semantics of the language, let us assume that D is a dictionary of geometry terms in English, and x is a document (written in English). If $L(x) = \{\text{isosceles, triangle, rectangles}\}$, then (isosceles, triangle) could be in $E_2(x)$, (isosceles, rectangles, triangle) could be in $E_3(x)$, but (rectangles, isosceles, triangle) could be avoided by not being included in $E_3(x)$. Using these notations, we can define the following classifications:

$f_n(x) = \{y \in \mathcal{W} \mid E_n(x) \subseteq E(y)\}$,
$g_n(x) = \{y \in \mathcal{W} \mid \cup_{n \in N^*} E_n(x) \subseteq E(y)\}$, and
$f(x) \;\; = \{y \in \mathcal{W} \mid E(x) \subseteq E(y)\}$, where $n \in N^*$.

Then we have

$g_1(x) = f_1(x) = \{y \in \mathcal{W} \mid L(x) \subseteq L(y)\}$, and $f_n \leq g_n \leq g_{n+1} \leq f$.

Moreover, for all $x \in \mathcal{W}$, there exists $n_0 \in N^*$ such that $f_n(x) = \mathcal{W}, \forall n \geq n_0$.

A directed complete partial order (dcpo) is a partially ordered set where each of its directed subsets has a supremum. A dcpo with a least element is called a pointed dcpo. Considering the first two parts of the same Proposition 2, it results that $(\mathcal{C}(\mathcal{W}), \leq)$ is a pointed dcpo. Since "*a partially ordered set D is a dcpo if and only if each chain in D has a supremum*" ([1], Prop. 2.1.15), it follows from the first and the third part of Proposition 2 that $(\mathcal{C}(\mathcal{W}), \geq)$ is also a pointed dcpo. In domain theory, a function between two dcpos is called Scott-continuous if it maps directed sets to directed sets, and preserves their suprema.

Definition 2. *A Scott-continuous function $T : (\mathcal{C}(\mathcal{W}), \leq) \rightarrow (\mathcal{C}(\mathcal{W}), \leq)$ is called a direct classification operator. A Scott-continuous function $T : (\mathcal{C}(\mathcal{W}), \geq) \rightarrow (\mathcal{C}(\mathcal{W}), \geq)$ is called an inverse classification operator. A web classification f is called stable with respect to a classification operator T if $T(f) = f$.*

It is possible to get more than one stable classification with respect to a given classification operator; in such a situation, the interesting resulting classifications are given by the least or by the largest stable classifications. The existence of these classifications is ensured by the fixed-point theorem for Scott-continuous functions ([1], Theorem 2.1.19). We describe the process of building a stable classification.

Theorem 1. *Let T be a direct classification operator.*

1. *There exists the least stable classification f with respect to T, and $f = \sqcup_{n \in N} T^n(\bot)$.*
2. *The assignment $st : [\mathcal{C}(\mathcal{W}) \to \mathcal{C}(\mathcal{W})] \to \mathcal{C}(\mathcal{W})$ defined by $st(T) = \sqcup_{n \in N} T^n(\bot)$ is Scott-continuous, where $[\mathcal{C}(\mathcal{W}) \to \mathcal{C}(\mathcal{W})]$ is the set of Scott-continuous functions over $\mathcal{C}(\mathcal{W})$.*

A similar result is valid also for inverse classification operators; in this case, the least stable classification f with respect to T is $f = \sqcap_{n \in N} T^n(\top)$.

A classification operator could be considered as deriving from a set of criteria used to realize a web classification. In this way, a classification operator represents the key part of a classification process. Let us consider, for instance, a direct classification operator based on keywords and their semantic meaning.

First we apply the operator to the trivial classification operator, namely the bottom element of $(\mathcal{C}(\mathcal{W}), \leq)$. In this trivial classification all the documents belong to a single class. Next, by applying the classification operator, the class is refined and we get a finer classification. Let us assume that in the second step we differentiate the documents using a single keyword. For instance, we select a class containing the keyword "geometry". In the next step we differentiate the documents by using two keywords ("geometry" and "conics"); the selected class is a subclass of the previous one containing the combination of words "geometry" and "conics". And so on... The process stops when the application of the classification operator does not select a proper subset of the previous class; namely when the classification becomes stable. Notice that the finest classification is the discrete classification; however it is not interesting at all.

Let T be a direct classification operator that satisfies the additional condition $T(f \sqcup g) = Tf \sqcup Tg, \forall f, g \in \mathcal{C}(\mathcal{W})$. We refer to such operators as strong classification operators. Since T is Scott-continuous, and $(\mathcal{C}(\mathcal{W}), \leq)$ is a complete lattice, it follows that T holds supremum for any set. Then, based on the proof of existence for the superior adjunct ([1], Prop. 3.1.13), T is an inferior adjunct. More precisely, the function $S : (\mathcal{C}(\mathcal{W}), \leq) \to (\mathcal{C}(\mathcal{W}), \leq)$ defined as $Sf = \sqcup(T^{-1}(\downarrow f))$ is the only superior adjunct of T. Point-wise, the classification Sf is defined by $(Sf)(x) = \cap_{g \in T^{-1}(\downarrow f)} g(x) = \cap_{T(g) \leq f} g(x)$ for all $x \in \mathcal{W}$. Consequently, the pair (T, S) is an isotone Galois connection between $(\mathcal{C}(\mathcal{W}), \leq)$ and $(\mathcal{C}(\mathcal{W}), \leq)$, meaning that $f \leq S(g) \Leftrightarrow T(f) \leq g, \forall f, g \in \mathcal{C}(\mathcal{W})$. For any subset $A \subseteq \mathcal{C}(\mathcal{W})$ we define the sets $L(A) = lb(S(A))$ and $U(A) = ub(T(A))$, where $lb(S(A))$ is the set of lower bounds for $S(A)$, and $ub(T(A))$ is the set of upper bounds for $T(A)$. Then, for $A, B \subseteq \mathcal{C}(\mathcal{W})$ we have $A \subseteq L(B) \Leftrightarrow \forall f \in A$, $\forall g \in B, f \leq S(g) \Leftrightarrow \forall g \in B, \forall f \in A, T(f) \leq g \Leftrightarrow B \subseteq U(A)$. Thus, the triple $((\mathcal{C}(\mathcal{W}), \leq), (\mathcal{C}(\mathcal{W}), \leq), I)$, where $I = \{(f, g) \in \mathcal{C}(\mathcal{W}) \times \mathcal{C}(\mathcal{W}) | f \in U(\{g\})\} = \{(f, g) \in \mathcal{C}(\mathcal{W}) \times \mathcal{C}(\mathcal{W}) | g \in L(\{f\})\}$, is a formal context such that $L(A) = A^\uparrow$

and $U(A) = B^{\downarrow}$, where $^{\uparrow}$ and $^{\downarrow}$ are derivation operators associated with the formal context [9]. $((\mathcal{C}(\mathcal{W}), \leq), (\mathcal{C}(\mathcal{W}), \leq), I)$ is called the formal context associated with the classification operator T. The properties of this context, its concepts and its associated context lattice will be treated in a future work.

As an example, consider $T : (\mathcal{C}(\mathcal{W}), \leq) \to (\mathcal{C}(\mathcal{W}), \leq)$ defined by $(Tf)(x) = f(x) \cap f_1(x)$ for all $x \in \mathcal{W}$, where f_1 is the classification defined in the example above. It is easy to see that Tf is a classification, and so T is well defined. Moreover, it is not difficult to show that T is isotone and preserves the upper bounds, and so T is Scott-continuous. Therefore T is a direct classification operator.

The set $(Tf)(x)$ contains the documents of class $f(x)$ having the same meaningful terms as x. If we consider any given classification g instead of f_1, and define $(Tf)(x) = f(x) \cap g(x)$, then we obtain an operator that refines the classification f considered previously as an argument. Namely, we get $f \leq Tf$, and $(Tf)(x)$ contains those documents which verify both conditions: $y \in f(x)$ and $y \in g(x)$.

3   Classifications Associated to a Topology

Alexandrov topologies are well-known in mathematics, and more details can be found in [2,11]. Here we present only some basic notions required to prove the relationship between these topologies and web classifications.

Let $f : \mathcal{W} \to \mathcal{P}(\mathcal{W})$ be a function. We define the relation $\leq_f = \{(x, y) \in \mathcal{W} \times \mathcal{W} \mid x \in f(y)\}$. f is reflective iff $\leq_f$ is reflexive, and f is hereditary iff $\leq_f$ is transitive. Also, if f is a classification, then f is unambiguous iff $\leq_f$ is antisymmetric. Therefore, defining a classification over $\mathcal{W}$ is equivalent to defining a preorder over the same space. Moreover, defining an unambiguous classification over $\mathcal{W}$ is equivalent to defining an order relation over $\mathcal{W}$. Each preorder corresponds one-to-one to an Alexandrov topology, and each partial order corresponds one-to-one to a T_0 Alexandrov topology [2]. In a T_0 topology, every pair of distinct points is topologically distinguishable, namely for any two distinct points there is an open set which contains precisely one of the points. We can establish similar correspondences between classifications and Alexandrov topologies. We use the notions of neighborhood, open and closed sets.

A topology over $\mathcal{W}$ is a set τ of subsets of $\mathcal{W}$ such that $\emptyset, \mathcal{W} \in \tau$, for each $\{D_i\}_{i \in I} \subseteq \tau$ we have $\cup_{i \in I} D_i \in \tau$, and if $D_1, D_2 \in \tau$, then $D_1 \cap D_2 \in \tau$. The sets of a topology τ are called τ-open sets [8]. A set $A \subseteq \mathcal{W}$ is τ-closed iff its complement $\mathcal{W} \setminus A$ is τ-open. A topology τ on $\mathcal{W}$ is called an *Alexandrov topology* if the family of the τ-closed sets forms a topology. The monotone operators over $(\mathcal{W}, \leq_f)$ are continuous functions with respect to the Alexandrov topology associated to f.

A topology can be defined either directly by means of its open sets, or by means of a neighborhood operator. A neighborhood operator is a function $\mathcal{V} : \mathcal{W} \to \mathcal{P}(\mathcal{P}(\mathcal{W}))$ such that the following conditions hold, for all $x \in \mathcal{W}$:

1. if $V \in \mathcal{V}(x)$, then $x \in V$;
2. if $V_1, V_2 \in \mathcal{V}(x)$, then $V_1 \cap V_2 \in \mathcal{V}(x)$;
3. if $V \in \mathcal{V}(x)$ and $V \subseteq U$, then $U \in \mathcal{V}(x)$;
4. for all $V \in \mathcal{V}(x)$, there is $U \in \mathcal{V}(x)$ such that $V \in \mathcal{V}(y)$ for all $y \in U$.

If τ is a topology on $\mathcal{W}$, then $\mathcal{V}_\tau : \mathcal{W} \to \mathcal{P}(\mathcal{P}(\mathcal{W}))$ defined by $\mathcal{V}_\tau(x) = \{V \subseteq \mathcal{W} \mid \exists D \in \tau \text{ such that } x \in D \subseteq V\}$ is a neighborhood operator on $\mathcal{W}$. If $\mathcal{V}$ is a neighborhood operator on $\mathcal{W}$, then $\tau_\mathcal{V} = \{D \subseteq \mathcal{W} \mid D \neq \emptyset \text{ and } D \in \mathcal{V}(x), \text{ for } \text{all } x \in D\} \cup \{\emptyset\}$ is a topology on $\mathcal{W}$. The notions of topology and neighborhood operator are equivalent because $\tau_{\mathcal{V}_\tau} = \tau$ and $\mathcal{V}_{\tau_\mathcal{V}} = \mathcal{V}$ (see [8]).

Let $f : \mathcal{W} \to \mathcal{P}(\mathcal{W})$ be a classification. Then the function $\mathcal{V}_f : \mathcal{W} \to \mathcal{P}(\mathcal{P}(\mathcal{W}))$ defined by $\mathcal{V}_f(x) = \{V \subseteq \mathcal{W} \mid f(x) \subseteq V\}$ for all $x \in \mathcal{W}$ is a neighborhood operator on $\mathcal{W}$. Let $\tau_f = \{D \subseteq \mathcal{W} \mid D \neq \emptyset \text{ and } D \in \mathcal{V}_f(x), \text{ for all } x \in D\} \cup \{\emptyset\}$ be the topology generated by $\mathcal{V}_f$. Since every document $x \in \mathcal{W}$ has a minimal neighborhood, it results that τ_f is an Alexandrov topology [2]. We denote by $\mathcal{A}(\mathcal{W})$ the set of Alexandrov topologies on the space $\mathcal{W}$, and by $\mathcal{A}_0(\mathcal{W})$ the set of T_0 Alexandrov topologies on $\mathcal{W}$.

Theorem 2

1. $\varphi : (\mathcal{C}(\mathcal{W}), \leq) \to (\mathcal{A}(\mathcal{W}), \subseteq)$ *defined by* $\varphi(f) = \tau_f$ *is bijective and monotone;*
2. $\varphi : (\mathcal{C}_0(\mathcal{W}), \leq) \to (\mathcal{A}_0(\mathcal{W}), \subseteq)$ *given by* $\varphi(f) = \tau_f$ *is bijective and monotone.*

Proof. According to the definition of τ_f, φ is well defined. Let $f, g \in \mathcal{C}(\mathcal{W})$ such that $f \leq g$. Since $g(x) \subseteq f(x)$ for all $x \in \mathcal{W}$, it follows that $\mathcal{V}_f(x) \subseteq \mathcal{V}_g(x)$ for all $x \in \mathcal{W}$. Then $\tau_f \subseteq \tau_g$, i.e. $\varphi(f) \subseteq \varphi(g)$, and so φ is monotone.

We now prove that $f(x) \in \tau_f$ for all $x \in \mathcal{W}$. According to the definition of the neighborhood operator, and because $f(x) \in \mathcal{V}_f(x)$, there exists $U \in \mathcal{V}_f(x)$ such that $f(x) \in \mathcal{V}_f(y)$ for all $y \in U$. Since $U \in \mathcal{V}_f(x)$, then $f(x) \subseteq U$. Hence $f(x) \in \mathcal{V}_f(y)$ for all $y \in f(x)$, and so $f(x) \in \tau_f$.

Finally, let $f, g \in \mathcal{C}(\mathcal{W})$ such that $\varphi(f) = \varphi(g)$, i.e. $\tau_f = \tau_g$. Considering $x \in \mathcal{W}$, from $f(x) \in \tau_f$ it follows that $f(x) \in \tau_g$, and then $f(x) \in \mathcal{V}_g(y)$ for all $y \in f(x)$. Then $g(y) \subseteq f(x)$ for all $y \in f(x)$; in particular $g(x) \subseteq f(x)$. In a similar way we get $f(x) \subseteq g(x)$, and so $f(x) = g(x)$, proving that φ is injective. We consider an arbitrary $\tau \in \mathcal{A}(\mathcal{W})$. For each $x \in \mathcal{W}$ we have a minimal neighborhood V_x, and define $f : \mathcal{W} \to \mathcal{P}(\mathcal{W})$ by $f(x) = V_x$. Since $x \in V_x$ it follows that f is reflective, and since $V_x \in \tau$ it follows that f is hereditary. According to the definition of f, the neighborhood system of x with respect to the topology τ is identical with $\mathcal{V}_f(x)$. Thus $\tau_f = \tau$, i.e. φ is also surjective. The second part is proved in a similar way.

Therefore, each classification corresponds to an Alexandrov topology on $\mathcal{W}$. The discrete classification corresponds to the discrete topology. Thus, the Alexandrov topology seems to be relevant for the study of classifications in web systems, including the World Wide Web.

It is not difficult to note that the minimal neighborhood of a point x with respect to an Alexandrov topology is the closure of $\{x\}$ in that topology. Also, for each topology τ over $\mathcal{W}$, the closure function $f_\tau : \mathcal{W} \to \mathcal{P}(\mathcal{W})$ given by $f_\tau(x) = \overline{\{x\}}$ is a web classification. This is not the only way to associate a classification to a topology. Since each topology over the web space leads to a web classification, we study how to build new topologies. A basis $\mathcal{B}$ for a topology τ over $\mathcal{W}$ is a nonempty collection of sets of τ such that every nonempty set in τ

can be written as a union of elements of $\mathcal{B}$. We say that a basis generates the topology; it is enough to know a basis in order to build the whole topology. More details and results on bases can be found in [8].

A collection $\mathcal{B}$ of subsets of $\mathcal{W}$ is a basis for a topology τ over $\mathcal{W}$ if and only if:

1. $\forall B_1, B_2 \in \mathcal{B}$ and $\forall x \in B_1 \cap B_2$, there is $B \in \mathcal{B}$ such that $x \in B \subseteq B_1 \cap B_2$;
2. $\mathcal{W} = \cup_{B \in \mathcal{B}} B$.

Given a set family $\mathcal{B} \subseteq \mathcal{W}$ satisfying these two conditions, we can build a family τ including $\emptyset$ and all the unions of sets from $\mathcal{B}$. This is a topology over $\mathcal{W}$, and $\mathcal{B}$ is a basis for τ. In this topology, for each $x \in \mathcal{W}$ we define

$$f_\mathcal{B}(x) = \{y \in \mathcal{W} \,|\, \forall B \in \mathcal{B}, y \in B \Rightarrow x \in B\} \tag{1}$$

This means that each family $\mathcal{B} \subseteq \mathcal{W}$ satisfying the above two conditions induces a classification given by (1), and so the problem is reduced to determining an appropriate family $\mathcal{B} \subseteq \mathcal{W}$.

For instance, such a family is provided by the graph structure of the web. We define the links by a binary relation $\hookrightarrow: \mathcal{W} \longrightarrow \mathcal{W}$ called *points-to*; by $a \hookrightarrow b$ we denote that a document a contains a link to a document b. $\hookrightarrow$ is extended to the smallest preorder $\rightsquigarrow$ containing $\hookrightarrow$. Thinking now in terms of web navigation, we define first a surfing trip over the web documents.

Definition 3. *Let $a, b \in \mathcal{W}$ and $n \geq 0$. A n-trip from document a to document b is a function $f : \{1, ..., n+1\} \to \mathcal{W}$ such that $a = f(1) \hookrightarrow f(2) \hookrightarrow ... \hookrightarrow f(n+1) = b$. In this case n is called the length of the trip, and is denoted by $\lambda(f)$. The image of f is denoted by $Im(f)$. 0-trips $f : \{1\} \to \{a\}$ with $f(1)=a$ are used as a notation for a document a.*

Definition 4. *Let $a, b \in \mathcal{W}$.*

1. *Let $k \geq 0$. We say that a is k-connected with b if there is a k-trip from a to b. We denote this by $a \rightsquigarrow_k b$.*
2. *We say that a is connected with b if there is a $k \geq 0$ such that a is k-connected with b. We denote this by $a \rightsquigarrow b$.*

We assume that $a \rightsquigarrow_0 a$, and so the relation denoted by $\rightsquigarrow$ becomes a preorder over $\mathcal{W}$ called *trip preorder*. Let $a \in \mathcal{W}$, and $A \subseteq \mathcal{W}$. Then we denote

$Out_a = \{x \in \mathcal{W} \,|\, a \rightsquigarrow x\}$,

$In_a = \{x \in \mathcal{W} \,|\, x \rightsquigarrow a\}$,

$I_a = \{(x, y) \in \mathcal{W} \times \mathcal{W} \,|\, y \rightsquigarrow a \Rightarrow x \rightsquigarrow a\}$, and

$O_a = \{(x, y) \in \mathcal{W} \times \mathcal{W} \,|\, a \rightsquigarrow y \Rightarrow a \rightsquigarrow x\}$.

Let us consider $\mathcal{B} = \{In_a | a \in \mathcal{W}\}$. It is not difficult to see that $\mathcal{B}$ satisfies the two required conditions for a basis, and

$$f_\mathcal{B}(x) = \{y \in \mathcal{W} \,|\, \forall a \in \mathcal{W}, y \rightsquigarrow a \Rightarrow x \rightsquigarrow a\} = \cap_{a \in \mathcal{W}} I_a[x],$$

where $I_a[x]$ is the section of I_a by x, i.e. $I_a[x] = \{y \in \mathcal{W} \,|\, (x, y) \in I_a\}$. Similarly, for $\mathcal{B} = \{Out_a | a \in \mathcal{W}\}$ we define $f_\mathcal{B}(x) = \{y \in \mathcal{W} \,|\, \forall a \in \mathcal{W}, a \rightsquigarrow y \Rightarrow a \rightsquigarrow x\} = \cap_{a \in \mathcal{W}} O_a[x]$. Consequently, I_a and O_a define certain similarities regarding the contents of the documents based on the common links. We refer to both of them as content-based surroundings (entourages in topology, according to the notion introduced by Bourbaki [8]).

Regarding the separation in topological spaces, if $(\mathcal{W}, \tau)$ is a T_1 topological space, then f_τ coincides with the discrete classification. This means that the classifications associated to T_1 topologies (in particular, Hausdorff spaces) are not too interesting. T_3 topological spaces correspond to more interesting classifications; separation in T_3 topological spaces involves documents which do not belong to a class. If $(\mathcal{W}, \tau)$ is a regular space, (i.e. given any closed set F and any point $x \notin F$, there exists disjoint neighborhoods U of x and V of F), then $x \notin f_\tau(y) \Rightarrow \exists D \in \tau$ such that $x \notin D$ and $f_\tau(y) \subseteq D$ (**). Our attempt is to characterize this type of separation by using distinctness operators.

4   Distinctness Operators

In web classification, the documents are compared according to some semantic criteria. This process is difficult, and the class membership of a document is determined according to a list of keywords, or (even worse) according to the links contained in the documents. In both situations it is not difficult to decide that two documents are in the same class in terms of their contents. It seems to be more difficult to decide that two documents are different, that they are not in the same class of the given classification. A distinctness relation would be important for web classification. Since $x \neq y$ means that $x \in \neg\{y\}$, we need a suitable complement operator $\neg$ in order to have an appropriate relation $\neq$. This is the reason why we define the notion of *distinctness operator* by using the complement operator $\neg$. The result is that each distinctness operator identifies a web classification. Moreover, each web classification leads to a particular distinctness operator verifying an additional condition.

In defining a classification, we use the fact that a document is part of its class. It is useful to know when a document is not part of a certain class. To get a formal distinction between various documents, and so to distinguish a document from that of a certain class, we use the distinctness relation and distinctness spaces. A point-set distinctness relation $\bowtie$ is defined between points and subsets of a given set satisfying certain axioms [3]. This relation can be extended to a set-set distinctness relation which is closely related to the notion of proximity. Broadly speaking, a proximity on a set $\mathcal{W}$ is a binary relation δ between subsets of $\mathcal{W}$, and $S\delta T$ expresses that S is near to T in some sense. A point-set distinctness relation $\bowtie$ induces an operator $-$ defined by $-S = \{x \mid x \bowtie S\}$ which behaves like complementation and so is called a distinctness complement. This operator plays an important role, and distinctness spaces can be axiomatized using only the distinctness complement. This is a good reason to study the distinctness complement, and to emphasize its relationship with the Alexandrov topology.

Let $- : \mathcal{P}(\mathcal{W}) \to \mathcal{P}(\mathcal{W})$ be a mapping defined over the family of subsets of $\mathcal{W}$, and consider the following axioms:

A_1. $-S \subseteq \neg S$,
A_2. $-S \subseteq \neg T \Rightarrow -S \subseteq -T$,
A_3. $-(S \cup T) = -S \cap -T$,
A_4. $x \in -S \Rightarrow \mathcal{W} = -\{x\} \cup -S$,
A_5. $x \in -S \Rightarrow \exists R \subseteq \mathcal{W}(x \in -R \wedge S \subseteq -\neg R)$.

Proposition 3. *If $-$ satisfies axioms A_1, A_2 and A_3, then*

1. $-\mathcal{W} = \emptyset$.
2. $S \subseteq T \Rightarrow -T \subseteq -S$.
3. *If A_4 holds, then $-\emptyset = \mathcal{W}$.*
4. *If A_5 holds, then $x \in -S \Rightarrow S \subseteq -\{x\}$.*

We define a binary relation $<>$ on $\mathcal{W}$ by $x <> y$ iff $x \in -\{y\}$.

Proposition 4. *If $-$ satisfies axioms A_1, A_2 and A_3, the following assertions are equivalent:*

1. $x <> y \Rightarrow y <> x$,
2. $x \in -S \Rightarrow \mathcal{W} = -\{x\} \cup -S$,
3. $x <> y \Rightarrow (z <> x \vee z <> y)$ for all $z \in \mathcal{W}$,
4. $x \in -S \Rightarrow S \subseteq -\{x\}$.

Let $f_- : \mathcal{W} \to \mathcal{P}(\mathcal{W})$ defined by
   $f_-(x) = \{y \in \mathcal{W} \mid \neg(y <> x)\}$, i.e. $f_-(x) = \neg - \{x\}$.

Proposition 5

1. *If A_1 holds, then f_- is reflective.*
2. *If A_2 holds, then f_- is hereditary.*
3. *If A_4 holds, then $y \notin f_-(x) \Rightarrow f_-(y) \cap f_-(x) = \emptyset$.*
4. *If A_5 holds, then $y \notin f_-(x) \Rightarrow \exists R \subseteq \mathcal{W}(y \in -R \wedge f_-(x) \subseteq R)$.*

According to this result, if axioms A_1 and A_2 are satisfied then f_- is a classification (induced by the operator $-$). If axioms A_3 and A_4 are also satisfied, the classification f_- fulfils the separation conditions described in Proposition 5 (parts 3 and 4). The property $y \notin f_-(x) \Rightarrow \exists R \subseteq \mathcal{W}(y \in -R \wedge f_-(x) \subseteq R)$ describes regularity, and is similar to the condition (**) mentioned at the end of the previous section. In fact, we associate a topology to the operator $-$ such that the regularity condition with respect to this topology coincides with condition (**).

Definition 5

1. *A distinctness operator on $\mathcal{W}$ is a mapping $- : \mathcal{P}(\mathcal{W}) \to \mathcal{P}(\mathcal{W})$ on the family of subsets of $\mathcal{W}$ which satisfies axioms A_1, A_2, A_3 and A_4.*
2. *A distinctness operator is regular iff it satisfies A_5.*

Therefore a regular distinctness operator on $\mathcal{W}$ is a mapping $- : \mathcal{P}(\mathcal{W}) \to \mathcal{P}(\mathcal{W})$ satisfying axioms A_1, A_2, A_3 and A_5. If $-$ is a distinctness operator on $\mathcal{W}$, then the relation $<>$ is an *inequality relation* on $\mathcal{W}$, namely it satisfies the properties:

1. $x <> y \Rightarrow x \neq y$,
2. $x <> y \Rightarrow y <> x$,

where $x \neq y$ *iff* $\neg(x = y)$, i.e. $\neq$ is the logical inequality. We refer to this relation $<>$ as the *inequality relation induced by* $-$.

Let $-$ be a regular distinctness operator on X, and let $\kappa : \mathcal{P}(\mathcal{W}) \to \mathcal{P}(\mathcal{W})$ defined by $\kappa(S) = \neg - S$.

Proposition 6. *The mapping κ is a Kuratowski operator on $\mathcal{W}$, i.e. it satisfies the following conditions*

1. $\kappa(\emptyset) = \emptyset$;
2. $S \subseteq \kappa(S)$, *for each subset* $S \subseteq \mathcal{W}$;
3. $\kappa(S \cup T) = \kappa(S) \cup \kappa(T)$, *for any two subsets* $S, T \subseteq \mathcal{W}$;
4. $\kappa(\kappa(S)) = \kappa(S)$, *for each subset* $S \subseteq \mathcal{W}$.

Since κ is a Kuratowski operator, the family $\tau_- = \{D \subseteq \mathcal{W} \mid \kappa(\neg D) = \neg D\}$ is a topology on $\mathcal{W}$. A subset $S \subseteq \mathcal{W}$ is τ_-- closed iff $-S = \neg S$. Also, if $\overline{S}$ is the closure of S with respect to τ_-, we have $\overline{S} = \kappa(S)$. Therefore $-S = \neg \overline{S}$ for every subset S of $\mathcal{W}$. Consequently $f_-(x) = \overline{\{x\}} = f_{\tau_-}(x), \forall x \in \mathcal{W}$. Thus $f_- = f_{\tau_-}$.

Proposition 7. *The topology τ_- is regular.*

Remark 1. The regularity of τ_- is strongly related to axiom A_5.

Proposition 8. *If $x, y \in \mathcal{W}$ such that $x <> y$, then they can be separated by two disjoint open sets (i.e., topology τ_- has a Hausdorff-like property with respect to the inequality $<>$ induced by $-$).*

Propositions 7 and 8 express that topological space $(\mathcal{W}, \tau_-)$ is a T_3 space (i.e., a regular space) with respect to the inequality $<>$ induced by $-$.

We show that there exists a bijection between regular distinctness operators and regular topologies. Let $\mathcal{A}$ be the family of regular distinctness operators, and $\mathcal{T}_3$ be the family of regular topologies.

Theorem 3. *The mapping $\Theta : \mathcal{A} \to \mathcal{T}_3$ defined by $\Theta(-) = \tau_-$ is a bijection.*

Proof. From the previous results, for each distinctness operator $-$ we have $\Theta(-) \in \mathcal{T}_3$, i.e. Θ is well defined. Let $-_1, -_2 \in \mathcal{A}$ such that $\Theta(-_1) = \Theta(-_2)$. We denote by $\overline{S}$ the closure of S related to $\Theta(-)$. Then, for each subset $S \subseteq \mathcal{W}$ we have $-_1 S = \neg \overline{S} = -_2 S$. Thus, Θ is injective. We should also prove that Θ is surjective. Let $\tau \in \mathcal{T}_3$ and the application $- : \mathcal{P}(\mathcal{W}) \to \mathcal{P}(\mathcal{W})$ defined by $-S = \neg \overline{S}$, where $\overline{S}$ is the closure of S related to τ. We prove that $-$ is a regular distinctness operator. Let S and T be two subsets of $\mathcal{W}$. Since $S \subseteq \overline{S}$, we have $-S \subseteq \neg S$, and so A_1 is true. If $-S \subseteq \neg T$, then $T \subseteq \overline{S}$. It follows that $\overline{T} \subseteq \overline{S}$, then $-S \subseteq -T$, and so A_2 is true. Since $-(S \cup T) = \neg \overline{S \cup T} = \neg(\overline{S} \cup \overline{T}) = \neg \overline{S} \cup \neg \overline{T} = -S \cap -T$, and so axiom A_3 is satisfied. In order to prove A_4 and A_5, let $x \in -S$, i.e. $x \in \neg \overline{S}$. Since τ is regular, there exist $Q, R \in \tau$ such that $x \in Q$, $\overline{S} \subseteq R$, and $Q \cap R = \emptyset$. Since $Q, R \in \tau$ and $Q \cap R = \emptyset$, we have $\overline{Q} \cap R = \emptyset$, and therefore

$\overline{\{x\}} \cap \overline{S} = \emptyset$; i.e. $\overline{\{x\}} \subseteq \neg \overline{S}$. It follows that $\mathcal{W} = \neg \overline{\{x\}} \cup \overline{\{x\}} \subseteq \neg \overline{\{x\}} \cup \neg \overline{S} \subseteq \mathcal{W}$; hence $\mathcal{W} = \neg \overline{\{x\}} \cup \neg \overline{S} = -\{x\} \cup -S$. Therefore, A_4 is satisfied. Since $Q, R \in \tau$ and $Q \cap R = \emptyset$, we also have $Q \cap \overline{R} = \emptyset$. So $x \in Q$ implies $x \notin \overline{R}$, and therefore $x \in -R$. Since $S \subseteq \overline{S} \subseteq R = --\neg R$ we have $S \subseteq --\neg R$. Therefore, A_5 is also satisfied. Thus we can conclude that $-$ is an regular distinctness operator.

Moreover, $\Theta(-) = \{D \subseteq X \mid D = --D\} = \{D \subseteq X \mid D = \neg(\neg D)\} = \{D \subseteq X \mid D = int(D)\} = \tau$, where $int(D)$ is the interior of D in the topology τ. Therefore, Θ is also surjective.

We denote $\Theta^{-1}(\tau)$ by $-_\tau$, where $-_\tau = \neg \overline{S}$ and $\overline{S}$ is the closure of S related to τ. Let $(\mathcal{W}, \mathcal{U})$ be a uniform space, i.e., a topological space with additional structure which is used to define uniform properties such as completeness, uniform continuity and uniform convergence. Also, consider τ to be a topology induced by $\mathcal{U}$. The topological space $(\mathcal{W}, \tau)$ is completely regular, i.e. given any closed set F and any point $x \notin F$, x and F can be separated by a continuous function. Since $(\mathcal{W}, \tau)$ is completely regular, it is also regular [8]. Thus $-_\tau S = \{x \in \mathcal{W} \mid \exists U \in \mathcal{U}$ such that $(x, y) \notin U, \forall y \in S\}$. In particular, if $\mathcal{U}$ is induced by a metric ρ, then $-_\tau S = \{x \in \mathcal{W} \mid \exists r > 0$ such that $\rho(x, y) \geq r, \forall y \in S\}$.

Let $-$ be a distinctness operator, and $<>$ the inequality induced by $-$.

Definition 6. *For a subset S of $\mathcal{W}$, the set $\sim_- S = \{x \in \mathcal{W} \mid \forall y \in S(x <> y)\}$ is called the complement of S induced by the distinctness operator $-$.*

When there is no risk of confusion, we denote $\sim_-$ by $\sim$. We have $\sim S = \{x \in \mathcal{W} \mid \forall y \in S(x \in -\{y\})\} = \{x \in \mathcal{W} \mid x \in \cap_{y \in S} -\{y\}\} = \cap_{y \in S} -\{y\} = \cap_{y \in S} \neg \overline{\{y\}} = \neg \cup_{y \in S} \overline{\{y\}}$. From now on we use the following notation:

$$\sim S = \cap_{x \in S} -\{x\} = \neg \cup_{x \in S} \overline{\{x\}}.$$

Notice that $-S \subseteq \sim S \subseteq \neg S$ for all $S \subseteq \mathcal{W}$, and $-\{x\} = \sim \{x\}$ for all $x \in \mathcal{W}$.

Proposition 9. *The operator $-$ satisfies A_4 iff $\sim -S = \neg - S$ for all $S \subseteq \mathcal{W}$.*

According to Proposition 4, the condition $\sim -S = \neg - S$ for all $S \subseteq \mathcal{W}$ is equivalent to the symmetry of relation $<>$.

Proposition 10. *The following assertions are equivalent for a distinctness operator $-$:*

1. $\sim S = \neg S$ for all $S \subseteq \mathcal{W}$,
2. $x <> y \Leftrightarrow \neg(x = y)$,
3. τ_- is T_1 (with respect to the logical inequality).

This result shows that T_1 topologies in the classical sense do not provide new inequality relations, and in this way they are not interesting from the point of view of differentiating web documents.

Proposition 11. *The following assertions are equivalent for a distinctness operator $-$:*

1. $-S =\sim S$ for all $S \subseteq \mathcal{W}$,
2. τ_- is an Alexandrov topology.

From Propositions 10 and 11 it follows that $-S = \neg S$ for all $S \subseteq \mathcal{W}$ iff τ_- is a T_1 Alexandrov topology (i.e., τ_- is the discrete topology on X). As mentioned before, in general we have $f_- = f_{\tau_-}$. If $-S =\sim S$ (***) is additionally satisfied for all $S \subseteq \mathcal{W}$, then $\tau_- = \tau_{f_-}$. As a consequence, there is a bijection between web classifications and the distinctness operators satisfying (***).

Theorem 4. *If $-$ is a regular distinctness operator, then $\sim$ is also a regular distinctness operator.*

Proof. Since $-S \subseteq\sim S \subseteq \neg S$, axiom A_1 is true. In order to prove A_2, let S and T be two subsets of $\mathcal{W}$ such that $\sim S \subseteq \neg T$. Then $T \subseteq \neg \sim S = \cup_{x \in S}\overline{\{x\}}$. Thus, $\forall y \in T$, there is $x \in S$ such that $y \in \overline{\{x\}}$. Then $\forall y \in T$ we have $\overline{\{y\}} \subseteq \cup_{x \in S}\overline{\{x\}}$, and so $\cup_{y \in T}\overline{\{y\}} \subseteq \cup_{x \in S}\overline{\{x\}}$. Hence, $\sim S \subseteq\sim T$ and A_2 is satisfied.

Since $\sim (S \cup T) = \cap_{x \in S \cup T}(-\{x\}) = \cap_{x \in S}(-\{x\}) \cap \cap_{x \in T}(-\{x\}) =\sim S \cap \sim T$; thus, A_3 is also satisfied.

Let $x \in\sim S$. Since $x \in -\{y\}$ for all $y \in S$, from axiom A_4 of the operator $-$ it follows that $\mathcal{W} = -\{x\} \cup -\{y\}$ for all $y \in S$. Then $\mathcal{W} = \cap_{y \in S}(-\{x\} \cup -\{y\}) = -\{x\} \cup \cap_{y \in S}(-\{y\}) = -\{x\}\cup \sim S \subseteq\sim \{x\}\cup \sim S \subseteq \mathcal{W}$, and therefore $\mathcal{W} =\sim \{x\}\cup \sim S$. Thus, A_4 is satisfied.

Let $x \in\sim S$. Since $x \in -\{y\}$ for all $y \in S$, from axiom A_5 of the operator $-$, it follows that for all $y \in S$ there exists $R_y \subseteq X$ such that $x \in -R_y$ and $\{y\} \subseteq -\neg R_y$. Thus, we have $x \in \cap_{y \in S}(-R_y) = \cap_{y \in S}(\neg \overline{R_y}) = \neg \cup_{y \in S} \overline{R_y} \subseteq \neg \cup_{y \in S} \cup_{z \in R_y}\overline{\{z\}} = \neg \cup_{z \in \cup_{y \in S} R_y}\overline{\{z\}} =\sim \cup_{y \in S}R_y$. If $R = \cup_{y \in S}R_y$, then $x \in\sim R$. Since $y \in -\neg R_y$ and $-\neg R_y \subseteq -\neg R \subseteq\sim \neg R$, we have $y \in\sim \neg R$ for all $y \in S$. Therefore, $S \subseteq\sim \neg R$ and so A_5 is satisfied.

Since $-\{x\} =\sim \{x\}$ for all $x \in \mathcal{W}$, we have $\sim_\sim S = \cap_{x \in S} \sim \{x\} = \cap_{x \in S} - \{x\} =\sim S$. From Proposition 11 it follows that $\tau_\sim$ is an Alexandrov topology. In fact $\tau_\sim$ is a regular Alexandrov topology. Since $-S \subseteq\sim S$ for all $S \subseteq \mathcal{W}$, we have $\tau_- \subseteq \tau_\sim$. Moreover, $\tau_- = \tau_\sim$ iff $-S =\sim S$ for all $S \subseteq \mathcal{W}$ iff τ_- is an Alexandrov topology. Also $\sim$ has the property: $\sim (\cup_{i \in I}S_i) = \cap_{i \in I} \sim S_i$ for any arbitrary family of subsets $\{S_i \mid i \in I\}$. This remark leads to the following definition:

Definition 7. *A (regular) distinctness complement is a (regular) distinctness operator $-$ having the property $-(\cup_{i \in I}S_i) = \cap_{i \in I} - S_i$ for any arbitrary family of subsets $\{S_i \mid i \in I\}$.*

According to Theorem 4, the complement induced by a regular distinctness operator is a regular distinctness complement. Based on a result similar to Theorem 3, we can prove that there exists a bijection between regular distinctness complements and regular Alexandrov topologies. Therefore to each regular distinctness operator we can associate a regular distinctness complement which can then be associated to a regular Alexandrov topology corresponding to a web classification. Thus axiom A_5

$x \in -S \Rightarrow \exists R \subseteq W \ (x \in -R \wedge S \subseteq --R)$ is reduced to
$x <> y \Rightarrow \exists R \subseteq W \ (x <> z, \forall z \in R \wedge y <> z', \forall z' \in \neg R)$.
In this way it is possible to provide *a new separation criteria* for web documents
in the process of their classification.

5   Conclusion

Using metric spaces, Sergey Brin has emphasized in [4] the importance of neighborhoods and the nearness relation. Citing [4], "...metric spaces are a very
general concept and can be applied to vectors (for example, under Euclidean
distance), as well as objects like strings and graphs which cannot be easily represented as vectors. Finding near neighbors in a metric space refers to selecting
the elements of a data set (a finite subset of the space) which are within a certain
distance of a given point." In this paper we extend such a view, considering the
formal results on which a practical application could then be built. The example
of Brin and the success of his company is encouraging.

Web searches seeking to identify web-objects that reside in the proximity of
other web-objects are significant for the search engines, where exact searches are
often meaningless. For example, querying the web to find images that contain a
sunset is usually done by providing an example image and then, identifying those
images that are similar to the example. The natural framework for executing
such searches is provided by metric spaces or, more generally, by dissimilarity
spaces. Nevertheless, all the existing indexing algorithms for proximity searching
consist in building an equivalence relation, so that at search time some classes
are discarded and the others are exhaustively searched [5]. Any classification
f induces a reflexive and transitive relation defined by $y \sim x$ iff $y \in f(x)$.
This relation has not the property of symmetry, and so it is not an equivalence
relation. Therefore the use of classifications represent a generalization suitable
to a topological approach without the use of metrics spaces.

A web system is represented as a graph in which the nodes are the pages
and the edges are web links. We define a topology induced by the edges of this
graph, and provide some results regarding the classification of web documents.
Classification of web documents is useful for the automated navigation of mobile
agents and search engines [10]. In classification we use the fact that a document is
part of its class. It is therefore useful to know when a document does not belong
to a certain class. In a constructive setting, this aspect is described properly by
the newly introduced notion of distinctness. To get a formal distinction between
various documents, and setting apart a document from a certain class, we use and
study the distinctness relation. There is a bijection between regular distinctness
operators and regular topologies. This means that regular distinctness operators
are associated to regular Alexandrov topologies, and so web classifications are
related to distinctness operators.

The paper represents a theoretical contribution dealing with the classification process by using three related approaches: graph theory, domain theory
and topology. These approach provides a high flexibility in both defining and

employing the classification process. In this paper we only suggest by some examples how these theoretical results can be used effectively and successfully in search engines; software implementations inspired by this theoretical approach represent a further work.

In many applications, formal topology does not represent actually the primary structure. However, important topological information may be lost, being filtered out by other aspects. Moreover, formal topology provides an elegant approach, and relates the qualitative and quantitative aspects of a problem. A previous topological approach of the web systems is given in [6], continued by some formal results related to searching and classification over the web in [7]. Alexandrov topologies play an important role because they correspond one-to-one to web classifications. Here we extend the approach by considering some distinctness operators and reveal their links to Alexandrov topologies (associated to web classifications). According to our knowledge, there are no similar topological approaches to the problem of web classification, and any distinctness relation over the web documents was not involved in the web classification.

Acknowledgements. We thank the anonymous reviewers for their comments. The work was supported by a grant of the Romanian Authority for Scientific Research, project number PN-II-ID-PCE-2011-3-0919.

References

1. Abramsky, S., Jung, A.: Domain Theory. In: Handbook of Logic in Computer Science, vol. 3, pp. 1–168. Oxford University Press (1995)
2. Arenas, F.G.: Alexandroff Spaces. Acta Math. Univ. Comenianae LXVIII, 17–25 (1999)
3. Bridges, D., Vîta, L.: Apartness Spaces as a Framework for Constructive Topology. Annals of Pure and Applied Logic 119, 61–83 (2003)
4. Brin, S.: Near Neighbor Search in Large Metric Spaces. In: Proceedings 21st Conference on Very Large Data Bases, pp. 574–584. ACM Press (1995)
5. Chavez, E., Navarro, G., Baeza-Yates, R., Marroquin, J.L.: Searching in Metric Spaces. ACM Computing Surveys 33, 273–321 (2001)
6. Ciobanu, G., Rusu, D.: Topological Spaces of the Web. In: Proceedings 14th Int'l World Wide Web Conference, pp. 1112–1114. ACM Press (2005)
7. Ciobanu, G., Rusu, D.: A Topological Approach of the Web Classification. In: Barkaoui, K., Cavalcanti, A., Cerone, A. (eds.) ICTAC 2006. LNCS, vol. 4281, pp. 80–92. Springer, Heidelberg (2006)
8. Engelking, R.: General Topology, 2nd edn. Sigma Series in Pure Mathematics, vol. 6. Heldermann (1989)
9. Ganter, B., Wille, R.: Formal Concept Analysis. Mathematical Foundations. Springer (1999)
10. Meghabghab, G., Kandel, A.: Search Engines, Link Analysis, and User's Web Behavior. Springer (2008)
11. Smyth, M.B.: Topology. In: Handbook of Logic in Computer Science, vol. 1, pp. 641–761. Oxford University Press (1992)

Specifying Well-Formed Part-Whole Relations in Coq

Richard Dapoigny and Patrick Barlatier

LISTIC/Polytech'Annecy-Chambéry
University of Savoie, Po. Box 80439, 74944 Annecy-le-vieux Cedex, France
richard.dapoigny@univ-savoie.fr

Abstract. In the domain of ontology design as well as in Conceptual Modeling, representing part-whole relations is a long-standing challenging problem. However, in most papers the focus has been on properties of the part-whole relation, rather than on its semantics. In the last decades, most approaches which have addressed the formal specification of the part-whole relation (i) rely on First Order Logic (FOL) which is unable to address multiple levels of granularity and (ii) do not support any typing mechanism useful for the extensional side of concepts and then, many difficulties remain especially about expressiveness. In mathematical logic and program checking, type theories have proved to be appealing but so far, they have not been applied in the formalization of ontological relations. To bridge this gap, we present an axiomatization of the part-whole relation which hold between typed terms. Relation structures in the dependently-typed framework rely on a constructive logic. We define in a precise way what relation structures and their meta-properties, are in term of type classes using the Coq language.

1 Introduction

Formal ontologies include at least foundational ontologies which deal with formal aspects of entities irrespective of their particular nature and domain ontologies. These foundational ontologies are necessary for the organization of the world into well-defined categories which are required for enforcing consistency of the domain ontology content. For that purpose, they should rely on formal foundations [1] and should include an expressive logic. Foundational ontologies are basically directed towards formal structures and relations in reality. Two (binary) structures have emerged as foundational relations [2, 3], namely the relation of subsumption (also called *is_a* relation) witnessing the generalization/specialization relation and the *part-whole* relation which governs the relation of part to whole shared by all material domains.

Assuming the view of an ontology as a representation of a conceptual system through a logical theory, we focus here on the *part-whole*[1] relation which is at the heart of most mereological theories. Mereologies are seen as formal theories of structural relations that hold in reality. A coherent formalization of *part-whole* relations is a long standing challenge which has received a great attention in computer science [2–8] (to cite a few) as well as in philosophy [9–11]. While some approaches such as [12] have used the singleton operator as primitive to take advantage of classical mereology for describing sets and classes, others distinguish classes from sets and have suggested a formalism of

---

[1] Also denoted *part-of* in the paper.

N. Hernandez et al. (Eds.): ICCS 2014, LNAI 8577, pp. 159–173, 2014.
DOI: 10.1007/978-3-319-08389-6_14, © Springer International Publishing Switzerland 2014

classes using first-order logic [2, 4, 13]. It has been pointed out that none of the models of biological structure described so far are able to satisfy an ontologically founded, expressive and uncontroversial account of *part-whole* relations [14, 15].

Using a dependent type theory based on (higher-order) constructive logic we are able to propose a very expressive specification of part-whole relations that is both ontologically well-founded and logically certified. As such, the theoretical approach that is developed in this paper is intended to serve as a tool conceptualizing ontologies. Using the Coq theorem prover as a core language, a running example from anatomy will demonstrate the strength of the ideas.

2   Basic Challenges for Part-Whole Relations

We follow the classical assumption of formal ontologies making a distinction between universals and particulars characterized by means of instantiation (unlike universals, particulars are entities that cannot have instances). We also admit the view of an ontology based on concepts and properties. We first recall some well-known challenging problems about the formalization of part-whole relations in ontologies.

The first point is that part-whole theories such as mereology postulates an extensional view by considering particulars. The wide-spread use of part-of relations between classes appears intuitive, but it contrasts with the classical approach to mereology [9] which focuses on particulars, rather than classes of particulars. It follows that using classes, the implicit reference to instances should be taken carefully into account. Assuming an ontology with classes and properties, two individuals may have different properties while being made of the same parts. Alternatively, two individuals may have the same identifying properties while being made of different parts. Therefore, a foundational question concerns the coexistence of part-whole relations using an extensional view with the (intensional) subsumption relation in the same framework. Sometimes, the extensionality and some mereological assumptions are too strong and the question arises of how to deal with individuals described not only by means of their parts, but also by means of their properties.

A rigorous system of formal definitions should clarify also the relations between the concepts of whole and set. One of the most important problem with set theory appears when it comes to dealing with the relations between different levels of granularity [5]. For instance, an organism is a totality of organs, but it is also a totality of cells, yet the corresponding sets have distinct members and are therefore different. It follows that a well-founded system should express in a realistic way the relations between any whole and its constituent parts at distinct levels of granularity. Along this line for example, how to generate all parts from the existence of a single whole?

How to capture the notion of a whole as a one-piece entity, as opposed to a scattered entity made up of several disconnected parts. Introducing the sum of parts x and y as $x + y$, it is generally admitted that a whole corresponds to the sum of its parts. However, if an object is identified by the sum of its parts, removing one of its parts will change the identity of this object [16]. A possible solution to this difficulty is to distinguish these formal sums from integral wholes. While the sum exists iff the constituent parts exist, an integral whole composed with the same parts requires not only the parts, but also what is termed an unifying condition supposed to hold between these parts [9].

The connection between part and whole on the one hand and dependence on the other hand is another challenge. In some cases, a whole can be regarded as being existentially dependent on its own constituent parts (essential parts). While some parts are essential to a whole, it is a common agreement that not all parts are essential.

Under what conditions is one particular a part of another particular is an old question. It entails the classification of distinct part-whole relations having appropriate properties [7, 16, 17]. These approaches advocate for a strong connection between the issue of transitivity of parthood and the type of the relation being considered. Nevertheless, there exists a common agreement that considers a basic (mereological) part-of relation which is regarded as transitive [7, 9, 10].

3 Using a Dependently-Typed Framework

3.1 Motivations

Using a dependently-typed theory offers some benefits over usual first-order logic approaches. Besides higher-order capabilities, in dependent type theory axioms and predicates are types and, as such they can be manipulated like any type is. In [18], the authors have shown the ability of the type-theoretical approach to cope with scalability on the SUMO foundational ontology. Dependent types are based on the notion of indexed families of types and provide a high expressiveness since they can represent subset types, relations or constraints as typed structures. They will be exploited for representing knowledge in an elegant and secure way. This last aspect is analyzed in [19] where the authors investigate typing applied to reasoning languages of the Semantic Web and point out that dependent types ensure normalization. The availability of powerful theorem provers (e.g., Coq) which can be used to check the well-formedness of user-defined typed structures is another advantage. These aspects motivate us to consider a type system with a strong proof theory.

As underlined in [20, 21], there is a significant advantage for using a dependently typed functional language such as Coq: the programming language of such a theorem prover combines powerful logical capabilities and reasonable expressive power. Using dependent types, terms can even represent mathematical demonstrations which can be further reused within a single language witnessing for a higher level of reasoning. In Computer Science, and specifically for program checking, dependent types allow to refine descriptions and maximize expressiveness with the purpose of providing certified code. We will demonstrate with code fragments, written in the support language Coq [22], that the language is able to satisfy most of all the constraints inherent in an expressive conceptual model for designing *part-whole* relations in ontologies.

This work is based on the unified language KDTL (Knowledge-based Dependently Typed Language) proposed in [23]. Assuming a core layer using a dependently-typed theory $\mathcal{T}_{CIC}{}^2$, the envelope layer will correspond to a higher-level theory $\mathcal{T}_{KDTL}$ which provides a set of primitives for knowledge representation and reasoning. The type-theoretical semantics based on the usual BHK interpretation[3] provides the language

---

[2] The Coq language is based on the Calculus of Inductive Constructions (CIC).

[3] The Brouwer-Heyting-Kolmogorov interpretation.

$\mathcal{L}_{CIC}$ for the core layer. The semantics of the envelope layer relies on a language $\mathcal{L}_{KDTL}$ which includes a system of judgments and inference rules.

Proposition 1. *Given the theories* $\mathcal{T}_{CIC} \subseteq \mathcal{T}_{KDTL}$ *formulated in the respective languages* $\mathcal{L}_{CIC} \subseteq \mathcal{L}_{KDTL}$, *then* $\mathcal{T}_{KDTL}$ *is a conservative extension of* $\mathcal{T}_{CIC}$ *if the part of* $\mathcal{T}_{KDTL}$ *consisting of* $\mathcal{L}_{CIC}$-*sentences is identical with* $\mathcal{T}_{CIC}$.

Proof. The core intuitionistic logic of CIC is assumed to be the root of the KDTL theory. CIC universes are respectively mapped on Type for data structures and on Prop for logic. CIC coercions are mapped to ontological inheritance and CIC automatic generation of Type Classes fields is mapped to their counter-part of mereological classes. Then, all well-formed types are mapped to well-formed concepts and properties.   □

It follows that a conservative extension of a consistent theory is consistent. If an ontology is formalized as a logical theory $\mathcal{T}_{KDTL}$, then $\mathcal{T}_{CIC}$ is seen as a sub-theory while $\mathcal{T}_{KDTL}$ is a conservative extension of this sub-theory.

3.2 A Quick Introduction to Coq

The Coq language is the support language for KDTL. The building blocks of the Calculus of Inductive Constructions (CIC) are terms and the basic relation is the typing relation. CIC with universes has given rise to the Coq language [22] which is a tool for developing mathematical specifications and proofs. As a specification language, Coq is both a higher order logic[4] (quantifiers may be applied on natural numbers, on functions of arbitrary types, on propositions, predicates, types, etc.) and a typed lambda-calculus enriched with an extension of primitive recursion (further details are given in [22]).

Coq is designed such that evaluation always terminates (decidability). All logical judgments are typing judgments. The type-checker checks the correctness of proofs, that is, it checks using proof search that a data structure complies to its specification. The proof engine also provides an interactive proof assistant to build proofs using specific programs called tactics. The language of the Coq theorem prover consists in a sequence of declarations and definitions. A declaration associates a name with a qualification. Qualifications can be either logical propositions which reside in the universe[5] *Prop*, mathematical collections which are in *Set* or abstract types which belong to a universe $Type\_i$ with $i \in N^6$. Construction of new categories relies on the scheme Definition $< definiendum > := < definiens >$ in which the latter must be a well-formed expression. Subtyping relations between (dependent) types are specified by defining coercion functions that are definable in the type theory. For any subtype A' of type A, there is a unique coercion $c : A' \to A$ and any object $x : A'$ can be regarded as an object $(c\ x)$ of type A. Coercions are implicit since one may use the object x instead of $(c\ x)$ whenever an object of type A is expected.

Basic data structures are list for which well-founded recursion is available via the Fixpoint operator and Type Classes (TCs). The use of Coq features such as implicit

---

[4] The core logic is constructive.

[5] Also called *sort*.

[6] Notice that universes are present in Coq but not accessible for the user.

arguments, coercions and overloading through TCs renders the formal text close to informal ontological descriptions and makes easy the verification of correct transcription of definitions and statements into the formal language. In particular, the use of very powerful primitives such as TCs [24, 25] is a key for describing data structures. TCs are just dependent inductive types with one constructor and some fields. These fields are eliminators corresponding to each constructor argument. Coq allows us to specify the rules inside TCs. Dependent types give new power to TCs while types and values are unified. TCs allow parametric arguments, inheritance and multiple fields [25]. Coq's TCs are first class, i.e., classes and their instances are designed as record types and registered as constants of these types. If parameters are marked as implicit (i.e., using curly brackets) then Coq will try to infer them (instance resolution) automatically using type inference. Canonical names for reusable components are achieved with single-field TCs containing a single component each, also referred to as *operational type classes* [25]. The instantiation relation (:) and the subsumption with coercions (: >) are already available in Coq and do not require any additional modeling structures.

Regarding the conceptual choice for characterizing the *part-whole* relation, class types will make up the core component for expressing ontological properties of a given concept. In such a way, the present modeling comes closer to the modeling of properties (roles) in DL (unlike Object-Oriented Modeling in which a concept is any instance of a class). As a consequence, we are able to suggest multiple definitions for a concept, each of them depending on a domain ontology or a context. The conceptual choice defended here assumes that (existential) dependence between ontological categories is expressed with parameter(s) while inheritance relies on TC coerced fields. Since TCs can be predicates, one has to prove the type class with instances (a predicate is provable if it has a proof). It follows that instances are nothing else than models proving the specifications of the TC w.r.t. semantics. We restrict here the scope to the formalization of the *part-whole* relation described with TCs.

3.3   The Ontological Interpretation

We assume that types are intensional entities and that there exists a mapping between concepts and types in which types represent concepts, but not all types describe concepts (e.g., they can describe relations). In object-oriented frameworks, classes (representation of universals) are defined according to their set of properties. We depart from this choice by introducing the type `kind` for which the set of properties is implicit while properties are types depending on a particular. Kinds, properties and formal (ontological) relations are the basic components of the core ontology.

A kind is considered as a canonical concept, assumption which is in line with the existence of "natural types", i.e., it can be identified as a type in isolation [26] and relates to what is called an "atomic concept" in DL. Using kinds instead of unary predicates for the ontological categories (i) gives the possibility to find an unintended application of n-ary predicates during the type checking (e.g., for non well-typed kinds) and (ii) offers a rich structural knowledge representation by means of partially ordered kinds. Kinds are assembled in a taxonomy which makes use the DOLCE taxonomy of particulars [27, 28][7]. A strong argument for using this taxonomy is that it does not classify universals

---

[7] Without its logical apparatus.

and leaves room for conceptual choices about universal structures. Alternatively, the *part-whole* relation is rather arranged into a partonomic (or partitive) hierarchy where entities considered as wholes are refined to their parts.

4 Formalizing the Part-Whole Relation

4.1 Basic Part-Whole Properties

The essential point is that mereology, the theory of parts and wholes, is better suitable than set theory to describe in realistic fashion the dependences between wholes and their constituent parts at distinct abstraction levels. Since some postulates of mereology are not entirely uncontroversial and if we expect to support both representations (*part-whole* and *is_a*) in the same language, (i) we restrict the *part-whole* theory to a minimal number of relations and terms (PartOf, ProperPartOf, Overlap, Product and Sum) and (ii) we do not consider controversial principles. We will show that KDTL enables us to remain neutral about the existence of any ultimate objects in reality from out of which all other objects will be constructed via the previous operations (bottom-up approach). Assuming an extensional hierarchy, the set of instances of kind, i.e., particulars, are partially ordered by the *part-whole* relation.

Definition 1. *(Part-whole Hierarchy) Let P the set of all proof objects built from the subsumption hierarchy and let $\sqsubseteq$, a binary relation on $P \times P$. A part-whole hierarchy $P_w = (P, \sqsubseteq)$ is a partial order iff the relation $\sqsubseteq$ is reflexive, antisymmetric and transitive.*

Other relations which do not possess these properties are called Association. We also introduce the *proper part-of* relation ($\sqsubset$) as an irreflexive and transitive relation, i.e., a strict partial order. Usually, in *part-whole* theories, other notions are defined in terms of the *part-whole* relation taken as primitive [10].

Definition 2. *(Overlap) Two particulars x and y overlap iff there exists a non-null particular z such that: $O(x\,y) \triangleq \exists z(z \sqsubseteq x \wedge z \sqsubseteq y)$*

Checking properties of overlap requires the usual first-order definition:

Definition **Overlap** $(x\,y : \mathsf{kind}) := \exists z,\, z \sqsubseteq x \wedge z \sqsubseteq y.$

Definition 3. *(Binary Sum) For any pair of entities x and y there exists a binary sum which overlaps exactly those things that overlap either x or y.*

$$Sum\,(x,\,y) \triangleq \exists z \forall w(O(w, z) \leftrightarrow (O(w, x) \vee O(w, y))).$$

In the extensional view, if $\{x, y\} \subseteq P_w$, there exists a $z = x + y$ such that x and y are part of z and any part of z overlaps (at least) either with x or y. We will show that hierarchies having this property is a complete lattice provided that the sum is unique. If we expect that the sum (as a least upper bound) is unique then it requires the closure of the *part-whole* relation [29].

Lemma 1
$$\forall xy(x \sqsubseteq y \rightarrow \forall z(O(z,x) \rightarrow O(z,y)))$$

The universal closure of the *part-whole* relation is derived in Coq from the axiom of transitivity (t_of_PartOf):

Lemma PWClosure : $\forall x\,y$:kind, $x \sqsubseteq y \rightarrow \forall z$, Overlap $z\,x \rightarrow$ Overlap $z\,y$.
Proof.
    intros $H1\,H2\,H3\,H4\,G1$.
    destruct $G1$.
    assert ($x \sqsubseteq H1 \rightarrow H1 \sqsubseteq H2 \rightarrow x \sqsubseteq H2$) by apply t_of_PartOf.
    intuition.
    red; exists x; split;[exact $H5$ | exact $H0$].
Qed.

Definition 4. *(Binary product) The mereological product of two kinds only exists when the two kinds overlap (mereologically), that is when they have some shared sub-part.*

$$Product\ (x,\ y) \triangleq \exists z \forall w((w \sqsubseteq z) \leftrightarrow (w \sqsubseteq x \wedge w \sqsubseteq y))$$

Notice that this can be either their overlap or the smaller kind inside the greatest. It is easy to see that the product is unique. The two binary operators Product and Sum in *part-whole* theories are respectively interpreted as meet and join, then the *part-whole* hierarchy is a lattice.

Proposition 2. *(Partonomic Lattice) The part-whole hierarchy with the (fictitious) null object adjoined, is a lattice* $(P_w, \sqsubseteq, Product, Sum)$ *if, given any two entities x and y there is a join (least upper bound), Sum $x\,y$, and a meet (greatest lower bound) Product $x\,y$ which is null iff x and y are disjoint.*

Proof. Assuming that any pair of suitably related objects have an upper bound, it is easy to see that the Sum operator is idempotent, commutative, and associative. The absorption law should sum an object x together with an object which is both part of x and part of another object y. It yields that the sum including x and a part of x will give x itself. In the same way, we can show that the absorption law which first applies the product and then the sum also holds. Alternatively, we can interpret the binary product of two kinds x and y within a given *part-whole* hierarchy as the largest kind shared in common by x and y, i.e., as their greatest lower bound with respect to $\sqsubseteq$. It can be shown again that such an operator would have the basic properties of a meet i.e., it is idempotent, commutative, associative and satisfies absorption [30].   □

All these results are reported in the TC PWLattice which includes all usual mathematical properties of a lattice.

Class PWLattice (k:Type)(pw:relation k)($meet\,join$:$k \rightarrow k \rightarrow k$) := {
    joinPred1 : $\forall x\,y$:k, $pw\,x\,(join\,x\,y) \wedge pw\,y\,(join\,x\,y)$;
    joinPred2 : $\forall x\,y\,z$:k, $pw\,x\,z \wedge pw\,y\,z \rightarrow pw\,(join\,x\,y)\,z$;
    meetPred1 : $\forall x\,y$:k, $pw\,(meet\,x\,y)\,x \wedge pw\,(meet\,x\,y)\,y$;
    meetPred2 : $\forall x\,y\,z$:k, $pw\,z\,x \wedge pw\,z\,y \rightarrow pw\,z\,(meet\,x\,y)$;
    meet_commutative :> Commutative $meet$;

```
meet_associative :> Associative meet;
meet_absorptive :> Absorptive meet join;
meet_idempotent :> Idempotent meet;
join_commutative :> Commutative join;
join_associative :> Associative join;
join_absorptive :> Absorptive join meet;
join_idempotent :> Idempotent join}.
```

The type class KDTL is the ontological account of foundational relations. For that purpose, it separates the *part-whole* relation from its partial-order properties and the Association relation without any property. An instance of PWLattice is also created as a last field meaning that any instance of the KDTL TC should respect (i) the partial order properties of the *part-whole* relation and (ii) all the properties of a lattice.

```
Class KDTL : Type := {
    kind :> Type;
    PartOf : relation kind;
    ProperPartOf : relation kind;
    Association : relation kind;
    Product : kind → kind → kind;
    Sum : kind → kind → kind;
    t_of_PartOf : Transitive (PartOf);
    ant_of_PartOf : Antisymmetric (PartOf);
    ref_of_PartOf : Reflexive (PartOf);
    t_of_PPartOf : Transitive ProperPartOf;
    irr_of_PPartOf : Irreflexive ProperPartOf;
    pwlatt : PWLattice kind PartOf Product Sum}.
```

An instance of the KDTL class will automatically infer the required parameters allowing their use in further definition. The KDTL class and the core taxonomy of DOLCE types of particulars is encapsulated within the KDTL_Lattice module type which is instantiated and reused in further modules used e.g., in domain ontologies.

4.2   Expressing Part-Whole (Meta)Properties

Definition 5. *(Mandatory part) A part is mandatory if the whole cannot exist without it.*

TCs obviously capture mandatory parts. If a part as a member of the TC does not exist, then the whole described by the TC no longer exists. Conversely, if a whole is proved, then any of its fields corresponds to a single instance. If one expects non-mandatory parts, a typical problem of the conceptualization with classes, then they cannot appear as fields of the related TC. The solution to this problem requires separate definitions of parts, in which case their instances are not automatically generated.

Definition 6. *(Essential part) A part is essential if it is mandatory and cannot change without destroying the whole.*

Essential parts require a temporal analysis. All instances of TCs are available at a given time-stamp and changing a part yields changing the conceptual description of the whole. For that purpose Coq has modules acting as a (conceptual) closure. As a consequence distinct conceptualizations can be enclosed in distinct modules. Since modules can be related in several subtle ways such as inheritance, typing, etc., then we get a means for representing distinct but related conceptualizations. Only the proved conceptualization at a given time-stamp will be available.

Definition 7. *(Exclusive part) A part is exclusive if it can be part of at most one whole.*

For exclusive parts which are described by a field inside a TC, we must guarantee that the corresponding object is not shared with another field in another TC. By default all instances of fields inside TC are structurally distinct from fields of the same type in other TCs, and therefore these instances are exclusive. If one expects to share (at least) an instance between TC, then it requires the mechanism of automatic instance inference between part-whole relations (see next section).

5   Constructing Part-Whole Hierarchies in Coq

5.1   Constructing the Taxonomy

As a running example, we provide an excerpt from an anatomical ontology (Anatomy of the Cardio-vascular System) with terms extracted from figure 1 (overlapping parts are highlighted). A module type `KDTL_Lattice` is created which include reusable basic type classes such as all meta-properties of relation types, lattice operators, the `PWLattice` type class and all kind declarations.

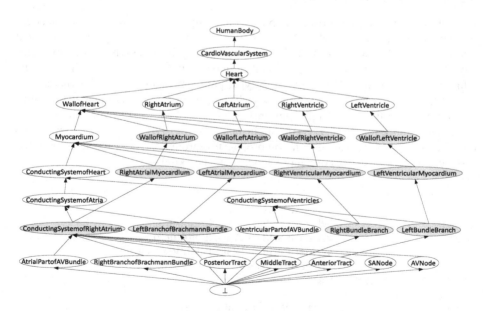

Fig. 1. An excerpt of the anatomical part-whole hierarchy

```
Class PT.
Class PD.
Class ED.
```
...
```
Class SubdivisionOfConductingSystemOfHeart.
Class AtrialPartofAVBundle.
Class RightBranchofBrachmannBundle.
Class PosteriorTract.
```
...

Then, all kinds are coerced to their subsumer in the subsumption hierarchy which supplements the DOLCE hierarchy (at the bottom) included in another module ECGSub defined as an instance of KDTL_Lattice.

Parameter $D2$: PD → PT. Coercion D2 : $PD \rightarrowtail PT$.
Parameter $D3$: ED → PT. Coercion D3 : $ED \rightarrowtail PT$.

...

Parameter $c1$: AtrialPartofAVBundle → SubdivisionOfConductingSystemOfHeart.
Coercion $c1$: $AtrialPartofAVBundle \rightarrowtail SubdivisionOfConductingSystemOfHeart$.
Parameter $c2$: RightBranchofBrachmannBundle → SubdivisionOfConductingSystemOfHeart.
Coercion $c2$: $RightBranchofBrachmannBundle \rightarrowtail SubdivisionOfConductingSystemOfHeart$.
Parameter $c3$: PosteriorTract → SubdivisionOfConductingSystemOfHeart.
Coercion $c3$: $PosteriorTract \rightarrowtail SubdivisionOfConductingSystemOfHeart$.

...

5.2 Constructing the Partonomy

Automatic Instance Generation. Coping with multiple levels of granularity that coexist in the Foundational Model of Anatomy (FMA) remains a challenge and, as pointed out in [31, 32], the FMA needs a thorough restructuring to address this issue satisfactorily. We will see now how this aspect is taken into account with Coq. First, in a module which instantiates the KDTL_Lattice type, parts are included as sub-structures of their respective wholes in a bottom-up way. Type classes provide automatized instance generation of parts. For example, any instance of WallOfHeart will automatically create an instance for Myocardium, an instance for WallOfRightAtrium, and so on. The process of instance generation propagates through TCs. At the same time, all instances are also instances of a more general TC, i.e., kind.

...
```
Class WallOfHeart := {
                    WHeartPart1 :> Myocardium;
                    WHeartPart2 :> WallOfRightAtrium;
                    WHeartPart3 :> WallOfLeftAtrium;
                    WHeartPart4 :> WallOfRightVentricle;
                    WHeartPart5 :> WallOfLeftVentricle}.
Parameter m25 : WallOfHeart → kind.
Coercion m25 : WallOfHeart ↣ kind.
Class RightAtrium := {RAPart1 :> WallOfRightAtrium}.
Parameter m26 : RightAtrium → kind.
Coercion m26 : RightAtrium ↣ kind.
```

```
Class LeftAtrium := {LAPart1 :> WallOfLeftAtrium}.
```
Parameter *m27* : LeftAtrium → kind.
Coercion *m27* : *LeftAtrium* ↣ *kind.*
```
Class RightVentricle := {RVPart1 :> WallOfRightVentricle}.
```
Parameter *m28* : RightVentricle → kind.
Coercion *m28* : *RightVentricle* ↣ *kind.*
```
Class LeftVentricle := {LVPart1 :> WallOfLeftVentricle}.
```
Parameter *m29* : LeftVentricle → kind.
Coercion *m29* : *LeftVentricle* ↣ *kind.*
```
Class Heart := {
```
$\qquad\qquad$ HeartPart1 :> WallOfHeart;

$\qquad\qquad$ HeartPart2 :> RightAtrium;

$\qquad\qquad$ HeartPart3 :> LeftAtrium;

$\qquad\qquad$ HeartPart4 :> RightVentricle;

$\qquad\qquad$ HeartPart5 :> LeftVentricle}.

Parameter *m30* : Heart → kind.
Coercion *m30* : *Heart* ↣ *kind.*

Then, *part-whole* relations are specified using dependent TCs labeled P_XX together with an operational TC Cpartof, that is a TC having a single field and no brackets [25]. Since the process of instance resolution only occurs between TC, then embedding the relation ⊏ inside a TC will ensure that instance resolution will propagate across P_XX structures. Furthermore, efficiency is ensured using a parameterized operational TC considered by Coq as an alias for the ⊏ relation.

Class Cpartof (*x y* : kind) := sdr1 :> *x* ⊏ *y*.

All TCs denoted P_XX provide available *part-whole* relations in the *part-whole* hierarchy on the basis of their internal arguments (*proper part-of*) or (*part-of*). A part may (as a limit case) be itself a whole which is expressed with an operational TC. By default, parts defined as arguments of *part-of* (or *proper part-of*) relations cannot exist outside the whole in which they appear i.e., they are mandatory parts. For that purpose they are expressed as substructures of TC, i.e., internal fields expressing the available *part-whole* relations. For example, since an instance of Heart is provided as argument, all its parts are available in P_Heart.

Class P_Heart (*x*:Heart) := {

$\qquad\qquad$ PHeart1 :> Cpartof HeartPart1 *x*;

$\qquad\qquad$ PHeart2 :> Cpartof HeartPart2 *x*;

$\qquad\qquad$ PHeart3 :> Cpartof HeartPart3 *x*;

$\qquad\qquad$ PHeart4 :> Cpartof HeartPart4 *x*;

$\qquad\qquad$ PHeart5 :> Cpartof HeartPart5 *x*}.

Let now consider the following test:

Class Test {*p*:AVNode}(*cl*:CardiovascularSystem): Prop.

It means that an instance of the class Test requires (i) that an instance of AVNode exists and (ii) that we provide an argument having the type CardiovascularSystem. Then, if we declare only the following instances:

Variable *hl* : Heart.
Variable *C* : P_Heart *hl*.

Variable *cv1* : CardiovascularSystem.
Variable *l1* : Test *cv1*.

then, the test succeeds which means that creating an instance of Heart will automatically create all instances that are below it in the lattice, and an instance of AVNode as well. This is exactly what we can expect of a partonomic lattice from a mereological point of view: if a whole exists, then its parts exist at all abstraction levels below it through transitivity. This is a strong argument for using TCs.

Shareable Parts. The necessary condition to be a shareable part requires the definition of fields in distinct *part-whole* structures having a common type. However, this condition is not sufficient and we must produce a proof that the relations refer to the same instances. For that purpose a whole W_1 will share a common instance with a whole W_2 if one of them admits an implicit argument having the type of the other whole. In such a way the common instance can be introduced without bundling, i.e., without giving explicit arguments. For example, proving that there are shareable parts between P_ConductingSystemofAtria and P_LeftAtrialMyocardium first requires the description of *part-whole* relations that hold with ConductingSystemof Atria. Then, the single *part-whole* relation standing within P_LeftAtrialMyocar dium should refer to the same instance than the instance of type LeftBranchofBrach mannBundle standing in P_ConductingSystemofAtria. For that purpose, the implicit argument '{P ConductingSystemofAtria} will automatically infer the missing instances. Overlaps can be checked in Coq using the Cpartof class witnessing their existence:

Class Overlap' (*x y* : kind){*z*:kind}{*v*:Cpartof *z x*}{*b*:Cpartof *z y*}.

All implicit instances between brackets (i.e., z, v and b) must exist in order to prove the Overlap' class, which boils down to apply definition 2. Finding a proof for Overlap' proves that the two TCs overlap by inferring the missing part.

```
Class P_ConductingSystemofAtria (x:ConductingSystemofAtria) := {
                    PConductingSystemofAtria1 :> Cpartof CSAtPart1 x;
                    PConductingSystemofAtria2 :> Cpartof CSAtPart2 x}.
Class P_LeftAtrialMyocardium (x:LeftAtrialMyocardium)
                    '{p:P_ConductingSystemofAtria} := {
                    PLeftAtrialMyocardium :> Cpartof CSAtPart2 x }.
...
Parameter csa1 : ConductingSystemofAtria.
Parameter lam1 : LeftAtrialMyocardium.
Parameter px : P_ConductingSystemofAtria csa1.
Parameter py : P_LeftAtrialMyocardium lam1.
Variable testOverlap1 : (Overlap' csa1 lam1).
```

The previous code sequence proves that, given an instance of a conducting system of atria, an instance of the left atrial myocardium and the two instances characterizing their relation instances (part-of), we can create a variable having the type Overlap' between the two instances. It is important to understand that we can create this type only if all the required instances have a common part which is proved using the class

Overlap' from definition 2. Instead of testing each first-order instance of the formula resulting from 2, instance resolution does the job automatically on all existing instances.

Notice that instance resolution creates automatically the appropriate instances, but instances can also be created by a real application (e.g., a database) and then only missing concrete values are created. Instance (proof) resolution with TC relies on a backtracking algorithm which uses constraint generation (max depth 100) and proof search among instances. The constraint generation algorithm is linear in the size of the rewritten term, while the customized proof search is achieved using a depth first algorithm with a linear average complexity (see [34] for more details).

6   Conclusion

In this paper, we present a study that provides a formal basis for the syntax and the semantics of *part-whole* constructs using syntactic structures of the Coq language. The type classes specifications provide expressive formal models of relationships. Thus, semantic analysis of these relationships can take place on these specifications using proof techniques provided for Coq. Any inconsistency or error discovered during the analysis provides feedback on the relationships.

Using a higher-order polymorphic type theory, i.e., the Calculus of Constructions with inductive types and universes, provides substantial benefits. First, higher-order is useful (i) to allow instances of categorization types to be types themselves, (ii) to abstract away from level distinctions and (iii) to directly support quantification over sets and general concepts. The major limitation holds in the difficulty for novice users to easily program in Coq. Further works include (i) the realization of an interface between Coq and the user to collect proof objects with requests and a tool for proving well-formed class diagrams and (ii) a deeper investigation on the conceptualization of the *part-whole* relation by replacing protothetic with type theory and using axioms.

References

1. Guarino, N., Welty, C.: A Formal Ontology of Properties. In: Dieng, R., Corby, O. (eds.) EKAW 2000. LNCS (LNAI), vol. 1937, pp. 97–112. Springer, Heidelberg (2000)
2. Smith, B., Rosse, C.: The Role of Foundational Relations in the Alignment of Biomedical Ontologies. In: Fieschi, M., et al. (eds.) MEDINFO 2004, pp. 444–448. IOS Press, Amsterdam (2004)
3. Guizzardi, G., Wagner, G.: What's in a Relationship: An Ontological Analysis. In: Li, Q., Spaccapietra, S., Yu, E., Olivé, A. (eds.) ER 2008. LNCS, vol. 5231, pp. 83–97. Springer, Heidelberg (2008)
4. Artale, A., Franconi, E., Guarino, N., Pazzi, L.: Part-whole relations in object-centered systems: An overview. Data and Knowledge Engineering 20(3), 347–383 (1996)
5. Bittner, T., Smith, B.: A Theory of Granular Partitions. In: Duckham, M., Goodchild, M.F., Worboys, M. (eds.) Foundations of Geographic Information Science, vol. 7, pp. 124–125 (2003)
6. Schwarz, U., Smith, B.: Ontological Relations. In: Munn, K., Smith, B. (eds.) Applied Ontology: An Introduction, Ch. 10, pp. 219–234. Ontos Verlag (2008)
7. Keet, C.M., Artale, A.: Representing and reasoning over a taxonomy of part-whole relations. Applied Ontology 3(1-2), 91–110 (2008)

8. Dapoigny, R., Barlatier, P.: Towards Ontological Correctness of Part-whole Relations with Dependent Types. In: Procs. of the Sixth Int. Conference (FOIS 2010), pp. 45–58 (2010)
9. Simons, P.: Parts: A Study in Ontology. Clarendon Press, Oxford (1987)
10. Varzi, A.C.: Parts, wholes, and part-whole relations: The prospects of mereotopology. Data and Knowledge Engineering 20(3), 259–286 (1996)
11. Johansson, I.: On the transitivity of the parthood relations. In: Hochberg, H., Mulligan, K. (eds.) Relations and Predicates, pp. 161–181. Ontos Verlag (2004)
12. Lewis, D.: Parts of Classes, pp. 79–93. Blackwell Publishers, Oxford (1991)
13. Opdahl, A.L., Henderson-Sellers, B., Barbier, F.: Ontological analysis of whole-part relationships in OO-models. Information and Software Technology 43, 387–399 (2001)
14. Aitken, J.S., Webber, B.L., Bard, J.B.: Part-of relations in anatomy ontologies: A proposal for RDFS and OWL formalisations. In: Procs. of the Pacific Symp. on Biocomputing, pp. 6–10 (2004)
15. Schulz, S., Stenzhorn, H., Boeker, M., Smith, B.: Strengths and limitations of formal ontologies in the biomedical domain. Elec. J. of Com., Inf. and Innovation in Health 3(1), 31–45 (2009)
16. Guizzardi, G.: On the Representation of Quantities and their Parts in Conceptual Modeling. In: Procs. of the Sixth Int. Conference (FOIS 2010), pp. 103–116. IOS Press (2010)
17. Gerstl, P., Pribbenow, S.: Midwinters, End Games, and Body Parts: a Classification of Part-whole Relations. Int. Journal of Human-Computer Studies 43, 865–889 (1995)
18. Angelov, K., Enache, R.: Typeful Ontologies with Direct Multilingual Verbalization. In: Rosner, M., Fuchs, N.E. (eds.) CNL 2010. LNCS, vol. 7175, pp. 1–20. Springer, Heidelberg (2012)
19. Cirstea, H., Coquery, E., Drabent, W., Fages, F., Kirchner, C., Maluszynski, J., Wack, B.: Types for Web Rule Languages: a preliminary study, technical report A04-R-560, PROTHEO - INRIA Lorraine - LORIA (2004)
20. Appel, A.W., Felty, A.P.: Dependent types ensure partial correctness of theorem provers. Journal of Functional Programming 14(1), 3–19 (2004)
21. Bertot, Y., Théry, L.: Dependent Types, Theorem Proving, and Applications for a Verifying Compiler. In: Meyer, B., Woodcock, J. (eds.) VSTTE 2005. LNCS, vol. 4171, pp. 173–181. Springer, Heidelberg (2008)
22. Bertot, Y., Castéran, P.: Interactive Theorem Proving and Program Development. Coq'Art: The Calculus of Inductive Constructions. An EATCS series. Springer (2004)
23. Dapoigny, R., Barlatier, P.: Modeling Ontological Structures with Type Classes in Coq. In: Pfeiffer, H.D., Ignatov, D.I., Poelmans, J., Gadiraju, N. (eds.) ICCS 2013. LNCS, vol. 7735, pp. 135–152. Springer, Heidelberg (2013)
24. Sozeau, M., Oury, N.: First-class type classes. In: Mohamed, O.A., Muñoz, C., Tahar, S. (eds.) TPHOLs 2008. LNCS, vol. 5170, pp. 278–293. Springer, Heidelberg (2008)
25. Spitters, B., van der Weegen, E.: Type classes for mathematics in type theory. Mathematical Structures in Computer Science 21(4), 795–825 (2011)
26. Sowa, J.F.: Using a lexicon of canonical graphs in a semantic interpreter. In: Relational Models of the Lexicon, pp. 113–137. Cambridge University Press (1988)
27. Masolo, C., Borgo, S., Gangemi, A., Guarino, N., Oltramari, A.: Ontology Library (D18), Laboratory for Applied Ontology-ISTC-CNR (2003)
28. Gangemi, A., Guarino, N., Masolo, C., Oltramari, A., Schneider, L.: Sweetening ontologies with DOLCE. In: Gómez-Pérez, A., Benjamins, V.R. (eds.) EKAW 2002. LNCS (LNAI), vol. 2473, pp. 166–181. Springer, Heidelberg (2002)
29. Smith, B.: Mereotopology: A Theory of Parts and Boundaries. Data and Knowledge Engineering 20, 287–303 (1996)

30. Varzi, A.C.: Mereology. In: Zalta, E.N. (ed.) The Stanford Encyclopedia of Philosophy (2012), `http://plato.stanford.edu/archives/win2012/entries/mereology`
31. Kumar, A., Smith, B., Novotny, D.: Biomedical Informatics and granularity. Comparative and Functional Genomics 5(6-7), 501–508 (2004)
32. Rector, A., Rogers, J., Bittner, T.: Granularity, scale and collectivity: When size does and does not matter. J. of Biomedical Informatics 39(3), 333–349 (2006)
33. Donnelly, M., Bittner, T., Rosse, C.: A formal theory for spatial representation and reasoning in biomedical ontologies. Artificial Intelligence in Medicine 36(1), 1–27 (2005)
34. Sozeau, M.: A New Look at Generalized Rewriting in Type Theory. Journal of Formalized Reasoning 2(1), 41–62 (2009)

Automatic Extraction of Semantic Relations by Using Web Statistical Information*

Valeria Borzì[1], Simone Faro[1], and Arianna Pavone[2]

[1] Dipartimento di Matematica e Informatica, Università di Catania
Viale Andrea Doria 6, I-95125 Catania, Italy
[2] Dipartimento di Scienze Umanistiche, Università di Catania
Piazza Dante 32, I-95124 Catania, Italy
faro@dmi.unict.it

Abstract. A semantic network is a graph which represents semantic relations between concepts, used in a lot of fields as a form of knowledge representation. This paper describes an automatic approach to identify semantic relations between concepts by using statistical information extracted from the Web. We automatically constructed an associative network starting from a lexicon. Moreover we applied these measures to the ESL semantic similarity test proving that our model is suitable for representing semantic correlations between terms obtaining an accuracy which is comparable with the state of the art.

1 Introduction

In recent years, with the increase of the information society, lexical knowledge, i.e. all the information that is known about words and all the relationships among them, is becoming a core research topic in order to understand and categorize all subjects of interest [14]. We need lexical knowledge to know how words are used in different ways to express different meanings [13].

An *associative network* is a labeled directed (or undirected) graph representing relational knowledge. Each vertex of the graph represents a concept and each edge (or link) represents a relation between concepts. Such structures are used to implement cognitive models representing key features of human memory.

Specifically, when two concepts, x and y, are thought simultaneously, they may become linked in memory. Subsequently, when one thinks about x, then y is likely to come to mind as well. Thus multiple links to a concept in memory make it easier to be retrieved because of many alternative routes to locate it.

A *semantic network* is an associative network where we introduce labels on the links between words [3], [14]. Labels represent the kind of relation between the two given concepts, such as "is-a", "part-of", "similar-to" and "related-to".

Aristotle firstly described some of the principles governing the role of associative networks and categories in memory, while the concept of semantic network

* This work has been supported by project PRISMA PON04a2 A/F funded by the Italian Ministry of University and Research within the PON 2007-2013 framework.

N. Hernandez et al. (Eds.): ICCS 2014, LNAI 8577, pp. 174–187, 2014.
DOI: 10.1007/978-3-319-08389-6_15, © Springer International Publishing Switzerland 2014

dates back to the 3rd century AD when the greek philosopher Porphyry, in his commentary on Aristotle's categories, drawn the oldest known semantic network, called *Porphyry's tree*. Despite its age, the Tree of Porphyry represents the common core of all modern type concept hierarchies.

The potential usefulness of large scale lexical knowledge networks can be attested by the number of projects and the amount of resources that have been dedicated to their construction [3], [14]. Creating such resources manually is a difficult task and it has to be repeated from ex novo for each new language. However there are a lot of important resources of this kind. Among the others the most relevant are WordNet, Wikipedia and BabelNet.

WordNet [3], is a lexical knowledge resource. It is a computational lexicon of the English language based on psycholinguistic principles. A concept in WordNet is represented as a synonym set (called *synset*), i.e. the set of words that share the same meaning. Synsets are related to each other by means of many lexical and semantic relations. Wikipedia instead is a multilingual Web-based encyclopedia. It is a collaborative open source medium edited by volunteers to provide a very large wide-coverage repository of encyclopedic knowledge. The text in Wikipedia is partially structured, various relations exist between the pages themselves. These include redirect pages (used to model synonymy), disambiguation pages (used to model homonymy and polysemy), internal links (used to model relations between terms) and categories. Finally, BabelNet [14] is a multilingual encyclopedic dictionary and a semantic network, currently covering 50 languages, created by linking Wikipedia network to WordNet, thus it includes lemmas which denote both lexicographic meanings and encyclopedic ones. However, a widely acknowledged problem with the above semantic networks is that they implements links which represent uniform distances between terms, while conceptual distances in real world relations between concepts could have a wide variability. As a consequence we find in Wikipedia, or in BabelNet, links between very close concepts but also links between terms that are conceptually distant, and no measure leading to distinguish between them.

In this paper we describe an automatic approach to identify semantic relatedness between concepts, by using statistical information extracted from the Web. We then use such semantic measure to construct a weighted associative network starting from the English WordNet lexicon, augmented with Wikipedia encyclopedic entities. From our preliminary experimental results it turns out that our presented approach can be efficiently used to identify semantic relatedness between concepts. The paper is organized as follows. In Section 2 we introduce the concept of semantic relatedness, which is particularly connected with our results, and present the most significant results on computing such measures. Then in Section 3 we introduce a new model The construction process is described in Section 4. In the next sections we present some experimental results (Section 5) and some examples (Section 6) in order to evaluate the effectiveness of the new presented model. We discuss our results and describe future works in Section 7.

2   Measuring Semantic Relatedness

Lexical semantic relatedness is a measure of how much two terms, words, or their senses, are semantically related. It has been well studied and categorized in linguistics.

Evaluating semantic relatedness using network representations is a problem with a long history in artificial intelligence and psychology, dating back to the spreading activation approach of Quillian [15] and Collins and Loftus [2]. It is important for many natural language processing or information retrieval applications. For instance, it has been used for spelling correction, word sense disambiguation, or coreference resolution. It has also been shown to help inducing information extraction patterns, performing semantic indexing for information retrieval, or assessing topic coherence.

Semantic relations between terms include typical relations such as

- synonymy: identity of senses as "automobile" and "car";
- antonymy: opposition of senses such as "fast" and "slow";
- hypernymy or hyponymy: such as "vehicle" and "car";
- meronymy or holonymy: part-whole relation such as "windshield" and "car".

Most of the recent research has focused on *semantic similarity* [17], [21,22], [6], [18], which represents a special case of semantic relatedness. For instance, antonyms are related, but not similar. Or, following Resnik [17], "car" and "bicycle" are more similar (as hyponyms of "vehicle") than "car" and "gasoline", though the latter pair may seem more related in the world.

Thus, while typical relations implying sense similarity are widely represented in lexicons like WordNet [3] and BabelNet [14], the latter types of relations are usually not always included in state-of-the-art ontologies, although they are relevant in conceptual connections between terms. Such relations include, for instance, the following

- synecdoche: a portion of something refers to the whole, as "information" for a "book" or as "cold" for the "winter";
- antonomasia: an epithet for a proper name, as "The Big Apple" for "New York" or as "The Conqueror" for "Caesar";
- trope: a figurative meaning for its literal use, "to bark" for "to shout".

Current approaches to address semantic relatedness can be categorized into three main categories: lexicon-based methods, corpus-based methods, and hybrid approaches.

In a *lexicon-based methods* the structure of a lexicon is used to measure semantic relatedness. Such approach consists in evaluating the distance between the nodes corresponding to the terms being compared: the shorter the path from one node to another, the more similar they are. Such approaches rely on the structure of the lexicon, such as the semantic shortest link path [11], the depth of the terms in the lexicon tree [23], the lexical chains between synsets and their relations [6], or on the type of the semantic edges [21]. Finally in [18] the authors use all 26 semantic relations found in WordNet in addition to information found in glosses to create an explicit semantic network.

However, a widely acknowledged problem with this approach is that it relies on the notion that links in the taxonomy represent uniform distances [18]. Unfortunately, this is difficult to define, much less to control. In real lexicons, there could be wide variability in the distance covered by a single relation link.

Differently, *corpus-based methods* use statistical information about words distribution extracted from a large corpus to compute semantic relatedness. For instance in [19,5] the authors used the statistical information from Wikipedia.

For the sake of completeness we mention also *hybrid methods* which use a combination of corpus-based and lexicon-based methods [7,1] to compute semantic relatedness between two terms.

3   A New Model for Directional Semantic Relatedness

In this section we formalize the model of semantic relatedness which has been used to construct our associative network. Unlike state-of-the-art networks, in our structure the edges represent a certain correlation between two terms and give a measure of such relations. Thus we obtain a weighted network where terms closely related have small distances while weakly correlated terms have a great distance. The distance between two nodes of the network is inversely proportional to their semantic correlation which we measure by an *attraction coefficient*. The closer is the semantic correlation between the two words, the greater is their attraction. In turn the semantic attraction between two different terms is a function of their *usage coefficient*, i.e. a numeric value which measures how much the corresponding term is used in a given language.

In what follows we formalize this concepts and give the mathematical definitions of the formulas we use for computing the semantic relatedness in our network.

The Usage Coefficient

All natural languages like English consist of a small number of very common words, a larger number of intermediate ones, and then an indefinitely large set of very rare terms. We define the *usage coefficient* (U.C.) of a lexical term x, for a given language $\mathcal{L}$, as a value indicating how much x is used in $\mathcal{L}$. Such coefficient has been classically computed as the frequency of the term x in large corpora as the Oxford English Corpus[1], the Brown Corpus of Standard American English[2] or Wikipedia[3].

In order to give a real estimate of the frequency of a given term we compute the U.C. of words as a function of the number of pages resulting in a Google

---

[1] http://www.oxforddictionaries.com
[2] http://www.essex.ac.uk/linguistics/external/clmt/w3c/corpus_ling/
  content/corpora/list/private/brown/brown.html
[3] http://en.wikipedia.org

search[4] for the term x. Specifically, for each term x contained in the English ontology, we performed a query on Google for x and use the number of page results for computing the U.C. of the term. We use the symbol $\rho(x)$ to indicate the U.C. of a term x.

Although the Google search engine does not guarantee the ability to return the exact number of results for any given search query[5] such value can be considered a good estimate of the actual number of results for the search request [16,8]. We observed an upper bound for the number of page results retrieved by Google, i.e. MAX_RESULTS = 25.270 millions of results.

The U.C. of a term x is then computed by

$$\rho(x) = \frac{\text{PAGE\_RESULTS}(x)}{\text{MAX\_RESULTS}}$$

In our search we activate automatic filtering feature in order to reduces undesirable results such as duplicate entries. Moreover we filter search results by language and we use the `allintext:` operator[6] in order to reduce the search to internal text of the web pages.

The Co-occurrence Usage Coefficient

Given a set of k terms, $\{x_1, x_2, \ldots, x_k\}$, of a given language $\mathcal{L}$, the *co-occurrence usage coefficient* (C.U.C.) of the terms x_i, is a value indicating how much such terms co-occur together in any context of the language. As before, we compute the C.U.C. as the number of pages resulting from a Google query for $x_1 \wedge x_2 \wedge \ldots \wedge x_k$, divided my the constant MAX_RESULTS.

We use the symbol $\rho(x_1 : x_2 : \ldots : x_k)$ to indicate the C.U.C. of the set $\{x_1, x_2, \ldots, x_k\}$.

By the definition given above it is trivial to observe that, for each $i = 1, \ldots, k$, the property $\rho(x_i) > \rho(x_1 : x_2 : \ldots : x_k)$ holds.

The Attraction Coefficient

A straightforward way to compute a similarity coefficient between two lexical terms is to use the *Jaccard similarity coefficient*, a statistic index introduced for comparing the similarity and diversity of sample sets. It is defined as the size of the intersection divided by the size of the union of the sample sets. More formally if $\rho(x)$ and $\rho(y)$ are the U.C. of terms x and y, respectively, and $\rho(x : y)$ is their co-occurrence coefficient, the Jaccard similarity coefficient of x and y can be computed by using the following formula

$$jacc(x : y) = \frac{\rho(x : y)}{\rho(x) + \rho(y) - \rho(x : y)}$$

---

[4] www.google.com

[5] http://www.google.com/support/enterprise/static/gsa/docs/admin/72/
gsa_doc_set/xml_reference/appendices.html

[6] http://www.googleguide.com/advanced_operators_reference.html

Such similarity coefficient has been used in [19], in combination with a lexicon-based approach, to measure the similarity relatedness of two terms. However it defines a symmetric semantic relation between x and y, thus assuming that $jacc(x : y) = jacc(y : x)$, which does not reflect the real world representation of associative networks where relations are, in general, represented by direct edges, i.e. the measure of the relation between x and y could be different from the measure of the relation between y and x.

Example 1. The terms "gasoline" and "car" are undoubtedly related in real world, thus if we think to gasoline the term "car" comes to mind with a great probability. However the contrary is not true, if we think to a car probably other terms come to mind with higher probability, like "road" or "parking". So we can say that "gasoline" is more related with "car" than viceversa.

In our model we define an unidirectional measure of semantic similarity between two terms. Specifically, the *attraction coefficient* (A.C.) of a lexical term x, on another term y of the same language, measures the semantic correlation of y towards x. In other words it is a numerical value evaluating how much the term x is conceptually related with term y (the contrary is not necessary).

More formally, let x and y two lexical terms, and let $\rho(x)$ and $\rho(y)$ the U.C. of x and y, respectively. Moreover let $\rho(x : y)$ be their co-occurrence coefficient. Then the attraction coefficient of y on x is defined by

$$\varphi(x \to y) = \frac{\rho(x : y)}{\rho(x)} \tag{1}$$

The following properties follow directly from the above definition and they are trivial to prove.

Property 1. If x and y are two lexical terms of $\mathcal{L}$, then the A.C. of y on x is a real number between 0 and 1. Formally

$$0 \leq \varphi(x \to y) \leq 1$$

Property 2. If x and y are two lexical terms of $\mathcal{L}$, and $\rho(x) > \rho(y)$ then the A.C. of x on y is greater than the A.C. of y on x. Formally

$$\rho(x) > \rho(y) \implies \varphi(y \to x) > \varphi(x \to y)$$

Due to Property 2 it turns out that lexical terms with a huge U.C. are more attractive than other terms with smaller coefficient. This is the case, for instance, of general terms as "love", "man", "science", "music" and "book".

Example 2. Consider the numerical values related to the terms "bark", "kennel", "dog" and "man", presented in the following table, where the U.C. are expressed in million of results.

U.C.		C.U.C.		A.C.		
ρ(bark)	$\sim 0{,}043$	ρ(bark : dog)	$\sim 0{,}012$	φ(bark $\rightarrow$ dog)	~ 0.28	(a)
ρ(kennel)	$\sim 0{,}026$	ρ(bark : man)	$\sim 0{,}015$	φ(dog $\rightarrow$ bark)	~ 0.01	(b)
ρ(dog)	$\sim 0{,}813$	ρ(kennel : dog)	$\sim 0{,}016$	φ(bark $\rightarrow$ man)	~ 0.35	(c)
ρ(man)	$\sim 1{,}000$	ρ(kennel : man)	$\sim 0{,}005$	φ(kennel $\rightarrow$ dog)	~ 0.62	(d)
		ρ(dog : man)	$\sim 0{,}372$	φ(kennel $\rightarrow$ man)	~ 0.19	(e)
				φ(dog $\rightarrow$ man)	~ 0.45	(f)

The term "bark" directly calls to mind the term "dog", since the bark is a prerogative of dogs, so that we can say that "bark" is semantically attracted by "dog" (a: 0.28). On the other hand, the contrary is not true since "dog" not necessarily calls to mind the term "bark" (b: 0.01), which is only one of the many inherent attitudes of a dog. Observe also that "bark" has a figurative meaning which can be applied to men, so it is semantically attracted also by the term "man" (d: 0.35). Differently the term "kennel" is strongly attracted by "dog" (d: 0.62) and is subject only to a feeble conceptual attraction by "man" (e: 0.19). The term "dog" is instead semantically attracted by the term "man" (f: 0.45), since the dog is the most popular domestic animal.

4   Building the Directed Semantic Graph

We construct our semantic network starting from state of the art lexicon resources and by enriching them with new information and new semantic relations induced by the relatedness model described above. Specifically we start from the English WordNet semantic network.

The algorithm for building the corresponding directed semantic graph is depicted in Figure 1 and is named BUILDNETWORK. It takes as input the set $\mathcal{L}$ of all terms of the lexicon and constructs a directed weighted graph where each term x of the lexicon is a node in the graph, and directed links between two nodes represent semantic relations between the corresponding terms. Each link is associated with a weight value representing the attraction coefficient between the two related terms. The construction is divided in two steps, a bootstrap process and an exploration process, as described below.

The Bootstrap Process. In the bootstrap process (see Figure 1, on the left) the algorithm initializes the usage coefficient $\rho(x)$ for each term x of the set $\mathcal{L}$ (lines 2-3). In addition, for each term x, the algorithm also initializes the set, $\Psi(x)$ of all terms y such that $\varphi(x \rightarrow y) \geq \delta$, for a given bound δ (lines 4-12). In our construction we set $\delta = 0.1$. Specifically the set $\Psi(x)$ initially consists in all terms y which are related to x in the lexicon (line 5). In addition $\Psi(x)$ is augmented with the set of all significant terms from its definition, excluding all those words (conjunctions, adverbs, pronouns) that will not be particularly useful in the construction of the semantic field of x (line 6). Then all terms in the set $\Psi(x)$ are investigated in order to compute the attraction coefficient $\varphi(x \rightarrow y)$ (lines 7-10). During this process the algorithm deletes from $\Psi(x)$ all term y such that $\varphi(x \rightarrow y) < \delta$ (lines 11-12).

```
BOOTSTRAP(L)                                    EXPLORE(x)
  1. for each x ∈ L do                            1. explored(x) ← 1
  2.      if ρ(x) = null do                        2. for each y ∈ Ψ(x) do
  3.           ρ(x) ← getUC(x)                      3.      if (explored(y) = 0) then
  4.      Ψ(x) ← ∅                                  4.           EXPLORE(y)
  5.      Ψ(x) ← Ψ(x) ∪ getRelated(x)              5.      for each z ∈ Ψ(y) do
  6.      Ψ(x) ← Ψ(x) ∪ getDefinition(x)           6.           φ(x → z) ← ρ(x : z)/ρ(x)
  7.      for each y ∈ Ψ(x) do                      7.           if (φ(x → z) < δ) then
  8.           if ρ(y) ← null do                    8.                Ψ(x) ← Ψ(x) ∪ {z}
  9.                ρ(y) ← getUC(y)
 10.           φ(x → y) ← ρ(x : y)/ρ(x)            BUILDNETWORK(L)
 11.           if (φ(x → y) < δ) then               1. bootstrap(L)
 12.                Ψ(x) ← Ψ(x) \ {y}               2. for each x ∈ L do
 13.      explored(x) ← 0                           3.      if (explored(x) = 0) then
                                                    4.           EXPLORE(x)
```

Fig. 1. The algorithm which construct the semantic directed network. The construction makes use of two procedures, the BOOTSTRAP procedure and an EXPLORE procedure.

The Exploration Process. The next step of the algorithm consists in exploring each node graph by setting a recursive process (see Figure 1, on the right). For each term x, the flag explored(x) allows the algorithm to keep track of nodes already analyzed (a value set to 1), and nodes not yet explored (a 0 value). During the exploration process of the node x, the algorithm try to increase the set $\Psi(x)$ by adding new related terms contained in the lexicon. To do that the algorithm firstly recursively explore all neighbors y of node x (lines 3-4), i.e. all terms in the set $\Psi(x)$, and then it tries to add new links from x to all the neighbor nodes of y (lines 5-8). In other words, if the term x is semantically attracted by the term y and the latter is attracted by the term z, then the algorithm tries a possible relation between x and z. Observe that If a new node z enters the set $\Psi(x)$ (line 8) then all its neighbors will be considered for inclusion in the set. This process continues until all terms have been explored.

5   First Experimental Results

To test our approach to semantic relatedness between two terms of the lexicon, we evaluated it on a synonym identification test. Although different tests are available on the net, as for instance the WordSimilarity-353 similarity test[7], the one we experimented with is the larger English as a Second Language (ESL) test, which was first used by Peter Turney in [22] as an evaluation of algorithms measuring the degree of similarity between words. Specifically the ESL test includes 50 synonym questions. Each question includes a sentence, providing context for

---

[7] http://www.cs.technion.ac.il/~gabr/resources/data/wordsim353/

the question, containing an initial word, and a set of options from which the most synonymous word must be selected. The following is an example question taken from ESL data set:

> To [firmly] refuse means to never change your mind and accept
> 1. steadfastly
> 2. reluctantly
> 3. sadly
> 4. hopefully

The results are measured in terms of accuracy. For each question with initial word x and option words $\{y_1, y_2, y_3, y_4\}$ we compute the attraction coefficients $\varphi(y_i \to x)$, for $i = 1 \ldots 4$ and put them in a decreasing order. Then we gave a decreasing score to each option word, from 4 to 1. Then the accuracy is computed as the sum of the scores obtained in the 50 questions compared with a full score result. The results of our approach, along with other approaches, on the 50 ESL questions are shown in Table 4. Our approach has achieved an accuracy of 84% on the ESL test, which is slightly better than the reported approaches in the literature. It should be noted that sometimes the difference between two approaches belonging to the same category are merely a difference in the data set used (Corpus or Lexicon) rather than a difference in the algorithms. Also, the ESL question set includes a sentence to give a context for the word, which some approaches (e.g. [22]) have used as an additional information source; we on the other hand, did not make use of the context information in our approach.

Table 1. Results with the ESL Data Set

Approach	Year	Category	Accuracy
Resnik [17]	1995	Hybrid	32.66%
Leacock and Chodorow [11]	1998	Lexicon	36.00%
Lin [12]	1998	Hybrid	36.00%
Jiang and Conrath [10]	1997	Hybrid	36.00%
Hirst and St-Onge [6]	1998	Lexicon	62.00%
Turney [22]	2001	Corpus	74.00%
Terra and Clarke [20]	2003	Corpus	80.00%
Jarmasz and Szpakowicz [9]	2003	Lexicon	82.00%
Tsatsaronis et al. [21]	2010	Lexicon	82.00%
Siblini and Kosseim [18]	2013	Lexicon	84.00%
Our Approach	2014	Corpus	84.00%

6 Some Examples

In this section we present some experimental evidences related with the structure of the semantic net which has been constructed at the date of the paper submission (January 16th, 2014). This is the reason why some terms are not depicted in the semantic nets, since they where not still added. In particular we briefly discuss portions of the semantic net connected with the terms "book" and "conquest". We present measures of relatedness between connected terms in both graphical and tabular forms. In Figure 2 and Figure 3 the diameter of a node representing a term x is proportional to its U.C. $\rho(x)$. Concentric circles represent distances from the main term, ranging from 1.0 (the innermost) to 0.3 (the outmost).

The Network around "book". The term "book" has a very large semantic network and attracts different related words, since its U.C. is very large. We can observe that both terms "book" and "information" got the same U.C. value. Moreover their A.C. is equal to 1. This means that the two terms often occur together. Thus their relation can be interpreted as a synecdoche, which is distinguished by metonymy because it is based on quantitative relationships, through the broadening of meaning. Therefore it is assumed that "book" is a medium which convey "information" in general The terms "magazine", "title" and "cover" are positioned very close to the center and they are therefore strongly related to "book". Furthermore the relationships between the various terms of the semantic network are bidirectional. Thus, for example, the term "book" is strictly related to "cover" by a relationship of metonymy, viceversa the term "cover" is strictly related to "book" but with a relation of hyponymy, thus "book" is the hyperonymy term and "cover" is the hyponym. In other cases we notice that the semantic connections are unidirectional, for example the relation between "book" and "publishing house", where the latter term is directly related to the book, but the book is not directly related to "publishing house".

The Network around "conquest". Table 3 shows the twenty closer lexical terms related with the term "conquest", while Figure 3 shows a graphical representation of the portion of the semantic network containing all terms related with "conquest". Typical relations of hyponymy and hypernyms can be found as "conquest" and "war", or "conquest" and "battle". A relation of metonymy can be read in the connection between "conquest" and "strategy".

Also, observing results shown in Table 3 we find a very interesting relation between "conquest" and "Caesar". Analyzing the results it is possible to notice that the two terms are strongly related, and also in this case we find a figure of speech, the antonomasia: the term "conquest" (as root of "conqueror") can be considered as representative of the term "Caesar", indeed. The relation between "conquest" and "attack" can be read as a metonymy, as cause-and-effect relation.

In addition, the relation between "conquest" and "freedom" can be read as a trope relation, since "conquest" here is used in its figurative meaning in place of "achievement". Similarly the same term is used, in connection with "love", with a figurative meaning in place of "seduction".

Table 2. The twenty lexical terms which have been found to be semantically closer to the term book. The number of results are expressed in millions of pages (Mr).

x	U.C.	C.U.C.	A.C. book	A.C. x	x	U.C.	C.U.C.	A.C. book	A.C. x
book	1,000								
information	1,000	1,000	1,00	1,00	copybook	0,001	0,000	0,52	-
magazine	0,895	0,830	0,93	0,83	ebook	0,192	0,091	0,47	-
cover	1,000	0,922	0,92	0,92	periodical	0,010	0,005	0,47	-
title	1,000	0,746	0,75	0,75	monograph	0,013	0,006	0,46	-
review	1,000	0,694	0,69	0,69	collection	1,000	0,444	0,44	0,44
school	1,000	0,624	0,62	0,62	press	1,000	0,421	0,42	0,42
publishing house	0,007	0,005	0,62	-	education	1,000	0,416	0,42	0,42
fiction	0,006	0,141	0,58	-	thriller	0,075	0,031	0,41	-
author	1,000	0,539	0,54	0,54	Gutenberg	0,008	0,003	0,37	-
word	1,000	0,535	0,54	0,54	reader	0,508	0,180	0,35	-

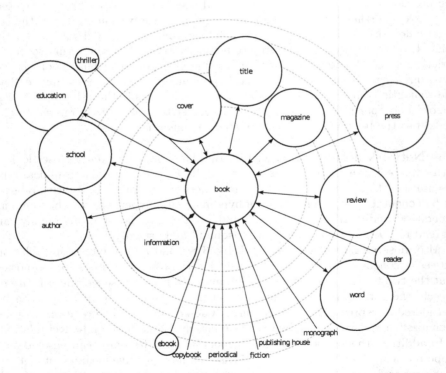

Fig. 2. A portion of the Semantic Net of the lexical term book. The diameter of a term x of the net if proportional to its U.C. $\rho(x)$. Concentric circles represent distances from the main term book.

Table 3. The twenty lexical terms which have been found to be semantically closer to the term conquest. The number of results are expressed in millions of pages (Mr).

x	U.C.	C.U.C.	A.C. conquest	A.C. x	x	U.C.	C.U.C.	A.C. conquest	A.C. x
conquest	0,029								
tyran	0,001	0,001	0,90	-	history	1,000	0,014	-	0,53
Athene	0,001	0,001	0,59	-	battle	0,579	0,009	-	0,30
military	0,026	0,009	0,34	-	right	1,000	0,014	-	0,49
Caesar	0,052	0,027	0,51	0,92	war	0,961	0,013	-	0,46
people	0,430	0,017	-	0,57	attack	0,497	0,007	-	0,24
empire	0,196	0,007	-	0,26	man	1,000	0,013	-	0,45
soldier	0,136	0,005	-	0,16	land	1,000	0,012	-	0,40
freedom	0,298	0,007	-	0,26	age	1,000	0,011	-	0,37
strategy	0,314	0,007	-	0,25	love	1,000	0,010	-	0,33
science	1,000	0,007	-	0,26	field	1,000	0,008	-	0,28

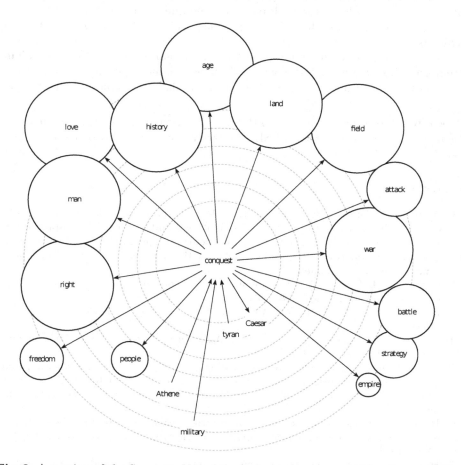

Fig. 3. A portion of the Semantic Net of the lexical term conquest. The diameter of a term x of the net if proportional to its U.C. $\rho(x)$. Concentric circles represent distances from the main term conquest.

7 Conclusions and Future Works

In this paper we described the construction of a semantic associative network for the English language. We start from the state-of-the-art semantic networks, as WordNet and Wikipedia, and enrich them with new informations measuring how much a term is used in practice. Then our algorithm explores the entire network in order to delete or add new semantic link according to a given model of directional semantic relatedness, based on statistical informations extracted from the Web. We then applied these measures to a real-world NLP task such as the ESL semantic similarity test. Our results show that our model is suitable for representing semantic correlations between terms obtaining an accuracy which is comparable with the state of the art.

Our algorithm is still exploring the network in order to complete the process of connecting all related terms. From our preliminary observations it turns out that several connections have been identified which do not appear in typical lexicons.

In future works we intend to construct a similar structure for the Italian language. Moreover we would like to perform additional experimental evaluation in order to test our model in field of semantic similarity or semantic relatedness.

Acknowledgements. We wish to thank Peter Turney for having provided the English as a Second Language (ESL) similarity test and for his precious suggestions.

References

1. Agirre, E., Alfonseca, E., Hall, K., Kravalova, J., Pasca, M., Soroa, A.: A study on similarity and relatedness using distributional and wordnet-based approaches. In: Proceedings of Human Language Technologies: The 2009 Annual Conference of the North American Chapter of the Association for Computational Linguistics, Boulder, pp. 19–27 (June 2009)
2. Collins, A., Loftus, E.: A spreading activation theory of semantic processing. Psychological Review 82, 407–428 (1975)
3. Fellbaum, C. (ed.): WordNet: An Electronic Database. MIT Press, Cambridge (1998)
4. Francis, W.N., Kucera, H.: Frequency Analysis of English Usage: Lexicon and Grammar. Houghton Mifflin (1982)
5. Gabrilovich, E., Markovitch, S.: Computing semantic relatedness using wikipedia based explicit semantic analysis. In: Proceedings of the 20th International Joint Conference on Artifical Intelligence (IJCAI 2007), Hyderabad, pp. 1606–1611 (January 2007)
6. Hirst, G., St-Onge, D.: Lexical chains as representations of context for the detection and correction of malapropisms. WordNet An electronic lexical database, 305–332 (April 1998)
7. Hughes, T., Ramage, D.: Lexical semantic relatedness with random graph walks. In: Proceedings of the Conference on Empirical Methods in Natural Language Processing - Conference on Computational Natural Language Learning (EMNLP-CoNLL), Prague, pp. 581–589 (June 2007)

8. Janetzko, D.: Objectivity, Reliability, and Validity of Search Engine Count Estimates. International Journal of Internet Science 3(1), 7–33 (2008)
9. Jarmasz, M., Szpakowicz, S.: Roget's thesaurus and semantic similarity. In: Proceedings of Recent Advances in Natural Language Processing, Borovets, pp. 212–219 (September 2003)
10. Jiang, J.J., Conrath, D.W.: Semantic similarity based on corpus statistics and lexical taxonomy. In: Proceedings of International Conference on Research in Computational Linguistics, Taipei, Taiwan, pp. 19–33 (August 1997)
11. Leacock, C., Chodorow, M.: Combining local context and wordnet similarity for word sense identification. WordNet: An Electronic Lexical Database 49(2), 265–283 (1998)
12. Lin, D.: An information-theoretic definition of similarity. In: Proceedings of the 15th International Conference on Machine Learning, Madison, vol. 1, pp. 296–304 (July 1998)
13. Navigli, R.: Word Sense Disambiguation: A survey. ACM Computing Surveys 41 (2009)
14. Navigli, R., Ponzetto, S.: BabelNet: The Automatic Construction, Evaluation and Application of a Wide-Coverage Multilingual Semantic Network. Artificial Intelligence 193 (2012)
15. Ross Quillian, M.: Semantic memory. In: Minsky, M. (ed.) Semantic Information Processing. MIT Press, Cambridge (1968)
16. Rayson, P., Charles, O., Auty, I.: Can Google count? Estimating search engine result consistency. In: Proceedings of the Seventh Web as Corpus Workshop, pp. 23–30 (2012)
17. Resnik, P.: Using information content to evaluate semantic similarity in a taxonomy. In: International Joint Conference for Artificial Intelligence, Montreal, pp. 448–453 (August 1995)
18. Siblini, R., Kosseim, L.: Using a Weighted Semantic Network for Lexical Semantic Relatedness. In: Proceedings of Recent Advances in Natural Language Processing (RANLP 2013), Hissar, Bulgaria (September 2013)
19. Strube, M., Ponzetto, S.P.: WikiRelate! Computing semantic relatedness using Wikipedia. In: Proceedings of the National Conference on Artificial Intelligence, Boston, vol. 21, p. 1419 (July 2006)
20. Terra, E., Clarke, C.L.A.: Frequency estimates for statistical word similarity measures. In: Proceedings of the 2003 Conference of the North American Chapter of the Association for Computational Linguistics on Human Language Technology, Edmonton, vol. 21, pp. 165–172 (May 2003)
21. Tsatsaronis, G., Varlamis, I., Vazirgiannis, M.: Text relatedness based on a word thesaurus. Journal of Artificial Intelligence Research 37(1), 1–40 (2010)
22. Turney, P.D.: Mining the Web for Synonyms: PMI-IR versus LSA on TOEFL. In: Flach, P.A., De Raedt, L. (eds.) ECML 2001. LNCS (LNAI), vol. 2167, pp. 491–502. Springer, Heidelberg (2001)
23. Wu, Z., Palmer, M.: Verbs semantics and lexical selection. In: Proceedings of the 32nd Annual Meeting on Association for Computational Linguistics, New Mexico, pp. 133–138 (June 1994)

Efficient Representation of Extensional Constraints

Cristian Frăsinaru and Florentin Olariu

Faculty of Computer Science, "A.I.Cuza" University, Iaşi, Romania
{acf,olariu}@info.uaic.ro

Abstract. Constraint programming is a paradigm where a given problem is formalized in terms of satisfying a set of restrictions. Difficult combinatorial problems, from virtually any area, can be modeled as constraint satisfaction problems (CSP) and then solved with state-of-the-art CSP solvers. Constraints are logical relations between various entities (variables) each taking values in a given domain. There are no limitations on how constraints should be specified – they can be seen as predicates that indicate partial information known about the problem. The *extensional* or *table constraint* is used to express directly the combination of values that are allowed for some variables. Many problems from industry or academic research are modeled using extensional constraints as they are created based on sets of existing data, represented as n-ary relations (sets of tuples).

In this paper, we describe an efficient data structure for implementing relations of any arity, that combines the concepts of trie (prefix tree) and support list, offering a quick query method. Because real-life problems may lead to bulky relations, containing a large number of tuples, it is essential to provide a trade-off between space and time complexity. In our representation, a relation is not tied to a specific extensional constraint and this allows multiple constraints to share the same data structure, in order to decrease the memory requirements. Our approach is generic and can be adapted to any method of solving constraint satisfaction problems, either systematic or hybrid, making it easy to be integrated within a CSP solver.

In order to test the efficiency of our data structure, we performed empirical analyses using various constraint satisfaction problems from known benchmarks.

1 Introduction

In many situations one has to find the solution of a problem described as a set of restrictions that apply to some entities. A *constraint satisfaction problem (CSP)* is a formal description that encapsulates all elements of such a scenario: a set of variables must be assigned values from specific domains in order to satisfy a set of constraints that restrict the space of possibilities [3], [14].

Once we have identified the variables, their domains and the constraints we have an abstract model of the problem that can be solved in an effective manner by specialized software systems called *constraint solvers*, such as ILOG [9], Choco [10], Minion [6], Omnics [4], etc. These solvers accept as input a representation of the problem either in a standard or a specific format and attempt to find a *solution* of the problem, that is a mapping between variables and values that satisfy all the constraints. Most of the times, this is not an easy task – the candidate problems for constraint solvers are usually difficult ones, for which dedicated algorithms are not known. In order to tackle this type

N. Hernandez et al. (Eds.): ICCS 2014, LNAI 8577, pp. 188–201, 2014.
DOI: 10.1007/978-3-319-08389-6_16, © Springer International Publishing Switzerland 2014

of problems and to be useful in practice, a solver must implement advanced algorithmic techniques and efficient data structures. As the number of practical applications of constraint programming is increasing in many areas like scheduling, industrial design, vehicle routing, configuration problems, business applications, combinatorial mathematics, etc. [15], [12] so is the demand of effective solvers capable of approaching difficult problems.

The search for a solution is usually implemented as a systematic exploration of the space in which solutions are defined. At each step, the solver makes a decision (assigning a value to a variable, for instance) and tries to move forward to complete the solution. If, at some point, this cannot be done anymore the backward step is performed, choosing another path in the search tree. However, the systematic exploration is not the only possibility as stochastic methods, local search or hybrid approaches can prove helpful in some situations [12].

Regardless of the algorithm used for the search, each time a decision is taken the solver will fire some sort of *filter-and-propagate* trigger [1], responsible with reducing the domains of the variables as a result of the current decision. A filter can be defined at the problem level, taking into consideration the connections between the constraints of the problem or it can be defined at the constraint level, taking into account the particularities of that specific constraint. Maintaining arc-consistency during the search [11] is one of the most widely used problem-level type of filter designed to reduce the number of incorrect decisions made by the solver. It is a *look ahead* mechanism that ensures that each decision has some sort of a continuation, therefore it does not lead to an immediate failure. As an example of constraint-level filter, global constraints like *AllDiff* [8] have specialized algorithms that will propagate a decision better than any generic method.

The *extensional* or *table constraint* is used to express directly the combination of values that are allowed for some variables. Many problems from industry are modeled using extensional constraints as they are created based on large sets of existing data. As noted in [7], it is important to represent that large volume of data using an efficient structure that offers a fast method for querying, since this will be used extensively by the propagation algorithm. The usage of *tries (prefix trees)* in the context of constraint programming is new and promising from the perspective of speeding up the propagation of table constraints. In this paper, we continue in the direction proposed in [7] and introduce an improved data structure, called *SupportTrie*, that combines the concepts of trie and support list. Empirical analyses on various problems taken from a well-known collection of benchmark files prove that we can speed the execution time up to 5 times, as against the default implementation of the trie.

The rest of the paper is organized as follows. Section 2 gives the basic definitions on constraint satisfaction problems. Section 3 presents the notion of extensional constraint and the manner in which they are usually implemented in a constraint solver. It is analyzed the relationship between extensional constraints and the sets of tuples that define them. Section 4 describes the naive representation of a relation, using support lists. Section 5 presents the trie data structure as a method of representing sets of tuples. Section 6 introduces our improved data structure *SupportTrie* and presents the algorithms for creating and querying it. Section 7 describes the experiments we conducted and the results obtained. Conclusions are given in Section 8.

2   Constraint Satisfaction Problems

A constraint satisfaction problem is represented as a *constraint network* $(X, \mathcal{D}, \mathcal{C})$ [3], where $X = \{x_1, ..., x_n\}$ is a set of variables, $\mathcal{D} = \{D_{x_1}, ..., D_{x_n}\}$ are finite domains of values that can be assigned to variables and $\mathcal{C} = \{C_1, ..., C_t\}$ represents the set of constraints. The *scope* of a constraint is the set of all variables involved in that constraint, for instance $scope(x + y = z) = \{x, y, z\}$. The *arity* of a constraint is the size of its scope. Constraints of arity two are named *binary constraints*. An *assignment* $(x_{i_1} = a_1, ..., x_{i_k} = a_k)$ is a mapping between some variables and values such that $a_j \in D(x_{i_j}) \, \forall j \in 1..k$. An assignment is *consistent* if it does not contradict any constraint. A *solution* is a consistent assignment that is also *complete*, i.e. it defines values for all variables of the problem. A network of constraints is *satisfiable* if it has at least one solution.

Intensional constraints are expressed using a functional predicate defined over the scope of the constraint that returns `true` whenever the arguments represent a valid combination of values, and `false` otherwise. *Extensional* constraints are expressed as a relation of allowed or forbidden tuples, a tuple being an element of the cartesian product of the corresponding domains of the variables that define the scope of the constraint. For example, the intensional constraint $C(x, y, z) : x + y = z$ defined over the variables x, y, z with values in the domain $D = \{0, 1\}$ states that in any solution of the problem the values $a, b, c \in \{0, 1\}$ assigned to x, y, z ($x = a, y = b, z = c$) satisfy the given expression, that is $a + b = c$. The same constraint can be expressed either in its *positive* extensional form, as the set of allowed combinations of values: $C^+(x, y, z) : \{(x = 0, y = 0, z = 0), (x = 0, y = 1, z = 1), (x = 1, y = 0, z = 1)\}$.

The solving process of a constraint network can be seen as a *transitional system*: $\sigma_0 \rightarrow \cdots \rightarrow \sigma_f$. In order to find a solution for a constraint satisfaction problem, the solving process will start from an initial state σ_0 that usually is a representation of the original problem and will take various decisions that move the system into another state. The process will finish whenever it encounters a state σ_f that contains a complete assignment satisfying all the constraints or inconsistency is detected by some means. During these transitions the domains of the variables may change. In a state σ of the solving process, we will denote the domain of a variable x as D_x^σ.

3   Extensional Constraints

Let $\{D_{i_1}, ..., D_{i_k}\}$ be a set of domains. A *relation* R of arity k is a subset of the cartesian product $D_{i_1} \times \cdots \times D_{i_k}$. An element of a relation is called a *tuple*. If $t = (a_1, ..., a_k)$ is a tuple we will denote as t_j or $t[j]$ the element a_j, $j = 1..k$.

An *extensional constraint* is defined as a triplet (R, S, τ) where $S = \{x_{i_1}, ..., x_{i_m}\}$ is a set of variables, R is a relation of arity m defined over the domains $\{D'_{i_1}, ..., D'_{i_m}\}$ where $D'_{i_j} \subseteq D_{i_j}, \forall j = 1..m$ and $\tau \in \{+, -\}$ specifies if the constraint is positive or negative. For a positive constraint, R represents the *support* set of tuples, for a negative constraint it represents the *conflict* set. We will denote such a constraint by R_S^+ or simply R_S, respectively R_S^-. We may omit the scope S if it can be inferred from the context. Of course, a negative constraint can be rewritten as a positive one and, because of that, in the rest of the paper we will consider only positive constraints.

It is not uncommon that the definition of a problem contains multiple constraints, with different scopes, which are defined over the same relation. For example, consider a problem where $X = \{x, y, z\}$, $C = \{x \neq y, x \neq z, y \neq z\}$ and $D_x = D_y = D_z = \{0, 1\}$. The extensional representation of the constraints is $C = \{R_{x,y}, R_{x,z}, R_{y,z}\}$ where $R = \{(0, 1), (1, 0)\}$ is the relation describing them.

An assignment can be viewed as an association between a set of variables $S = \{x_{i_1}, \ldots, x_{i_k}\}$ and a tuple $t = (a_1, \ldots, a_k)$, i.e. a mapping between variables and values: $\bar{a} = (x_{i_1} = a_1, \ldots, x_{i_k} = a_k)$. Given a constraint R_S, each tuple $t \in R$ induces an assignment: considering the set of variables S, each position of the tuple has a corresponding variable. We denote $tuple(\bar{a})$ the tuple of the assignment $\bar{a}$, $scope(\bar{a})$ its variables, $\bar{a}[x]$ the value of the variable x and $size(\bar{a}) = |scope(\bar{a})|$. An assignment $\bar{a}$ is R_S-complete if $scope(\bar{a}) = S$.

Let R_S be a table constraint and $Y = \{x_{i'_1}, \ldots, x_{i'_p}\} \subseteq S$. The *projection of a tuple* $t = (a_{i_1}, \ldots, a_{i_k}) \in R$ on the set Y is defined as $t[Y] = (a_{i'_1}, \ldots, a_{i'_p})$, that is, in $t[Y]$ are retained only the values corresponding to the variables in Y. If $Y = \{y\}$, then we denote $t[\{y\}]$ simply as $t[y]$.

Let σ be a state created during the solving of a problem. The *selection of a constraint* R_S over the state σ is a new constraint $\sigma(R_S)$ with the same scope S, created from the relation R by removing all tuples that contain values no longer present in the corresponding domains of the variables. Therefore, $\sigma(R_S) = R'_S$ where:

$$R' = R - \{t \in R | \exists y \in S \; t[y] \notin D_y^\sigma\}$$

For example, let $R = \{(0, 0), (0, 1), (1, 0), (1, 1)\}$ be a relation, $C = \{R_{x,y}, R_{x,z}, R_{y,z}\}$ the constraints and the original domains of the variables $D_x^{\sigma_0} = D_y^{\sigma_0} = D_z^{\sigma_0} = \{0, 1\}$. Assume that, in the next state σ_1, the domain of x becomes $D_x^{\sigma_1} = \{0\}$. Then $\sigma_1(R_{x,y}) = \sigma_1(R_{x,z}) = \{(0, 0), (0, 1)\}$, $\sigma_1(R_{y,z}) = \sigma_0(R_{y,z})$.

A tuple t is *valid* in a state σ, with respect to a constraint R_S, if $t[S] \in \sigma(R_S)$. An assignment $\bar{a}$ is valid if $tuple(\bar{a})$ is valid. An assignment is consistent if it is valid on all constraints of the problem.

A solver must be able to verify quickly the validity of a given tuple at any point of the solving process. It is not efficient to compute the selected relations of each constraint during the solving process. First of all, some constraints may share the same initial relation, since relations may be quite large. Creating a relation for every constraint will certainly have a negative impact on the memory required for their representation. Secondly, if the solver must go back to a previous state, all relations must be restored as they were in that state. That means that the memory required will depend on the number of the states, which is definitely not a good idea. Therefore, we need an effective solution to compute the selection of a constraint in a given state, whenever it is necessary.

In order to define the logic of an extensional constraint $C = R_S$, one has to implement a method $eval_C(\sigma)$ which will evaluate the given state σ, analyze if this state is feasible from the perspective of the constraint C and identify the values that can be removed from their current domains because they do not belong to any valid tuple. This revision is part of the filter-and-propagate mechanism which will move the solving process to the next state. Regardless of the algorithm used for implementing the evaluation method, in order to verify if an assignment is consistent we must query the relations involved and take into consideration the current domains of the variables. The implementation of a relation R must offer the method $valid_R(\overline{a}, \sigma, S)$ to verify if an assignment is valid in the state σ for the constraint R_S.

Let us consider the following example: $X = \{x, y, z\}$, $D_x = D_y = D_z = \{0, 1, 2\}$, $C = \{C_1 = (R_1, \{x, y\}), C_2 = (R_2, \{x, z\}), C_3 = (R_3, \{y, z\})\}$ where $R_1 = \{(0, 0), (1, 1)\}$, $R_2 = \{(0, 0), (2, 2)\}$, $R_3 = \{(1, 1), (2, 2)\}$.

Evaluating the initial state, the constraint C_1 will remove the value 2 from the domains D_x and D_y since there is no valid tuple where $x = 2$ or $y = 2$:

$$valid_{R_1}((x = 2), \sigma_0, \{x, y\}) = false, valid_{R_1}((y = 2), \sigma_0, \{x, y\}) = false.$$

Note that in this state $valid_{R_3}((y = 2), \sigma_0, \{y, z\}) = true$. In the next state σ_1, $D_y^{\sigma_1}$ no longer contains the value 2 so $valid_{R_3}((y = 2), \sigma_1, \{y, z\}) = false$.

Therefore, the method $valid_R$ will return true if the relation R contains a tuple t that matches the argument $\overline{a}$ and all values of t are still in their corresponding domains:

$$valid_R(\overline{a}, \sigma, S) \Leftrightarrow \exists t \in R \; t[scope(\overline{a})] = tuple(\overline{a}) \wedge \forall x \in S \; t[x] \in D_x^{\sigma}$$

Representing a binary relation can be done efficiently using its *support matrix* where each row is actually an array of bits indicating what values are supported by the value corresponding to the column. In some cases, relations with higher arity can also be implemented using multidimensional support arrays but, obviously, this is not a solution for the general case. In this paper, our goal is to describe an efficient data structure for implementing relations of any arity.

4   Support Lists

Let $R \subseteq D_1 \times \cdots \times D_k$ be a relation containing m tuples $\{t_1, ..., t_m\}$ of arity $k \geq 2$. We will assume that the tuples are lexicographically sorted. We will denote $index_R(t)$ the position of the tuple t in this representation of R.

The naive implementation of the $valid_R$ method is to iterate through this list and verify the validity of each tuple. With proper data structures, checking if a value belongs to a domain can be done in constant time. (In the states created during the solving process the domains of the variables will not hold actual values but bits indicating if a certain value is still present in the domain (bit set) or not (bit clear) – this technique is called *domain trailing*.

The time complexity of the naive approach is $O(mk)$, which is not good especially because the number of the tuples in a relation is usually large. Recall that this method is going to be invoked a large number of times so keeping its complexity low is essential. Considering that the space required to represent any element of the problem's domains or the value of an index is $O(1)$, the space complexity of this representation is also $O(mk)$.

The original paper on GAC-schema [2] introduces a simple method to improve the search for a valid tuple. For each *(position,value)* pair (i, a) a list $tupleSupport(i, a)$ is created, containing all tuples t that have the value a on the i-th position ($i \in 1..k, a \in D_i$). When looking for a tuple, the method $valid_R(\overline{a}, \sigma, S)$, where $S = \{x_{i_1}, ..., x_{i_k}\}$, should only consider the support list of minimum size:

$$min\{|tupleSupport(i, \overline{a}[x_i])|, \quad x_i \in scope(\overline{a}), i = 1..k\}.$$

Note that the assignment $\overline{a}$ might not specify values for all variables from S.

Using this simple optimization, the time complexity of the $valid_R$ method is now $O(pk)$, where p is the maximum size of a support list. In the worst case, it is possible that $p = m$, however the average complexity certainly decreases. Since each tuple is present in k support lists and there are kd lists, where d is the maximum size of a domain, the space complexity increases to $O(mk) + O(k^2d)$.

5   Tries (Prefix Trees)

Assuming that the domains of a problem consist of characters, symbols from some alphabet, a relation of arity k can be viewed as a set of strings each of length k. Introduced in [5], a *trie* or *prefix tree* (the term comes from 'retrieval') is an efficient data structure for representing and querying large sets of strings, such as a dictionary. The root of the tree is an empty node (no label) and all other nodes are labeled with some character from the alphabet. Each level of the tree corresponds to a position between 1 and the maximum length of the strings and there can be no two nodes on the same level labeled with the same symbol. Level 0 is represented by the root. A string $a_1a_2 \ldots a_k$ belongs to the dictionary if and only if there is a path of length k from the root to a leaf node such that the node at level i is labeled with $a_i, \forall i = 1..k$.

It is easy to see how tries can be used to represent relations. The nodes will be labeled with values from the domains of the variables and each path of the tree, from the root to a leaf, will correspond to a tuple.

Let $R \subseteq D_1 \times \cdots \times D_k$ be a relation with m tuples of arity $k \geq 2$ and $d = max_{i=1..k}|D_i|$. A prefix tree representing R will have at most $O(d^k)$ nodes. We can assume that representing a node takes $O(1)$ space, so the complexity is $O(d^k)$. Note that if R is complete and all domains are of size d, the number of tuples is $m = d^k$, so the space complexity of the trie is k times better than in the case of naive representation, using a list of tuples, which is $O(mk)$.

Let us analyze now the time complexity of $valid_R(\overline{a}, \sigma, S)$, that is the method that verifies if an assignment $\overline{a}$ is valid in a state σ, regarding the extensional constraint R_S.

Checking if a tuple $t = (a_1, \ldots, a_k)$ belongs to the relation can be accomplished in $O(dk)$. The algorithm is straightforward: start with the root node as the current node; when the current node is at level i descend to the child that is labeled a_{i+1} and make it the current node; if such node does not exist, signal that the tuple does not belong to the relation; stop when the current node is a leaf.

That means that checking the validity of an R_S–complete assignment $\overline{a}$ can be accomplished in $O(dk)$ by checking if $tuple(\overline{a})$ belongs to the relation and verifying that $\overline{a}[x] \in D_x^\sigma \ \forall x \in scope(\overline{a})$. Recall that testing if a domain still contains a value in a state σ is done in $O(1)$.

In [7] the authors explored the use of tries for implementing table constraints in the context of GAC-schema [2] algorithm. Using a large experimental analysis, it is noted that significant improvements have been obtained using this data structure over other approaches, such as support lists with various optimizations. However, their algorithm is strictly related to the context and has some particularities. Each extensional constraint has its own relation and for each variable x in the scope of a constraint, a trie with variable x at the first level is created. In our opinion, that approach puts a lot of pressure on the memory requirements, especially for problems with many variables and many constraints that share a small numbers of relations.

Let us consider a simple crossword puzzle generation problem. We must fill a crossword puzzle grid 7×7 such that each row and column represents a word from a predefined 7–letter dictionary. There are many variants that can be used to model this problem as a CSP instance. We consider the model in which there is a variable for each cell of the grid, having the domain $\{A, B, \ldots, Z\}$. For every row and column there is a table constraint imposing that they represent a word from the dictionary. All 14 constraints share the same relation and, since in English there are more than 12.000 words formed with 7 letters, this is actually the number of tuples in the relation. It does not seem practical to duplicate the relation for every constraint and use a trie for each of its variables; the memory required multiplies by 98.

6   SupportTries (Tries with Support Lists)

In this paper we present an improved data structure for representing a relation, called *SupportTrie*, that combines the concepts of trie and support list and offers a fast algorithm for testing the validity of an assignment. A relation is not tied to a specific constraint and this allows multiple constraints to share the same data structure, in order to decrease the memory requirements. Our approach is generic and can be adapted to any method of solving constraint satisfaction problems.

Let $R \subseteq D_1 \times \cdots \times D_k$ be a relation with m tuples $\{t_1, \ldots, t_m\}$ of arity $k \geq 2$, lexicographically sorted.

A trie is actually a tree formed of *nodes* labeled with values from the domains D_i. The root forms level 0 of the tree, the i-th level of the tree contains values from D_i, $i = 1..k$. We denote $value(node)$ the label of the node, $parent(node)$ its parent, $children(node)$ its children and $level(node) \in 0..k$ the level in which the node appears in the tree. If the node is a leaf, we denote $tuple(node)$ the tuple represented by the path from the root to that node.

If a node is internal, we denote $first(node)$ and $last(node)$ the first, respectively the last index in $\{1, \ldots, m\}$ such that for any $leaf$ of the subtree rooted in $node$ we have $index_R(tuple(leaf)) \in [first(node), last(node)]$, where $index_R(t)$ denotes the position of the tuple t in the representation of R as a list.

In addition, we create a data structure called *support* such that $support(level, value)$ represents a list containing the nodes from the specified level of the trie labeled with the specified value. Unlike the support lists introduced in Section 4, these contain references to nodes of the trie and not to tuples of the relation. We denote $supportHead$ and $supportTail$ the first and the last elements of a support list and $nextSibling(node)$ the next node that has the same parent as $node$ in the $support$ list containing them.

6.1   Creating the Trie and the Support Lists

Creating the trie and its support lists is done incrementally. Whenever a new tuple is read from the input source, we have to identify a path that represents a prefix of that tuple and create new nodes in the tree for the rest of the values in the tuple. For each created node, we update the *support* data structure. The time complexity of this method is $O(dk)$, where k is the arity of the relation and $d = max_{i=1..k}|D_i|$. Therefore, the overall time complexity of creating *trie* and *support* data structures is $O(mdk)$, where m is the number of tuples in the relation.

The space complexity of our data structures is $O(d^k)$ required for the trie and another $O(kdp)$ required for the support lists, where p is the maximum size of a support list. Note that the support lists are smaller for the upper level of the trie and they grow larger towards the lower levels. The size of a support list at level i, for a specific value, does not exceed d^{i-1} so the space complexity for the support lists is actually: $O(kd\sum_{i=1}^{k} d^{i-1}) = O(kd^k)$.

The overhead of using more memory than in the case of the standard trie is compensated by a faster algorithm for checking the validity of a given assignment during the solving process.

6.2   Checking the Validity of an Assignment

The method we describe next is $valid_R(\overline{a}, \sigma, S)$, responsible for verifying if an assignment $\overline{a}$ is valid in the state σ for the constraint R_S.

We assumed that the tuples are lexicographically sorted. If D is the ordered domain of a variable, we denote $first(D)$ and $last(D)$ the first, respectively the last element of the domain. Note that each level $i > 0$ of the trie corresponds to the variable x_i from $S = \{x_1, \ldots, x_k\}$, $k = arity(R)$.

The idea of the algorithm is to restrict the search space as much as possible, taking into account the variables of the given assignment and the current domains.

If the assignment is $\overline{a} = (y_1 = b_1, \ldots, y_j = b_j)$, the best place to start is with the variable y_i that is closest to the root, since the number of nodes is lower at the top of the tree. Once we selected that *level*, we iterate through all the nodes that offer support for the value b_i, that is the list $support(level, b_i)$. For every node in the list, we must check the validity of the path from the node up to the root and then search for a valid path to some leaf (a path is valid if the partial assignment it represents is also valid).

Before we present the algorithm, let's consider an example. The relation from Fig. 1 contains 10 tuples of arity 4, indexed from 1 to 10. We will denote a certain node in the trie by the path from the root to it, for example $[aca]$ is the second node labeled with a at level 3. Consider a constraint satisfaction problem with variables $X = \{x_1, \ldots, x_n\}$, all having as initial domain $D = \{a, b, c, d\}$. Consider $S = \{x_1, x_2, x_3, x_4\}$ and σ a state of the solving process where domains are $D_{x_1}^\sigma = \{a, c\}$, $D_{x_2}^\sigma = D_{x_3}^\sigma = D_{x_4}^\sigma = \{a, b\}$. We will describe a trace of the invocation $valid_R((x_4 = a, x_3 = a), \sigma, S)$.

The starting level of the trie is 3 (dotted), corresponding to the higher level of the assignment's variables $\{x_3, x_4\}$. First of all, since $d \notin D_{x_1}^\sigma$ the whole subtree rooted in $[d]$ will be ignored by the search algorithm. This is implemented by lowering the maximum search index from 10 down to 8. The minimum search index is 1 since $a \in D_{x_1}^\sigma$. Next, the algorithm will iterate through the support list $support(3, a)$, containing all nodes at level 3 labeled with a. The node $[aaa]$ is not good because, looking down, it has only the continuation $x_4 = b$. The node $[aca]$ is also not good because, looking up, it leads to $x_2 = c$ and $c \notin D_{x_2}^\sigma$.

The next node in the support list is $[baa]$. This is also not good, because it leads up to $x_1 = b$ and $b \notin D^{\sigma}_{x_1}$. In this case, we can ignore the whole subtree rooted in $[b]$ and jump to a node from the support list that has its first index greater than $last([b]) = 6$. So, the next node is $[caa]$, which is also not good because it has only the continuation $x_4 = b$. The next node from the support list $[cba]$ is good and it leads to a valid tuple $(x_1 = c, x_2 = b, x_3 = a, x_4 = a)$, so the method returns $true$.

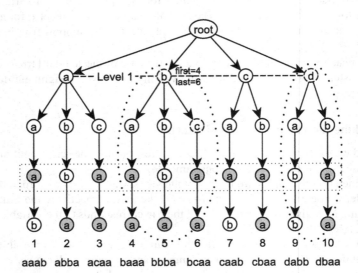

Fig. 1. Trie representing a relation with 10 tuples of arity 4

The $valid_R(\bar{a}, \sigma, S)$ method

Require: $\bar{a}$ an assignment (not empty), $S = \{x_1, \ldots, x_k\}$ a set of variables with $k = arity(R)$ and $scope(\bar{a}) \subseteq S$, σ a state of the solving process

Ensure: Returns $true$ if there is a tuple $t \in R$ such that $t[scope(\bar{a})] = tuple(\bar{a})$ and $\forall x \in S \ t[x] \in D^{\sigma}_x$, otherwise returns $false$.

{**Init:** Reduce the search to a subtree with tuple indices between $minPos$ and $maxPos$}

{**Init$_1$:** Take into consideration the current domains}

$minPos = min_{i=1..k} supportHead(i, first(D^{\sigma}_{x_i}))$

$maxPos = max_{i=1..k} supportTail(i, last(D^{\sigma}_{x_i}))$

{**Init$_2$:** Take into consideration the assignment $\bar{a}$}

$level = 0, value = null$

for all $i = k..1$ **do**

    if $x_i \notin scope(\bar{a})$ **then**

        continue

    end if

    if $support(i, \bar{a}[x_i]) = null$ **then**

        return false

end if
$\quad level = i, value = \overline{a}[x_i]$
end for
$\{level < k$ and $value \neq null$, otherwise $\overline{a} = \emptyset\}$
$minPos = min(minPos, supportHead(level, value))$
$maxPos = max(maxPos, supportTail(level, value))$
$supportNode = supportHead(level, value)$

$\{$**Search:** Iterate through the support list$\}$
$supportLevel = level(supportNode), nextPos = minPos$
while $supportNode \neq null$ **do**
$\quad$**if** $last(supportNode) < nextPos$ **then**
$\quad\quad supportNode = nextSibling(supportNode)$
$\quad\quad$**continue**
$\quad$**end if**
$\quad$**if** $first(supportNode) > maxPos$ **then**
$\quad\quad$**return false**
$\quad$**end if**
$\quad\{$**Search Up:** Move up in the tree and check the validity of the partial tuple from the root to the current support node$\}$
$\quad upNode = supportNode, validTuple = true$
$\quad$**for all** $j = supportLevel - 1..1$ **do**
$\quad\quad upNode = parent(upNode)$
$\quad\quad$**if** $value(upNode) \notin D^{\sigma}_{x_j}$ **then**
$\quad\quad\quad validTuple = false, nextPos = last(upNode) + 1$
$\quad\quad\quad$**break**
$\quad\quad$**end if**
$\quad$**end for**
$\quad$**if** $validTuple$ **then**
$\quad\quad$**if** $supportLevel = arity$ **then**
$\quad\quad\quad$**return true**
$\quad\quad$**end if**
$\quad\quad\{$**Search Down:** Move down in the tree and find a valid partial tuple from the current support node to a leaf$\}$
$\quad\quad$**if** $validNode(\overline{a}, \sigma, S, supportNode)$ **then**
$\quad\quad\quad$**return true**
$\quad\quad$**end if**
$\quad$**end if**
$\quad supportNode = nextSibling(supportNode)$
end while
return false

In the previous method, before the invocation of $validNode$ we identify a node $parent$ such that the projection of the assignment $\overline{a}$ on the set of variables $\{x_1, \ldots, x_i\}$ is valid, where $i = level(parent)$. We must now attempt to complete the response tuple by searching downward.

The recursive method $validNode_R(\overline{a}, \sigma, S, parent)$ checks if the subtree rooted in *parent* contains a leaf *node* such that $tuple(node)$ is the response tuple. To do that, we must verify all paths from *parent* to the possible leaves and for each of these paths check the validity of the resulting tuple. Note that an invocation of the form $validNode_R(\overline{a}, \sigma, S, root)$ performs the default, full exploration of the trie.

The $validNode_R(\overline{a}, \sigma, S, parent)$ method

Require: $\overline{a}$ an assignment (not empty), $S = \{x_1, \ldots, x_k\}$ a set of variables with $k = arity(R)$ and $scope(\overline{a}) \subseteq S$, σ a state of the solving process, *parent* a node in the trie;

Ensure: Returns *true* if there is a leaf node $node \in trie$ representing a tuple t such that if we denote $i = level(node)$, $S' = \{x_{i+1}, \ldots, x_k\}$ we have $t[scope(\overline{a}) \cap S'] = tuple(\overline{a})[S']$ and $\forall x \in S' \ t[x] \in D_x^\sigma$, otherwise returns *false*.

$value = value(parent)$, $level = level(parent)$, $x = x_{level}$
{**Case 1.** The variable x is not present in the assignment}
if $x \notin scope(\overline{a})$ **then**
   {descend in all subtrees}
   for all *node* child of *parent* **do**
      if $value(node) \notin D_x^\sigma$ **then**
         continue
      end if
      if $level = arity$ or $validNode(\overline{a}, \sigma, S, node)$ **then**
         return true
      end if
   end for
   return false
end if

{**Case 2.** The variable x is present in the assignment}
{descend only in the corresponding subtree}
$node = null$
for all *node*, child of *parent* **do**
   if $value(node) = value$ **then**
      break
   end if
end for
return $node \neq null$ and ($level = arity$ or $validNode(\overline{a}, \sigma, S, node)$)

The method can be further improved in the case of a *monotonic* sequence of states, that is for any two consecutive states created in the solving process σ and σ' and for any variable x we have: $D_x^{\sigma'} \subseteq D_x^\sigma$. This scenario occurs especially in the case of systematic solvers, during the forward phase, when values are being pruned from their domains as a result of the filter-and-propagate algorithm. Each state can be labeled with a *version* identifier such that $version(\sigma') > version(\sigma)$ for any two consecutive states σ and σ'. Whenever a node in the trie is found to be invalid either because its value no longer belongs to the corresponding domain or because none of its children is valid, the version of the current state can be attached to that node. Later on, the search algorithm will avoid checking subtrees that are rooted in a node marked as invalid with a version id smaller than the current one.

7   Experiments

In order to make experiments we have used our CSP solver, called Omnics [4]. The solver is written in the Java programming language and it accepts problems either encoded in the XCSP format [13] or created using its own application programming interface (API). The tests were run on an Intel Core i7 at 2.4GHz. The Java Virtual Machine was given 1024 MB of memory (-Xmx1024m -Xms1024m).

The XCSP instances used for testing were taken from a well-known collection of benchmark files maintained by Christophe Lecoutre and available at http://www.cril.univ-artois.fr/~lecoutre/benchmarks.html

The instances prefixed with *rand*, located in the "Random Instances" category, represent problems involving positive table constraints, each having a corresponding nonbinary relation with a large number of tuples.

We tested 10 instances of unsatisfiable problems, prefixed with *rand-10-20-10*. Each variable has the domain $\{0, 1, \ldots, 9\}$ and each relation has exactly 10.000 tuples of arity 10. The table below presents the general parameters of the problems and the running times obtained for each instance in three solving scenarios: *List*: the case in which *support lists* were used for representing relations (Section 4); *Trie*: the standard implementation of a *trie* (Section 5); *SupportTrie*: the case when our data structure was used, that combines the concept of trie and support list (Section 6). Time is expressed in seconds. Timeout was set to five minutes.

rand-10-20-10				*List*	*Trie*	*SupportTrie*
Number of variables	20		rand-10-20-10-11	9.53	3.37	2.31
Number of constraints	5		rand-10-20-10-12	9.19	1.32	1.26
Number of relations	5		rand-10-20-10-13	11.44	1.64	1.52
Domain size	10		rand-10-20-10-17	15.34	0.80	0.75
Relation size	10.000		rand-10-20-10-18	1.15	0.32	0.36
Arity	10		rand-10-20-10-19	10.38	8.42	4.27

The first set of random problems are easy to solve, with very few backtracks. Despite having to create the additional support lists, our data structure is still outperforming the other implementations.

The instances prefixed with *tsp*, located in the "Patterned Instances" category, refer to the traveling salesman problem (TSP), a well-known NP-hard combinatorial problem.

The table below presents the general parameters of the problems and the results obtained. Note that the number of constraints is 10 times the number of relations. This is a common scenario in patterned problems, due to existing symmetry. Even if the trie has only 3 levels (the arity of the tuples is 3), the execution time was reduced by half. This is in fact due to the large number of elements in the domains of the variables.

tsp-20				$List$	$Trie$	$SupportTrie$
Number of variables	61		tsp-20-142	-	51.22	25.63
Number of constraints	230		tsp-20-193	-	236.40	105.68
Number of relations	23		tsp-20-366	-	28.25	17.14
Domain size	1001		tsp-20-453	282.34	21.65	11.89
Relation size	10.000 − 46.000		tsp-20-76	76.14	6.87	3.97
Arity	3		tsp-20-901	104.52	9.37	4.66

The instances prefixed with *crossword*, located in the "Real-World Instances" category, refer to crossword puzzles taken from various sources, such as the Herald Tribune (Spring, 1999).

The table below presents the general parameters of the problems and the results obtained. Note that in this case the number of relations is 1 or 2 and the number of constraints is around 20 – in fact, most of the constraints share the same relation. For this type of problem the improvement is significant, our data structure is up to 5 times faster than the next best one.

crossword-lex-vg				$List$	$Trie$	$SupportTrie$
Number of variables	∼ 100		crossword-lex-vg7-7	-	-	143.00
Number of constraints	∼ 20		crossword-lex-vg7-8	-	-	125.74
Number of relations	1 − 2		crossword-lex-vg7-9	207.58	110.79	31.12
Domain size	26		crossword-lex-vg8-8	-	175.55	54.70
Relation size	∼ 4000		crossword-lex-vg8-9	134.99	84.03	21.08
Arity	7 − 12		crossword-lex-vg9-9	75.89	54.77	10.28

8 Conclusions

This paper describes an efficient data structure for representing relations of any arity, that combines the concepts of trie (prefix tree) and support list. The purpose of this is to minimize the time spent by an extensional constraint in order to verify if a given assignment is valid at some point of solving a CSP problem. Actually, for extensional constraint problems, querying the relations is the most time-consuming operation.

We have continued in the direction proposed in [7], confirmed the good experimental results obtained by the authors and presented a novel data structure that improves even further the use of tries.

In our implementation, there is a *one-to-many* relationship between the data structure that represents the set of tuples and the constraints of the problem. This allows multiple constraints to share the same relation, decreasing the memory requirements. As we have seen in our experiments, there are many types of problems that benefit from this separation.

We did not assume any method for solving CSP problems – this process is seen as a transitional system that moves from one state to another until a solution is found or inconsistency is detected. Therefore, our approach is generic and can be integrated into any CSP solver, regardless of its implementation.

The set of experiments we performed show that our data structure makes significant improvements, measured in execution time, over the simple representation of relations as support lists or standard tries.

Extensional constraints are widely applicable as many problems from industry or academic research are modeled using them. We believe that this area of research offers potential for other novel data structures to be analyzed, in order to efficiently represent large sets of tuples in the context of constraint programming.

References

1. Bessiere, C.: Constraint Propagation. In: Rossi, F., van Beek, P., Walsh, T. (eds.) Handbook of Constraint Programming, ch. 7. Elsevier (2006)
2. Bessiere, C., Regin, J.C.: Arc consistency for general constraint networks: Preliminary results. In: IJCAI (1), pp. 398–404 (1997)
3. Dechter, R.: Constraint processing. Elsevier Morgan Kaufmann (2003)
4. Frasinaru, C.: Omnics solver (2013), http://profs.info.uaic.ro/~acf/omnics
5. Fredkin, E.: Trie memory. Commun. ACM 3(9), 490–499 (1960), http://doi.acm.org/10.1145/367390.367400
6. Gent, I.P.: Minion solver (2007), http://minion.sourceforge.net/index.html
7. Gent, I.P., Jefferson, C., Miguel, I., Nightingale, P.: Data structures for generalised arc consistency for extensional constraints. In: AAAI, pp. 191–197. AAAI Press (2007), http://dblp.uni-trier.de/db/conf/aaai/aaai2007.html#GentJMN07
8. van Hoeve, W.J., Katriel, I.: Global constraints. In: Rossi, F., van Beek, P., Walsh, T. (eds.) Handbook of Constraint Programming, ch. 7. Elsevier (2006)
9. IBM: ILOG CPLEX Solver (2009), http://www-01.ibm.com/software/commerce/optimization/cplex-optimizer
10. Laburthe, F., Jussien, N.: Choco solver (2012), http://www.emn.fr/z-info/choco-solver
11. Mackworth, A.: Consistency in networks of relations. Artificial Intelligence 8(1), 99–118 (1977)
12. Rossi, F., van Beek, P., Walsh, T. (eds.): Handbook of Constraint Programming. Elsevier (2006)
13. Roussel, O., Lecoutre, C.: XML representation of constraint networks: Format XCSP 2.1. CoRR abs/0902.2362 (2009), http://dblp.uni-trier.de/db/journals/corr/corr0902.html#abs-0902-2362
14. Tsang, E.P.K.: Foundations of constraint satisfaction. Computation in cognitive science. Academic Press (1993)
15. Wallace, M.: Practical applications of constraint programming. Constraints 1, 139–168 (1996)

On the Usability of Random Indexing
in Patent Retrieval

Mihai Lupu*

Institute of Software Technology and Interactive Systems
Vienna University of Technology, Austria
lupu@ifs.tuwien.ac.at

Abstract. Statistical semantics methods are fairly controversial in the
IR community, mostly because of their instability and difficulty to debug.
At the same time, they are extremely tempting, in the same way perhaps,
as Artificial Intelligence was in the 60s. Then, it took a few decades for
the hype to pass and for us to learn the real utility and limits of the great
technologies developed earlier. This paper takes an exhaustive view of
the performance and utility of a particular statistical semantics method,
Random Indexing, in the context of difficult texts. After over a year of
CPU time in experiments, we provide a global view of the behaviour of
the method on a particularly challenging test collection based on patent
data. In the end, we observe interesting patterns emerging in the semantic
space created by the method, which we hypothesize to be the cause of
the behaviour observed in the experiments.

1 Introduction

The IR community has shown in recent years an increasing amount of attention
to domain specific tasks such as legal, medical or patent search. In these particular domains, three major conferences in our field, SIGIR, CIKM and ECIR,
have published 10 papers in each 2008 and 2009, 12 in 2010 and 22 in 2011. We
add to this the different evaluation campaigns organized specifically for them
(TREC-CHEM, Legal TREC, CLEF-IP, NTCIR, ImageCLEF) and see that it
is indeed worth taking a look at what makes these domains different and how
some of the "old" methods apply to them; or why they do not apply.

What makes these particular domains interesting is not only their perceived
utility—after all, web search is used by millions more—but rather the challenges
they present. We call them *difficult* and define this difficulty in two ways. First,
the documents of these domains are not easily understood by an educated, native speaker of the language in which they were written. Second, they exhibit
term and document frequencies which are statistically different from natural
language [16].

At the same time, statistical semantics methods have been considered, with
more or less success in IR, for almost three decades. Like AI before them, they are

---

* Mihai Lupu is partially supported by the following projects: MUCKE (FWF),
KHRESMOI (EC FP7).

N. Hernandez et al. (Eds.): ICCS 2014, LNAI 8577, pp. 202–216, 2014.
DOI: 10.1007/978-3-319-08389-6_17, © Springer International Publishing Switzerland 2014

higher or lower on our research agenda, function of many different factors. One of these factors should be our understanding of their applicability in different domains and use cases. This is what we set out to do in this work, for the patent domain mentioned above.

We therefore look at statistical semantic methods, of which Latent Semantic Indexing [8] has been the poster-child for the past two decades. Due to the size of the collections and the known scalability issues of this principal statistical semantics method, we will work with Random Indexing [20,5], a relatively recent and promising approximation of LSI.

For our experiments we focus on the so called "Prior Art" patent scenario. The modelled task is finding documents that describe inventions similar to the one presented in a patent application to a particular patent office. We choose this domain because of the queries used in this scenario, namely full documents. A statistical semantics method should be able to extract core concepts of the query document and compare them to the concepts in the test collection. This is much more of a proper test for latent semantics than the usual short query web use-case. In this latter scenario, it is intuitively harder to justify how the method learns something about the query itself.

1.1 Statistical Semantics

There is something fundamentally attractive to the concept of statistical semantics, to the idea of computers simulating human behaviour, or, in this case, human understanding. The original LSI came first as a mathematical best-approximation of the original term-document matrix, and was only later analysed as a functional model for cognitive activities [14]. Conversely, Random Indexing starts from philosophical ideas of meaning. In their 2001 article, Karlgren and Sahlgren [13] start from Ludwig Wittgenstein's 1953 *Philosophical Investigations* and define knowing the meaning of a word as knowing how to use it correctly in context. Note that the definition refers to the use of meanings, rather than the meaning of meaning.

Random Indexing works essentially in two steps: first, an N-dimensional random label is assigned to each word type in the data. This contains only a small random number k of +1 or -1 values among many 0s. Second, for every term occurrence, a context vector is created by summing together the vectors of the terms in its context. This way (and as we will see, with potential repetition of the two steps) we obtain a vector considered to represent the context of the term, and therefore, its meaning.

All statistical semantics methods (and we include here the works of Schütze [25] and Lund [15]) are therefore based on the philosophical wisdom that the meaning of words can be represented by the sum of contexts in which they appear. The difference between all of them is given by the interpretation and implementation of the "sum" and "context".

Such methods have been shown to display functional behaviour comparable to users speaking English as a foreign language [14,13]. The next step is to see how well these methods perform in a search environment that requires expert users. We have low expectations regarding the quality of the results, but at the same time, such a study needs to be done, and it is to some extent surprising that we could not find something in the existing literature.

1.2   Patent Data

Together with scientific articles, patent data represents the main source of technical information in the world. It covers all technical, scientific and engineering fields. Patents have a substantial amount of meta-data indicating authors, owners, classifications, dates, related patents. They are probably the only world-wide collection of documents where content follows a (minimal) set of guidelines or practices. In general, each document (be it an application or a granted patent) contains an abstract, a description of the invention, and a set of claims for intellectual property protection.

This all would make it one of the easiest collections to work with, but the fact remains that understanding a patent document is often extremely difficult. This is due to the technical nature of the publication, but also to the writing style and the mixture of different genres within the same document [18]. While the Description section generally resembles a scientific article, the Claims section reads more similar to a contract, with frequent uses of the same term (i.e. high term frequency) and numerous references to "aforementioned" entities. We defer the presentation of corpus statistics for the later sections of this article, where their effects will be analysed.

Recently, Adams [2] described many of the problems that the textual part of the patent documents presents to the searcher. Partly they are the results of the long tradition of the patent system. For example, in digitizing legacy patents, the quality of the OCR system may not always have been a cause of concern and action. Therefore, many unique terms appear in the body of the patent, seriously damaging the capability of a system to learn anything from the text. Additionally, the patent corpus is said to suffer from extensive usage of the so-called "patentese" [3]. Like "legalese" in e-discovery, it brings about complex and unnatural sentences, where the basic assumptions of IR models do not hold.

It is with this textual part that we work in this study, a text that presents a significant challenge to Vector Space Model (VSM) techniques, and particularly to statistical semantics.

1.3   Motivation

To a large extent, we depart in this study from the assumption that statistical semantics will not work with this kind of data. This is based on knowledge about long standing issues of high dimensional similarities and prior work described in Section 5. At the same time, it is profoundly beneficial to understand why this is. Ultimately, if meaning is defined as proper use in context; context is nothing more than the sequence of letters that make up words and documents; and if the documents that we are working with are in fact intelligible to some people, we should find a method for the computer to identify what humans find meaningful in each document.

The other questions are: With such a huge amount of documents and terms, can we begin to comprehend what is going on inside a mathematical model such as those at the base of statistical semantics? Is there some way to visualize what the system generates ?

This paper contributes to our understanding of the utility of Random Indexing, through the most extensive empirical study to date. In the process, we identified a simple procedure for determining the optimal number of seeds per each dimension and collection. The procedure works before the indexing phase and therefore eliminates one parameter to fine-tune. The paper also provides a visual understanding of the semantic space, allowing us to peer into the space created by the statistical semantics method.

We structure the presentation as follows: Section 2 describes the methodology we used in analysing Random Indexing. Section 3 covers a selection of the experiments done in this study. We then analyse these results and propose some hypothesis for their behaviour in Section 4. Related work is presented in Section 5 and we conclude with a summary of the results in Section 6.

2   Methodology

For our experiments we used a recent patent retrieval test collection, CLEF-IP 2011 [19]. It contains just over 3 million patent documents, in three languages: English, French and German. For the purposes of this study, we only used the English parts[1] of the collection. For each document, we indexed the English abstract (ab), description (desc) and claims (clm)- the three fields that together represent the full textual body of the patent document.

CLEF-IP 2011 provides 1351 topics in English, of a total of 4000 topics. Each topic is a patent application document and the task is to find those documents which the patent examiner also found in the process of determining the novelty of the invention described in the patent application. Therefore, the relevance judgements consist of patent numbers, identified by the issuing authority code and a unique document number. Since the topics and the documents in the collection are actually the same type of objects, we indexed them together, making sure that none of the topic documents appear as results of any topic.

Information retrieval in general has a lot of parameters to fine-tune and random indexing brings yet another set. We start our experiments with a burst of experiments at an arguably low dimensionality (100), which is relatively fast computed. We take some decisions with respect to other parameters based on this. Then, as we move on through the experiments, we will come back to some of the decisions taken and run experiments to check that the initial observations still hold, given all the other parameter changes.

For running experiments, we used the SemanticVectors library [27]. The library takes an existing Lucene index and creates semantic vectors based on term frequencies in the source index. Therefore, the choices taken in the initial

---

[1] We say *parts* rather than *documents*, because some documents contain different sections in different languages.

Lucene indexing process, affect the end results. Since the objective of this study is to analyse the performance of the statistical semantics methods, we kept the pre-processing at a minimum. Full details of the performed experiments, including Lucene preprocessing parameters, are available at `http://mihailupu.net/ri4patents/`

3   Experimental Results

3.1   Collection Statistics

It has been often said, and there are many anecdotal examples, that patent documents are more "difficult" than other documents used in IR tests, with little proof so far. We start by looking at the collection term frequencies. It is well known that term frequencies follow a Zipfian distribution, and that the plot of frequencies per ranks on logarithmic axes should be, approximately, a line. Figure 1a shows this plot for our corpus and for the American National Corpus (http://americannationalcorpus.org/). Although the figures vary, the plot is made such that the two extremes (top left corner and lower right corner) match. They appear to be similar, but we should not forget that these are logarithmic plots. What matters here is the second derivative of the curves: the rate at which the frequency changes. Compared to a corpus of "regular" American English, the patent corpus decreases the frequency much faster. This means that it is rich both in high frequency and low frequency terms, but poor in average frequency terms.

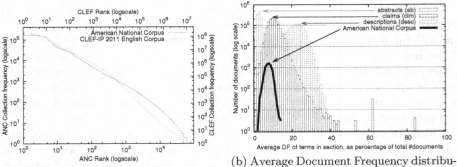

(a) Term frequencies in patents vs standard american English

(b) Average Document Frequency distribution, per sections(s)

Fig. 1. Collection statistics

The second aspect to observe is the specificity of the terms used in this corpus. Given the relationship between term specificity and document frequency (DF) [11], we can infer something about the specificity of each section by looking at the average DF of the terms within. Figure 1b shows the distribution of the average document frequency of terms in the three sections of a patent document, as a percentage of the total number of sections of that type. For comparison, the figure also shows the average DF in documents of the Open American National Corpus. Extreme cases are visible in the patent data, such as those 4 documents whose *average* DF is 80% of the total number of documents. Even without these extremes, most Description sections have an average DF over 25% of the size of the corpus.

3.2   Initial Experiment Burst

We start with a set of parameter values which have previously been shown to provide good results, together with control values, to cover potential unexpected behaviour in this new corpus. The main objective of this burst is to identify the fields of the patent documents, or the combination thereof, most likely to provide good retrieval results. We varied the use of a stemmer, the fields in the index, and the minimum and maximum collection term frequency.

The first observation we made was that using a Porter stemmer decreases the effectiveness. Using a stemmer is generally considered best practice in IR, but in this case, the statistical semantics are able to first, identify variants of the same term and second, distinguish different meanings for different variants of the same term. We show an example for each of the two cases, to illustrate what we mean. For the first point, Listing 1.1 shows the 10 most similar terms to the term `coatings`.

Listing 1.1. Top 10 similar terms to "coatings"

```
1.0: coatings
0.9999339: rubs
0.9999338: coating
0.9999328: acrylics
0.9999271: vinyls
0.9999268: cratering
0.9999251: distinctness
0.9999246: blistering
0.9999235: pompano
0.9999234: cyanamid
```

As can be seen, the singular form appears very high in the list, surrounded by different types of coatings.

Regarding the second point, we take the example of "crystal" and "crystals". The first is generally used in optics, while the second in chemistry. Listings 1.2 and 1.3 show the two top-10 similar word lists. Note how Listing 1.2 identifies optical devices (and misspellings at positions 2 and 3) , while Listing 1.3 refers mostly to the process of crystallisation (and it even manages to identify a poorly tokenized version of the term, at position 5 and 10).

Listing 1.2. Top 10 similar terms to "crystal"

```
1.0: crystal
0.9999378: cyrstal
0.9999305: crytal
0.9999022: nicol// a type of prism
0.9999014: jjap
0.9999006: nicols
0.9998996: nematic// a type of liquid crystal
0.9998943: uniaxial//minerals that form crystals used in
    optics
0.9998894: cb15//a particular liquid crystal
0.9998887: anisotropy
```

Listing 1.3. Top 10 similar terms to "crystals"

```
1.0: crystals
0.9998632: supersaturation
0.9998519: crystallizing
0.9998281: supersaturated
0.9998213: crys
0.9998193: purer
0.9998166: soda
0.9998120: crystallize
0.9998105: crystallizers
0.9998081: tals
```

In the following subsections, we will take a more in depth look at the different parameters and their values. We will work only with the top three combinations of fields identified in this experiment burst.

3.3   Number of Seeds

We start with the only parameter that is really method-specific. Random Indexing relies on the Johnson-Lindenstrauss lemma and particularly on the proof they provided in their 1984 article [10]. There, they show that if one chooses at random a rank n orthogonal projection, then, with positive probability, the projection restricted to the set of vectors of interest, will satisfy the condition in the Lemma. RI relies on the observation that, in a high dimensional space, a random set of vectors is always almost orthogonal.

However, this says nothing about how this random set of vectors should be created. Random Indexing relies on the results of Achlioptas [1] and uses a variant of a very simple distribution, which assigns values of -1 or 1 to a small set of coordinates in an otherwise null vector. These non-null coordinates are referred to as *seeds*.

There is apriorically little guidance as to what a good value for the number of seeds should be. Intuitively, it should depend on the dimensionality of the new space and the number of vectors we need to project (i.e. the size of the projection matrix). Random Indexing works by first assigning random vectors to the documents and calculating the term vectors as a weighted sum of the documents in which they appear. The following equation shows this first step,

$$T'_{m,k} = T_{m,n} \times D_{n,k} \tag{1}$$

where m is the number of terms, n is the number of documents, k is the reduced dimension, $T_{m,n}$ is the original term-document matrix, $D_{n,k}$ is the set of random vectors assigned to documents (one vector per line) and $T'_{m,k}$ is the resulting reduced dimensionality term vectors. The almost orthogonality of the new semantic space is given not by the lines in $D_{n,k}$, but by the columns. Given the random procedure, it is extremely difficult to calculate what would make a good seed for each dimensionality. Instead, it is relatively easy to numerically observe this. In fact, what we need to estimate is the error introduced in the calculation of the product of two vectors by the lack of orthogonality of the basis vectors. These basis vectors being now the columns of the newly generated matrix, we can calculate the error by calculating the cosines of the angles between them. Note that the number of seeds depends heavily on how exactly they are generated. This simple numerical observation is however independent of this and therefore applicable to any seed generation method.

In an orthogonal system this would of course be zero, and we would then suspect that the higher this value, the worst the performance of the system. The approximate results are presented in Table 2[2]. Comparing these values with the results obtained by the system using these values and shown in Table 1, confirms the utility of this method of determining the optimal number of seeds.

Table 1. The effect of the number of seeds

Fields	Dimensionality:100, Min term freq:1, Max term freq:100k				
	2	3	4	5	10
ab-desc	24.7(23.6,25.9)	28.6(27.4,29.8)	25.4(24.2,26.6)	28.5(27.3,29.7)	25.6(24.5,26.8)
desc-clm	26.3(25.1,27.5)	26.8(25.6,28)	26.8(25.6,28)	26.4(25.2,27.6)	27.8(26.6,29)
ab-desc-clm	26.4(25.2,27.6)	28.4(27.2,29.7)	27.1(26,28.4)	28.7(27.5,29.9)	25.3(24.1,26.4)

Table 2. The approximate sum of cosine values between random vectors

Dimensionality	Number of seeds				
	2	3	4	5	10
100	50	33	50	40	50
300	150	100	150	120	150
600	300	200	300	240	300
1000	500	347	503	411	506

---

[2] For space, we do not present confidence intervals here, but note that they do not overlap.

In a series of exhaustive experiments, we have varied the dimensionality of the space, the maximum and minimum frequency, as well as the minimum document frequency. We do not list here the full results because none of the combinations obtained results above a baseline provided by Lucene. However, it is important to try to understand why the promise of latent semantics has not been fulfilled in the case of the CLEF-IP 2011 test collection.

4   Discussion

In our experiments we have observed how Random Indexing is impervious to practically all IR engineering best-practice and only reacts to parameters that are specific to the method itself, such as the number of seeds or dimensionality of the reduced, semantic space. We have also seen that the results obtained are far below a relatively standard method, as that implemented in Lucene 3.5.0. The low results are that much more puzzling, as manual inspection of the term space shows that RI is able to correctly identify semantically related terms, as well as misspellings and even wrong tokenizations. We have shown examples of this in Section 3.2. Given all this, and a previous study showing that in LSI identifying related terms is at the heart of what makes spectral retrieval work in practice [4], we are left rightfully wondering why does it not work in this case.

To try to understand this, we need to visualize the semantic space created by the method. Figure 2 shows bitmap representations of the document and term vectors. Each line in these figures is a series of pixels whose RGB values are proportional to the coordinates of that particular vector in the semantic space. As RGB values are only between 0 and 255, we used Equation 2 to map the coordinates to the visible range.

$$colour[i] = \frac{1}{5} \left(log \left(|v[i]|(e^5 - 1) + 1 \right) \right)^3 \cdot 255 \qquad (2)$$

Here, we take the absolute value of the vector v at position i, map it first from the $[0, 1]$ interval to $[1, e^5]$, spread out the values with logarithm at power 3 and finally mapping the result to $[0, 255]$. As RGB has three components, we use blue for negative values of $v[i]$, red for positive values, while keeping the green component constant. The figures are converted to a high-contrast black-and-white representation suitable for printing. The full colour pictures are available on our website (anonymized).

There are a couple of immediate observations we can make based on Figure 2. First, we notice that the document vector matrix is full of spurious black lines. Second, the term vectors appear much more random than the document vectors.

The black lines in the document space are due to the nature of the collection. Some documents do not have all sections in English. When this happens, the document vector is NaN and appears on the plot as a black line. This however does not affect the retrieval performance, because NaN are always at the end of the similarity list. We plot the document space without these vectors in Figure 3. This allows us to observe other issues with the collection. For instance, we

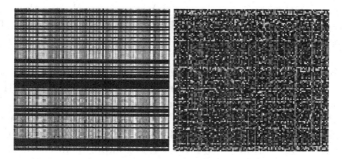

Fig. 2. Snapshot of document vectors (left) and term vectors (right) in the semantic space

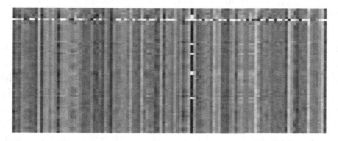

Fig. 3. Snapshot of document vectors in the semantic space, excluding null vectors

observe a visible horizontal line at the top of this figure. Upon examination of the document it represents, we found that the document was labeled as English, but contained German text instead.

Looking now at Figures 2 and 3, we can hypothesize that the reason for which Random Indexing is able to perform remarkably in terms of identifying similar terms but is not so good in finding similar documents is that the document space shows some extreme correlations between its documents. The vertical lines observed in Figure 3 indicate that practically all documents are mapped to the same small area of the space. This also explains why adding more dimensionalities to the space does not help, as observed in our experiments.

There are two possible causes for this behaviour. First, the input to the method is flawed. That is, the key components that make a statistical semantic method work - *context* and *sum* - are not properly defined. Second, the method itself does not work. The following two sections explore these directions.

4.1   Input Alternatives

As seen before, the vectors generated by the method appear to present high correlations between each other. This can only be due to the way they are created, and, in particular, the definition of the *context* and *sum* concepts mentioned in the definition of statistics semantics. So far, these have been *context*='document field' and *sum*='arithmetic sum'. In this subsection, we change these definitions to observe any changes. We start with the definition of context and then follow up with that of the sum.

Window-Based Term Vectors. In one of the main instances of Random Indexing [13], but also in early work on statistical semantic [26], the method defines context as a window around every term instead of the whole document or section. Proponents of this method argue that the term is better defined by its immediate context rather than all the terms in the document. In terms of document retrieval performance however, our experiments show a significant decrease when using this method. Furthermore, because now we are creating many more contexts, the computational resources needed for this version of RI increase substantially. We conclude that for the purpose of document retrieval defining context as a window around each term is counter productive.

Term Weighting. The SemanticVectors library also implements the LogEntropy method proposed by Martin and Berry [17]. We additionally implemented a term weighing scheme based on the Lucene model[3]. The lucene-like formula is similar in the sense that it uses the same formula for TF-IDF as Lucene, but discards the boosting and query-match factors, which are not available in our case. While both methods have been demonstrated to improve similarity detection, in their respective contexts, in our case, both weighting methods degraded the performance of the system.

4.2   Method Variations

The other possible cause of the high clustering observed in the document space may simply be the method itself. RI works by first generating a set of random vectors for documents, then creating term vectors based on the documents containing each term, and then iterating once again to make sure the latent semantics emerge distinct from the randomness in the initial set of vectors. This makes sense if the objective is to identify term similarity, but may result in self-reinforcing features of the semantic space, by the time the document vectors are created in the second iteration. For the purpose of document retrieval it makes more sense to start with random term vectors. This method is also present in the SemanticVectors library, but until now was only used in finding indirect relations between terms [7]. Figure 4 shows the document semantic space created by this method. The results improved by 10% over the best set of parameters found so far, while the image shows indeed less clustering of the documents in one part of the space. There are still however visible vertical lines, although this time they are less frequent and only visible in the colour version of the image.

The reason for this coagulation of the document vectors in areas of the semantic space can only be explained through the extremely skewed distribution

---

[3] http://lucene.apache.org/core/old_versioned_docs/versions/3_5_0/api/all/org/apache/lucene/search/Similarity.html

Fig. 4. Snapshot of document vectors in the reflective semantic space, excluding null vectors

of terms in documents, as observed in Section 3.1. Ultimately, the method relies on a small iteration multiplying the term-document matrix with a set of random vectors, but also with itself 3, 4 times—in the reflective and standard version, respectively. This means that, despite normalisation at every step, the imbalances in the original data make it such that the method fails in identifying meaningful correlations[4]. In this sense, we are somewhat surprised that the two term weighting methods we tried in this study degraded the results. Although we have made progress in understanding the inner workings of Random Indexing, this goes to show that there is still more work to do in adapting it to a particular domain.

5   Related Work

To the best of our knowledge this is the first study of this magnitude on the performance of Random Indexing in a domain specific IR context. Nevertheless, statistical semantic methods are certainly not new in IR, but their understanding is still limited. LSA gets the lion's share of the attention among published works. Bradford [6] created a survey of prior applications of LSA, covering over 40 publications up to 2008, showing the optimal dimensionality in different cases. His study also shows that, in general, the size of the collection to which this was applied was extremely small, with a maximum of 348k documents, minimum: 82, and an average in the low thousands of documents. More recently, Garron and Kontostathis [9] have analysed the performance of LSI at the TREC Legal Learning track, but the results presented there are very difficult to generalize, due to the nature of the track (i.e. the use of only a handful of topics).

   Random Indexing is more recent and its impact on information retrieval tasks has not really been assessed. It is this gap that this paper aims to fill. Instead, most studies focus on information or knowledge extraction tasks. In particular,

---

[4] Note that our full set of experiments includes testing RI with only one iteration, but performance degraded and, for space reasons, we do not list them here.

it has been extensively used in biomedical texts [12], but also in opinion mining [23]. In an IR context, Sahlgren [21,22] has used Random Indexing as a query extension method, with limited success.

Statistical semantic methods are attractive because they claim to be able to disambiguate text. It is unclear however to what extent they have really been tested on ambiguous text so far. The issue of lack of ambiguity in test collections has been approached by Sanderson before [24]. Patent test collections present a different kind of ambiguity.

6   Conclusion

The results presented in this study confirm our starting hypothesis that although random indexing is intellectually interesting and in some cases useful, for the task of patent retrieval it does not outperform the standard Vector Space Model (VSM). Figure 5 shows the best results obtained via Random Indexing compared to the selected baselines. The x-axis lists the runs submitted to the 2011 CLEF-IP campaign, while the y-axis the values obtained by each run in terms of *bpref* - one of the metrics used in CLEF-IP. The message is not at all that we should disrepute such methods. On the contrary, what we should take away from this is that there is significant space for improvement and that more work has to be done in this field. Random Indexing requires a deep understanding of its cognitive foundations, the mathematics modelling these foundations, and—equally important—the implementation details, to make sure that when one deals with huge amounts of terms and documents, the limits of the machines' precision are not inadvertently overran.

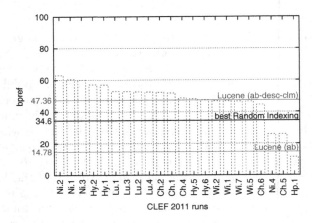

Fig. 5. Random Indexing compared with the baselines

Through this study, we observed how Random Indexing is unable to overcome the very skewed term and document frequencies present in the patent domain. We have shown that the method is able to automate the engineering methods on which other IR models rely, such as the use of a stemmer. However, it is unable to outperform an established IR system such as Lucene. Based on visual observation of the semantic space, we suggest that this is due to its higher sensitivity to imbalances in the term-document matrix. The way to correct these imbalances remains a challenge to this date, as previously suggested term weighting schemes are found to actually degrade performance.

Random Indexing appears able to identify concepts, as demonstrated in related works and by observation of the term similarities. For this to work in an IR task, the method has to be able to combine these concepts properly into a query vector, and it is here that we identify the most potential for future development in this area.

References

1. Achlioptas, D.: Database-friendly random projections. In: Proc. of PODS (2001)
2. Adams, S.: The text, the full text and nothing but the text: Part 1 - standards for creating textual information in patent documents and general search implications. WPI Journal 32(1), 22–29 (2010)
3. Atkinson, K.H.: Towards a more rational patent search paradigm. In: Proc. of PaIR (2008)
4. Bast, H., Majumdar, D.: Why spectral retrieval works. In: Proc. of SIGIR (2005)
5. Bingham, E., Mannila, H.: Random projection in dimensionality reduction: applications to image and text data. In: Proc. of KDD (2001)
6. Bradford, R.B.: An empirical study of required dimensionality for large-scale latent semantic indexing applications. In: Proc. of CIKM (2008)
7. Cohen, T., Schvaneveldt, R., Widdows, D.: Reflective random indexing and indirect inference: A scalable method for discovery of implicit connections. Journal of Biomedical Informatics 43(2) (2010)
8. Furnas, G.W., Dumais, S.T., Landauer, T.K., Harshman, R.A., Streeter, L.A., Lochbaum, K.E.: Information Retrieval using Singular Value Decomposition Model of Latent Semantic Structure. In: Proc. of SIGIR (1988)
9. Garron, A., Kontostathis, A.: Applying latent semantic indexing on the trec 2010 legal dataset. In: Text Retrieval Conference, TREC (2010)
10. Johnson, W.B., Lindenstrauss, J.: Extensions to lipschiz mapping into hilbert space. Contemporary Mathematics 26 (1984)
11. Joho, H., Sanderson, M.: Document frequency and term specificity. In: Large Scale Semantic Access to Content (Text, Image, Video, & Sound), RIAO (2007)
12. Jonnalagadda, S., Cohen, T., Wu, S., Gonzalez, G.: Enhancing clinical concept extraction with distributional semantics. Journal of Biomedical Informatics 45(1), 129–140 (2012)
13. Karlgren, J., Sahlgren, M.: From words to understanding. In: Uesaka, Y., Kanerva, P., Ashton, H. (eds.) Foundations of Real-World Intelligence (2001)
14. Landauer, T.K., Dumais, S.T.: A solution to Plato's problem: The latent semantic analysis theory of acquisition, induction, and representation of knowledge. Psychological Review, 211–240 (1997)
15. Lund, K., Burgess, C.: Producing high-dimensional semantic spaces from lexical co-occurrence. Behavior Research Methods 28 (1996)
16. Lupu, M., Hanbury, A.: Patent Retrieval. Foundations and Trends in Information Retrieval 7(1) (2013)

17. Martin, D., Berry, M.: Mathematical Foundations Behind Latent Semantic Analysis. In: Handbook of Latent Semantic Analysis (2007)
18. Oostdijk, N., D'hondt, E., van Halteren, H., Verberne, S.: Genre and domain in patent texts. In: Proc. of PaIR (2010)
19. Piroi, F., Lupu, M., Hanbury, A., Zenz, V.: Clef-ip 2011: Retrieval in the intellectual property domain. In: CLEF (Notebook Papers/Labs/Workshop) (2011)
20. Sahlgren, M.: An introduction to random indexing. Technical report, SICS, Swedish Institute of Computer Science (2005)
21. Sahlgren, M., Hansen, P., Karlgren, J.: English-Japanese cross-lingual query expansion using random indexing of aligned bilingual text data. In: Proc. of NTCIR (2002)
22. Sahlgren, M., Karlgren, J.: Vector-based semantic analysis using random indexing for cross-lingual query expansion. In: Peters, C., Braschler, M., Gonzalo, J., Kluck, M. (eds.) CLEF 2001. LNCS, vol. 2406, pp. 169–176. Springer, Heidelberg (2002)
23. Sahlgren, M., Karlgren, J.: Terminology mining in social media. In: Proc. of CIKM (2009)
24. Sanderson, M.: Ambiguous queries: test collections need more sense. In: Proc. of SIGIR (2008)
25. Schütze, H.: Dimensions of meaning. In: Proceedings of the Supercomputing 1992 (1992)
26. Schütze, H., Pederse, J.O.: A cooccurrence-based thesaurus and two applications to information retrieval. Information Processing & Management 33(3) (1997)
27. Widdows, D., Cohen, T.: The semantic vectors package: New algorithms and public tools for distributional semantics. In: Proc. of ICSC (2010)

Teaching Syllogistics Using Conceptual Graphs

Peter Øhrstrøm[1], Ulrik Sandborg-Petersen[1],
Steinar Thorvaldsen[2], and Thomas Ploug[1]

[1] Department of Communication and Psychology, Aalborg University,
9000 Aalborg, Denmark
{poe,ulrikp,ploug}@hum.aau.dk
[2] Department of Education, University of Tromsø, 9037 Tromsø, Norway
steinar.thorvaldsen@uit.no

Abstract. It has for centuries been commonly believed that syllogistic reasoning is an essential part of human rationality. For this reason, Aristotelian syllogistics has since the rise of the European university been a standard component of logic teaching. During the medieval period syllogistic validity was presented in terms of a number of artificial words designed to summarize the deductive structure of this basic system. The present paper is a continuation of earlier studies involving practical experiments with informatics students using a student-facing Java-Applet running in the student's browser, implemented using the Prolog programming language as embodied in a Java implementation called Prolog+CG. The aim of the present paper is to study some interesting conceptual aspects of syllogistic reasoning and to investigate whether CG formalism can be helpful in order to obtain a better understanding of syllogistic reasoning in general and the system of Aristotelian syllogisms conceived as a deductive (axiomatic) structure in particular. Some prototypes of tools for basic logic teaching have been developed using Prolog+CG, and various preliminary tests of the tools have been carried out.

Keywords: Syllogistics, argumentation, conceptual graphs, deduction, logic teaching.

1 Introduction

Aristotelian syllogistic has been an essential part of almost all courses in basic logic since the rise of the European university in the 11th century; cf. [1] and [4]. This kind of reasoning has commonly been believed to constitute a crucial component of the ideal of basic human argumentation as it is carried out in social life. Clearly, most of us make frequent mistakes in the way we are treating syllogistic forms in our discussions in daily life. However, normally we are able to recognize and understand what went wrong when we are confronted with our errors. For this reason, we very much treasure and appreciate the logical ideal of correctness which has been expressed in syllogistic theory as it has been known since Aristotle.

It is an important aspect of Aristotelian syllogisms that the valid syllogisms can be reduced or explained in terms of some even more basic principles. In fact, the valid

N. Hernandez et al. (Eds.): ICCS 2014, LNAI 8577, pp. 217–230, 2014.
DOI: 10.1007/978-3-319-08389-6_18, © Springer International Publishing Switzerland 2014

syllogisms may be understood as an axiomatic system. Actually, the system of Aristotelian syllogisms may be understood as the very first presentation of an axiomatic system in logical literature. This view of the syllogisms as a deductive system was in fact incorporated in the standard medieval presentation of the syllogisms. In modern logic there is also a strong emphasis on the importance of deduction. John Sowa has suggested that deductive reasoning is represented in terms of so-called conceptual graphs cf. [5]. This formalism may be presented using the so-called Conceptual Graph Interchange Format (CGIF); cf. [7]. In the present paper we explore the use of this option in logic teaching. The main questions are:

a)   How should the understanding of the syllogisms as a deductive system be incorporated in modern logic teaching?

b)   Can a better understanding of the syllogisms as deductive system give rise to a more correct syllogistic reasoning in practice?

The present study is a continuation of earlier studies and practical experiments involving informatics students and other students from the humanities; cf. [9] and [10]. The experiments have been carried out using the so-called Syllog system, i.e. a Java-Applet running in the student's browser, implemented using the Prolog programming language as embodied in a Java implementation called Prolog+CG; cf. [2], [3], [6] and [8]. Syllog generates syllogisms at random, and the user is supposed to evaluate them using the system.

In section 2 of this paper we present the theory of Aristotelian syllogistics as a deductive system in the classical manner as well as a more modern way in terms of conceptual graphs. In section 3 we describe the general experimental setup as well as the practical experiments which have been carried out. In section 4 we discuss the results of the experiments. Section 5 is based on a qualitative interview with a group of students, and it deals with a potential improvement of the experimental setup which can lead to the development of a tool which may be used in logic teaching in order to give the students a better understanding of the nature of deductive structures. In section 6 it will be discussed what we may conclude from the present study and also what further developments and investigations the present study may suggest.

2   Aristotelian Syllogisms as Deductive Structures

In terms of modern logic the classical (medieval) syllogistics may be presented as a fragment of first order predicate calculus. A classical syllogism corresponds to an implication of the following kind:

$$(p \wedge q) \supset r$$

where each of the propositions p, q, and r matches one of the following four forms

$a(U, V)$ (read: "All U are V")
$e(U, V)$ (read: "No U are V")
$i(U, V)$ (read: "Some U are V")
$o(U, V)$ (read: "Some U are not V")

These four functors were suggested by the medieval logicians referring to the vowels in the words "affirmo" (latin for "I confirm") and "nego" (latin for "I deny"), respectively. Given that U and V are interpreted as predicates, we may express the functors in terms of the usual first order predicate calculus in the following way:

$$a(U, V) \leftrightarrow \forall x: (U(x) \supset V(x))$$
$$e(U, V) \leftrightarrow \forall x: (U(x) \supset \sim V(x))$$
$$i(U, V) \leftrightarrow \exists x: (U(x) \wedge V(x))$$
$$o(U, V) \leftrightarrow \exists x: (U(x) \wedge \sim V(x))$$

These four basic propositions are related in terms of the negation in the following manner:

$$i(U, V) \leftrightarrow \sim e(U, V)$$
$$o(U, V) \leftrightarrow \sim a(U, V)$$

The classical syllogisms occur in four different figures:

$$(u(M, P) \wedge v(S, M)) \supset w(S, P) \quad \text{(1st figure)}$$
$$(u(P, M) \wedge v(S, M)) \supset w(S, P) \quad \text{(2nd figure)}$$
$$(u(M, P) \wedge v(M, S)) \supset w(S, P) \quad \text{(3rd figure)}$$
$$(u(P, M) \wedge v(M, S)) \supset w(S, P) \quad \text{(4th figure)}$$

where $u, v, w \in \{a, e, i, o\}$ and where M, S, P are variables corresponding to "the middle term", "the subject" and "the predicate" (of the conclusion). The premises are mentioned according to the convention that the grammatical subject (here S) of the conclusion should appear in the second of the premises.

In this way 256 different syllogisms can be constructed. According to classical (Aristotelian) syllogistics, however, only 24 of them are valid. The medieval logicians named the valid syllogisms according to the vowels, $\{a, e, i, o\}$, involved. In this way the following artificial names were constructed (see [1]):

1st figure: barbara, celarent, darii, ferio, barbarix, feraxo
2nd figure: cesare, camestres, festino, baroco, camestrop, cesarox
3rd figure: darapti, disamis, datisi, felapton, bocardo, ferison
4th figure: bramantip, camenes, dimaris, fesapo, fresison, camenop

If needed, we may add the figure number to the name (e.g. baroco2, fesapo4). In these names some of the consonants signify the logical relations between the valid syllogisms, and they also indicate which rules of inference should be used in order to obtain the syllogism in question from syllogisms which were considered to be fundamental: barbara, celarent, darii, ferio. – In fact, the system of syllogisms may in this way be seen as an axiomatic system with these four axioms and the following inference rules (see [1] and [3]):

x-rule-1: $a(U,V) \rightarrow i(U,V)$ $\forall x: (U(x) \supset V(x)) \rightarrow \exists x: (U(x) \wedge V(x))$
x-rule-2: $e(U,V) \rightarrow o(U,V)$ $\forall x: (U(x) \supset {\sim}V(x)) \rightarrow \exists x: (U(x) \wedge {\sim}V(x))$
s-rule-1: $i(U,V) \leftrightarrow i(V,U)$ $\exists x: (U(x) \wedge V(x)) \leftrightarrow \exists x: (V(x) \wedge U(x))$
s-rule-2: $e(U,V) \leftrightarrow e(V,U)$ $\forall x: (U(x) \supset {\sim}V(x)) \leftrightarrow \forall x: (V(x) \supset {\sim}U(x))$
m-rule: $(p \wedge q) \supset r \leftrightarrow (q \wedge p) \supset r$
c-rule-1: $(p \wedge q) \supset r \leftrightarrow ({\sim}r \wedge q) \supset {\sim}p$
c-rule-2: $(p \wedge q) \supset r \leftrightarrow (p \wedge {\sim}r) \supset {\sim}q$

In addition, p-rule-1 may be defined as a combination of x-rule-1 and s-rule-1, whereas p-rule-2 may be defined as a combination of x-rule-2 and s-rule-2. It should also be noted that x-rule-1/2 is used from left to right in the conclusion, whereas it is used from right to left in the premises.

The use of the axioms in the deduction is rather simple, since only one axiom is involved in each deduction. From barbara1 we may derive barbarix1 (by x-rule-1), baroco2 (by c-rule-2), bocardo3 (by c-rule-1), bramantip4 (by m-rule and p-rule-1).

From celarent1 we may derive cesare2 (by s-rule-2), camestres2 (by m-rule and two applications of s-rule-2), camestrop2 (by m-rule, s-rule-2 and p-rule-2), cesarox2 (by s-rule-2 and x-rule-2), camenes4 (by m-rule and s-rule-2), camenop4 (by m-rule and p-rule-2). From darii1 we may derive darapti3 (by p-rule-1), disamis3 (by m-rule and two applications of s-rule-1), datisi3 (by s-rule-1), dimaris4 (by m-rule and s-rule-1). From ferio1 we may derive feraxo1 (by x-rule-1), festino2 (by s-rule-2), felapton3 (by p-rule-1), ferison3 (by s-rule-1), fesapo4 (by s-rule-2 and p-rule-1), fresison4 (by s-rule-2 and s-rule-1).

There is obviously a great deal of structural beauty built into this system of artificial names. It should be noted that the names of a syllogism may in fact be read as an abbreviation of its proof. For instance, the proof of camestrop2 from celarent1 may be presented in the following manner:

$(e(M, P) \wedge a(S, M)) \supset e(S, P)$	(celarent1)
$(a(S, M) \wedge e(M, P)) \supset e(S, P)$	(use of m-rule)
$(a(S, M) \wedge e(P,M)) \supset e(S, P)$	(use of s-rule-2)
$(a(S, M) \wedge e(P,M)) \supset o(P,S)$	(use of p-rule-2; camestrop2; QED)

Note that in this demonstration of camestrop2, P plays the role of the grammatical subject in the conclusion. For this reason the order of the premises should according to the medieval standard be presented as it is done here.

In logic teaching this nice medieval system may obviously be used in order to illustrate the idea of a logical proof. In this way, the system of the 24 valid syllogisms may be reduced to a system of four axioms and seven rules of inference. From a modern perspective we have to say that this reduction does not go far enough. On the other hand, it should be admitted that the system is elegant, and that the idea of using the proof of a theorem as its name is rather remarkable.

Given that we want to produce an even more convincing reduction of the system of syllogisms than the one which was suggested in medieval logic, we may apply conceptual graphs represented in the conceptual graph interchange format (CGIF); cf.

[7]. In terms of this formalism the four basic propositions syllogistics may be represented in the following manner, where "–attr->" stands for "has the attribute of":

$a(U, V) \leftrightarrow$ [All: *x] [If: (?x –attr-> U) [Then: (?x-attr-> V)]]
$e(U, V) \leftrightarrow$ [All: *x] [If: (?x –attr-> U) [Then: ~(?x-attr-> V)]]
$i(U, V) \leftrightarrow$ [*x] (?x –attr-> U) (?x-attr-> V)
$o(U, V) \leftrightarrow$ [*x] (?x –attr-> U) ~(?x-attr-> V)

In the following, we take the liberty of ignoring '*' and '?' in the CGIFs. - The CG formalism is to some extent integrated in the PROLOG+CG. In fact, expressions like, [x] –attr-> [U], are allowed according to the syntax of this system. For instance, if we want to represent "All Greeks are Mortal" in terms of CGs one might consider the following PROLOG clause:

[x] -attr-> [Mortal] :- [x] -attr-> [Greeks].

This representation will, however, be too simplistic in our case. We need more freedom of expression concerning the proof system. It may be argued that the representation of the syllogisms as implications should be represented in terms of sequents, i.e.

$p, q \vdash r$

This representation indicates that r is derivable from the combination of p and q. What derivability means in this context then has to be specified in terms of some rules of inference. The following rules are suggested to cover the deductive system of the Aristotelian syllogisms:

(TR)   [All: x] [If: $G_1(x)$ [Then: $G_2(x)$]]
        [All: x] [If: $G_2(x)$ [Then: $G_3(x)$]]
        Therefore:
        [All: x] [If: $G_1(x)$ [Then: $G_3(x)$]]

(SUB)   [All: x] [If: $G_1(x)$ [Then: $G_2(x)$]]
        [x] $G_1(x)$ $G_3(x)$
        Therefore:
        [x] $G_2(x)$ $G_3(x)$

(CON)   [All: x] [If: $G_1(x)$ [Then: $G_2(x)$]]
        Therefore:
        [All: x] [If: $\sim G_2(x)$ [Then: $\sim G_1(x)$]]

(MUT)   [x] $G_1(x)$ $G_2(x)$
        Therefore:
        [x] $G_2(x)$ $G_1(x)$

(EX)   [All: x] [If: $G_1(x)$ [Then: $G_2(x)$]]
        Therefore:
        [x] $G_1(x)$ $G_2(x)$

The usual understanding of negation and double negation is also assumed i.e. for any graph, G, it holds that: $\sim\sim G \leftrightarrow G$. Given these rules any of the 24 valid syllogisms may be demonstrated. Consider, for instance, the syllogism camestrop2 mentioned above. In this case we have to prove:

        [x] (x –attr-> P) ~(x-attr-> S)
Given

        [All: x] [If: (x –attr-> S) [Then: (x-attr-> M)]]
        [All: x] [If: (x –attr-> P) [Then: ~(x-attr-> M)]]

The proof can be carried out in the following manner:

    1.   [All: x] [If: (x –attr-> S) [Then: (x-attr-> M)]]   (premise)
    2.   [All: x] [If: (x –attr-> P) [Then: ~(x-attr-> M)]]   (premise)
    3.   [All: x] [If: (x –attr-> M) [Then: ~(x-attr-> P)]]   (from 2 & CON)
    4.   [All: x] [If: (x –attr-> S) [Then: ~(x-attr-> P)]]   (from 1, 3 & TR)
    5.   [All: x] [If: (x –attr-> P) [Then: ~(x-attr-> S)]]   (from 4 & CON)
    6.   [x] (x –attr-> P) ~(x-attr-> S)   (from 5 & EX; QED)

Note that here not only camestrop2 has been demonstrated. 1-5 actually also constitute a proof of camestres2.

The above demonstration is in fact a relatively complicated proof. In many of the other cases, the numbers of steps in the proofs are much smaller. Consider for instance darii1:

    1.   [All: x] [If: (x –attr-> M) [Then: (x-attr-> P)]]   (premise)
    2.   [x] (x –attr-> S) (x-attr-> M)   (premise)
    3.   [x] (x –attr-> S) (x-attr-> P)   (from 1, 2 & SUB; QED)

Using this deductive approach to the syllogisms, it will also be possible to show important results concerning the invalidity of certain syllogistic arguments. For instance, by going through the rules of inference listed above it is evident that if both premises are existential, then nothing follows. The same holds if both premises are negative i.e. o-propositions or e-propositions.

The use of (EX) corresponding to the medieval x-rule has sometimes been seen as controversial and the 9 syllogisms which depend on this rule have consequently been seen as "questioned". Clearly (EX) has to be rejected, if we accepted that the statement "all S are P" can be true even if the concept S corresponds to the empty set.

3   The Experiments in the Present Study

From the earlier studies, [9] and [10], it can be concluded that we have very strong evidence for the following claims regarding the way syllogisms are handled by students before they have been introduced to the theories of syllogisms:

a) In many cases students can make correct distinctions between valid and invalid syllogisms. Their ability to do so is significantly higher than the level of guessing.

b) It is easier for the students to find the correct validity in case of a valid syllogism than in case of an invalid syllogism, i.e. there are more errors in spotting the invalidity of invalid syllogistic arguments than in spotting the validity of valid syllogisms.

c) If the conclusion of the syllogistic argument is true (i.e. corresponds with the real world) then the students are likely to assume that the argument is valid.

d) The students are more likely to accept the validity of the 15 "unquestioned" syllogisms than the validity of the 9 "questioned" syllogisms.

e) In [10] it has been studied how gamified quizzing can be used in logic teaching. This study shows that at least in some cases the responses from the students indicate that the use of gamification had some positive effects on the motivation to learn.

The earlier studies also deal with the question of possible improvements in the score caused by the logic teaching. Here the main conclusion can be stated in the following manner:

f) There is no or just a very small improvement in the score if it is measured after the presentation of the medieval theory of syllogisms in general.

In the earlier studies we did not include all 232 invalid syllogistic arguments, but we selected 24 invalid syllogistic arguments assumed to be somewhat "tempting". This approach has been challenged. In consequence we have in the present study included all 232 invalid syllogisms. The present version of the system, Syllog, first selects at random whether the next syllogism to present on the screen should be valid or invalid. If it is supposed to be valid then one of the 24 valid syllogisms is selected at random. If it is supposed to be invalid then one of the 232 invalid syllogisms is selected at random. In each case the user is asked to evaluate the argument presented on the screen.

The 2nd year students of Humanistic Informatics at Aalborg University (in Aalborg and in Copenhagen) were on different dates in February 2014 asked to run Syllog in groups of 2-4 before they were introduced to syllogistic logic. All test results have been logged by Syllog. The score was computed as

$$Score = correctanswers/answercount$$

The statistical analyses of the scoring data were performed using standard methods from descriptive statistics and statistical testing. The two sample t-test is applied to detect increased score, and to look for significant differences between results from the pre-test and the post-test. The chi-square test is applied to detect differences in how students handle invalid and valid syllogisms.

In table 1 we compare the results from the pre-test (i.e. before the lectures on syllogisms) in Aalborg using the old dataset from earlier studies, with the test in Aalborg using the new dataset, where all invalid syllogisms are represented. There is no significant difference in how student handle the two datasets, and the average exercise score of the old dataset is 0.564, compared to 0.557 in the new data.

Table 1. The 2x2 tables summarizing counts of how often previous and present students in Aalborg replied correctly to invalid syllogisms in the pre-test. The p-value is by the chi-square test.

	Correct reply? Pre-test	
	No	Yes
Old dataset (24 invalid syllogisms)	**538**	**697**
New dataset (232 invalid syllogisms)	**627**	**789**
p-value	0.74	

This means that in this context it makes no essential difference to include all 232 invalid syllogistic arguments in Syllog. The procedure used in the earlier studies turns out to be fully satisfactory, when it comes to measuring the ability to do valid syllogistic reasoning.

The students were also asked to run the program after the lectures on syllogistic logic. The lectures dealt with the medieval theory of syllogism as well as a brief introduction (about 30 min.) to the treatment of syllogisms in terms of conceptual graphs (CGIF) focusing on the CG formalization of the syllogistic propositions and the rules of inference mentioned above.

In the pre-test the students were just presented with the arbitrary syllogism in its verbal form, e.g.

All teachers are mothers.
All teachers are females.
Ergo: Some females are mothers.

In the post-test, the students were also given the translations of the premises and the conclusion into conceptual graphs (CGIF). With the above example that would mean that the following translated are presented on the screen:

[All: x] [If: (x –attr-> [teachers]) [Then: (x-attr-> [mothers])]]
[All: x] [If: (x –attr-> [teachers]) [Then: (x-attr-> [females])]]
Ergo: [x] [(x –attr-> [females]) (x -attr-> [mothers])]

In addition, the system offered the possible deductions from the premises using (EX). This means that the following derivations were shown on the screen as well:

[x] [(x −attr-> [teachers]) (x -attr-> [mothers])]
[x] [(x −attr-> [teachers]) (x -attr-> [females])]

The challenge for students would then be to determine whether (TR), (SUB), (CON) and (MUT) can be used in order to derive the conclusion from the premises. In this case it should be noted that the conclusion obviously follows by (SUB) from the second premise and the first of the above derivations.

The pre-test were performed by 110 groups in Aalborg and 52 groups in Copenhagen. The post-test were performed by 49 groups in Aalborg, and 36 groups in Copenhagen. This big dropout in Alborg creates a bias for our analyses that must be handled with care. In fact we have decided to ignore the Aalborg Post-test and to concentrate on the results from Copenhagen in this case. All results are, however, shown in Table 2.

Table 2. The above table are summarizing the pre-test versus post-test results of the students average group score in Aalborg and Copenhagen. The p-values are by the two sample t-test (one sided not assuming equal variance). The analyses gives no statistical evidence against the presumption that student will handle syllogism equally well before and after the lecture on syllogisms and conceptual graphs.

		Mean Score	SD Score	p-value Pre vs. Post
Aalborg	Pre (N=110)	0.633	0.191	0.265
	Post (N=49)	0.598	0.176	
Copenhagen	Pre (N=52)	0.590	0.259	0.096
	Post (N=36)	0.677	0.240	

Table 2 shows inconsistent results with respect to progress between the pre-test and the post-test. The achieved scores are observed to be between 0.59 and 0.68, and are all significantly higher than random guessing with an expected score of 0.5 (p-values 0.01 or lower by the one sample t-test).

The below figure 1 shows the distribution of the scores in the post-test. It seems that there is a certain tendency of "two or more tops" in the distribution.

We also compared how much time the student groups spent thinking on each task of the syllogism exercise. The results are shown in Table 3, which shows that the students spent significantly more time to think about each task after the lecture had been given.

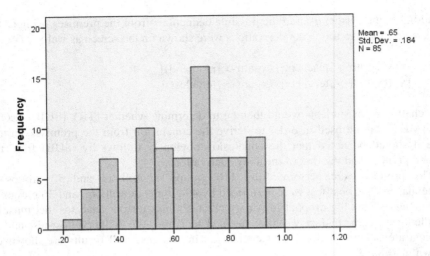

Fig. 1. The distribution of scores in the post-test for the participating groups (N=85)

Table 3. Table are summarizing the pre-test versus post-test results of students time spent on each task of the exercise. The p-values are by the two sample t-test (two sided not assuming equal variance). Some statistical evidence are found against the presumption that student will handle syllogism equally fast before and after the lecture.

		Mean time (s)	SD time (s)	p-value Pre vs. Post
Aalborg	Pre (N=110)	30.9	16.6	0.040
	Post (N=49)	39.2	25.4	
Copenhagen	Pre (N=52)	49.8	34.7	0.098
	Post (N=36)	61.3	32.3	

4   Discussion of the Experimental Results

It is evident that the students use more time on each syllogistic argument in the post-test than in the pre-test. This suggests that the presented CGs make them reflect. There is, however, no significant improvement of the average score when the CGs are listed along with the textual presentation of the argument. It should, however, be noted that Figure 1 shows certain tendency towards "two or more tops". This may indicate that there could be a group of students who are, in fact, able to benefit from the CGs although the majority of students do not benefit from the display of CGs.

The fact that there is no significant improvement of the average score when the CGs are listed as well may not be too surprising given the results from the earlier studies which have been summarised in e) above. In fact, it seems to be very difficult to obtain any improvement in the score through logic teaching. One possibility is that the logical part of the human mind is only trainable within a quite narrow limit, when it comes to practical skills in everyday reasoning. Even if this turns out to be the case, it does not necessarily mean that logic teaching is useless. As one of the students stated during the discussion, based on the insights from the teaching the students may better understand what is wrong in the invalid syllogistic arguments and why the valid arguments hold. In this way, although their reasoning skills in daily life are pretty much the same after the logic lectures, the students may in fact from the logic classes have gained a much deeper understanding of the very notion of validity. On the other hand, it may still be possible to use the CGs in a more convincing manner in order to give the students an even deeper understanding of what valid reasoning is. For this reason, it might still be a good idea to look for designs and tools different from the present post-test through which the students can benefit from the structural qualities of conceptual graphs in an optimal manner. One problem seems to be that the present design with all the translations and derivations mentioned above is too overloaded and complicated, which means that it has become hard to read and difficult to use.

5   Can Conceptual Graphs be Useful in Logic Teaching?

Would it be possible to improve the experimental setup used in the Post-test in order to create a design or even a tool which can effectively support the students in their understanding of syllogistic reasoning? In order to answer this question a qualitative interview with a group of students has been carried out. We conducted a semi-structured focus group interview with 7 students following the posttest at the end of the course on logics. The group of students consisted of 5 women and 2 men aged from 21 to 26.

The students were initially asked about their understanding of the basic concepts in logic. They all revealed a strong understanding of the basic concepts such as validity and soundness of arguments, and distinguished between propositional arguments and syllogistic reasoning with quantifiers. More specifically, they adequately defined syllogistic reasoning as reasoning containing two premises and a conclusion all constituted by one of the four basic forms $a(U,V)$, $e(U,V)$, $i(U,V)$ and $o(U,V)$ with (U,V) being constituted by a subject (S), predicate (P) or middle term (M). The students were not immediately able to reconstruct the conceptual graphs equivalent to the four basic forms in a correct manner.

Following these initial questions the students were presented with the screendump from the test in which they had all participated i.e. with the translations into CGIFs and the derivations presented above.

Looking at the screen the students were asked about a) their ability to read the conceptual graphs, b) their understanding of the rules applied in the example, and c) and their ability to decide the validity of the relevant syllogism. Their answers revealed three interesting trends elaborated on below:

1) The students find the formalism and the deduction rules of conceptual graphs abstract.
2) The students use the natural language formulation of arguments to support them in interpreting the conceptual graphs and not vice versa.
3) The students found the decision procedure to be applied in order to determine the validity of syllogisms to differ markedly from alternative decision procedures such as Venn-diagrams.

228 P. Øhrstrøm et al.

Concerning item 1 the students note that the vocabulary of conceptual graphs contains other elements [All: x]","[x]", "-attr->", "If ... Then ...", "~") than the classic four forms listed above. The students argue that the reading of the conceptual graphs is less intuitive in the sense that it translates differently into natural language than the basic four forms. Thus the *a(S,P)* simply reads "All S are P" whereas the equivalent conceptual graph

[All: x] [If: (x –attr-> [S]) [Then: (x-attr-> [P])]]

reads "For all x, if x has the attribute S, then x has the attribute P". The students are not convinced that anything is gained by going to this deeper lever of analysis.

Three students specifically mention the problem of reading conceptual graphs corresponding to the "some"-quantified basic forms, i.e. *i(U,V)* and *o(U,V)*. They find it strange and somewhat surprising that there is no specific conjunction included in the conceptual graph formalism. Seen from a natural language point of view they find that something must be missing in the formulation:

[x](([x] -attr-> [teachers]) ([x] -attr -> [women]))

In general, the students did not find the translation of such formalisms into natural language intuitively convincing.

In the conducted test the students were presented with the syllogism in natural language as well as in the form of conceptual graphs. The purpose of the interview was in part to determine if the conceptual graphs aided the students in their reasoning about the validity of the syllogism. All students noted that they did look at conceptual graphs but that they used the natural language formulation of each of the premises and conclusion in the syllogism to interpret the conceptual graphs and to make the deductions required in order to understand the different formulations of each of the premises and conclusion as conceptual graphs. In short, the natural language formulation of the syllogisms is the fix-point for the reasoning of the students, and they all confirm that they made their evaluations based on the arguments in natural language.

All students found the process of determining the validity of syllogistic arguments by means of conceptual graphs to differ markedly from alternative procedures such as Venn-diagrams. The students univocally state that while Venn-diagrams present a simple mechanic procedure for determining the validity of syllogisms, the process of determining the validity by means of conceptual graphs is markedly different by requiring reformulations of premises and then the ability to determine if the conclusion can be derived by the application of the derivation rules. The students seem to make two points here. First, that the procedure is different in nature than other procedures, and therefore "more difficult" to master. Second, that the procedure involves "more steps" and therefore is less simple than, for example, the use of Venn-diagrams.

The interview reveals that the students find natural language formulations and the classic formulations of syllogisms more intuitive in their evaluation of validity. All students noted, however, that they had received very little training in the formalism of conceptual graphs.

In further studies and experiments on logic teaching new methods should be employed. It may not be satisfactory just to let the students see the formal representations. They should also be allowed to operate on the structures according to the relevant rules. Maybe it would be a good idea to use gamification as suggested in [10]. We may develop a game-like system which allows the students to use the deductive rules, (TR), (SUB), (CON), (MUT) and (EX), on the premises of the syllogism in question in order to decide whether or not the syllogism is provable. The various conceptual graphs may even be presented in a graphical manner and not just in linear form. It might even be possible for the students to draw graphs, which may be significant for a deeper understanding of the structures.

As we have seen, the time needed to evaluate the syllogisms is significantly greater in the post-tests than in the pre-tests. It may be assumed that the use of conceptual graphs gives rise to more reflection and maybe even a better understanding of the notion of validity. An assumption like this may actually be tested performing new statistical tests. In addition to asking the students whether a given syllogism is valid or not, it might be interesting to ask them to evaluate the degree of confidence in the answer given.

6   Conclusions

As we have seen, it is possible to create systems implemented in PROLOG+CG which to some extent can measure the outcome of logic teaching. Such tools can clearly give the logic teacher a lot of important information which will turn out to be useful for the teacher when he or she is planning a logic course. We have also seen that it is possible to create tools which can incorporate the formalism of conceptual graphs (CGIF). However, if such tools should improve the reasoning skills of the students, a new design of the system has to be developed and tested. The qualitative interviews suggest that in order for conceptual graphs to be useful in logic teaching the CG formalism has to be presented more carefully than in the present experiment. An extensive practice will be needed for the students to acquire the skills necessary to read, reason with and evaluate syllogisms in terms of CGs. This will require several lectures - and not just a single lecture. Furthermore, it might be a good idea to develop new and more appealing ways of presenting the conceptual structures. In particular, it might be a good idea to present the structures graphically.

The students have a first-hand experience of the fact that conceptual graphs represent a fundamentally different approach to the evaluation of validity. In fact, it seems that they are aware of the fact that is it straightforward to present deductive aspects in terms of the CG formalism. Consequently, the inclusion of conceptual graphs in the curriculum for courses on logic could add to the understanding of the deductive structures and proof procedures. In order for such teaching to bring this understanding it would – judging from the interview – have to put emphasis on teaching the students a specific decision procedure for determining whether a syllogism can be proved or not. It will not be sufficient - as in the present experiment - just to exemplify how the rules of inference may be used. A more extensive (and if possible) complete proof procedure should be presented. Such a procedure can in fact be developed for the Aristotelian syllogisms in terms of a game-like system, which will allow the user to prove valid syllogisms applying the rules of inference presented above. In this way, the game may in a very practical manner illustrate what it means prove a theorem in a deductive system. In addition, the game may in a very clear manner illustrate that 15 of the valid syllogisms can be proved without the use of the (EX) rule, whereas the proofs of 9 of the syllogisms which are valid from an Aristotelian point of view will have to involve the use of the (EX) rule along with the other rules of inference mentioned above.

References

1. Aristotle: Prior Analytics, Jenkinson A.J. (transl.). The Internet Classics Archive (1994-2000), http://classics.mit.edu//Aristotle/prior.html
2. Kabbaj, A., Janta-Polczynski, M.: From PROLOG++ to PROLOG+CG: A CG object-oriented logic programming language. In: Ganter, B., Mineau, G.W. (eds.) ICCS 2000. LNCS (LNAI), vol. 1867, pp. 540–554. Springer, Heidelberg (2000)
3. Kabbaj, A., Moulin, B., Gancet, J., Nadeau, D., Rouleau, O.: Uses, improvements, and extensions of Prolog+CG: Case studies. In: Delugach, H.S., Stumme, G. (eds.) ICCS 2001. LNCS (LNAI), vol. 2120, pp. 346–359. Springer, Heidelberg (2001)
4. Parry, W.T., Hacker, E.A.: Aristotelian Logic. State University of New York Press (1991)
5. Petersen, U., Schärfe, H., Øhrstrøm, P.: Online Course in Knowledge Representation using Conceptual Graphs, Aalborg University (2005), http://cg.huminf.aau.dk
6. Petersen, U.: Prolog+CG: A Maintainer's Perspective. In: de Moor, A., Polovina, S., Delugach, H. (eds.) Proceedings of the First Conceptual Structures Interoperability Workshop (CS-TIW 2006). Aalborg University Press (2006)
7. Sowa, J.F.: Conceptual Graph Summary, http://www.jfsowa.com/cg/cgif.htm (accessed February 2014)
8. Øhrstrøm, P., Sandborg-Petersen, U., Ploug, T.: Syllog - A Tool for Logic Teaching. In: Proceedings of Artificial Intelligence Workshops 2010 (AIW 2010), Mimos Berhad, pp. 42–55 (2010)
9. Øhrstrøm, P., Sandborg-Petersen, U., Thorvaldsen, S., Ploug, T.: Classical Syllogisms in Logic Teaching. In: Pfeiffer, H.D., Ignatov, D.I., Poelmans, J., Gadiraju, N. (eds.) ICCS 2013. LNCS (LNAI), vol. 7735, pp. 31–43. Springer, Heidelberg (2013)
10. Øhrstrøm, P., Sandborg-Petersen, U., Thorvaldsen, S., Ploug, T.: Teaching Logic through Web-Based and Gamified Quizzing of Formal Arguments. In: Hernández-Leo, D., Ley, T., Klamma, R., Harrer, A. (eds.) EC-TEL 2013. LNCS, vol. 8095, pp. 410–423. Springer, Heidelberg (2013)

Principles of Regulated Activation Networks

Alexandre Miguel Pinto and Leandro Barroso

CISUC - University of Coimbra, Portugal
ampinto@dei.uc.pt, lafb@student.dei.uc.pt

Abstract. In a context of growing awareness and prevalence of mental disorders, cognitive modeling has emerged as an important contribution to the study of the mind and its processes. Computational models have proved to be indispensable tools for precise and systematic simulations of cognitive processes, and have a potential application in the diagnosis and treatment of such pathologies.

We propose a connectionist cognitive model that incorporates regulatory mechanisms, called Regulated Activation Networks (RANs), that will be applied to the modeling of psychological phenomena. This paper summarizes the current early stages of the development of the RANs model. The objectives, principles and approach taken are described, as well as the architecture of the RANs model, some preliminary results and plans for future work.

Keywords: Cognitive model, connectionism, conceptual spaces, psychological phenomena.

1 Introduction

Context: Every year, a third of the European Union population suffers from mental disorders [8]. In a global scale the numbers are also worrying: the World Health Organization predicts that depression will be the leading cause of disease burden by 2030 [16]. This alarming trend is one of the reasons for the growing importance of the cognitive science research field, which focuses on the study of the mind and its processes, especially information representation, processing and transformation. Particularly, the development of computational models which provide algorithmic specificity, conceptual clarity and precision, has allowed the realization of simulations that can either be useful to test and validate psychological theories or to generate new hypotheses about how the mind works. This has turned them into indispensable tools in the study of the human mind.

Motivation: We aim at modeling mental processes with our RANs tool and, hopefully, use it to aid in the diagnosis and treatment of some mental pathologies such as depression, schizophrenia and autism. To achieve this the RANs model will need to include the parameters that are enough to mimic the neurological features responsible for simulating and regulating both healthy and pathological cognitive processes. We envisage a scenario where, given a particular instance of the RANs model, we allow it to learn the values of those regulatory parameters from the inputs (e.g., words) given to, and outputs (e.g., other words) obtained from, a specific human user – with different users we train different instances of the RANs. As some users may be previously diagnosed as

N. Hernandez et al. (Eds.): ICCS 2014, LNAI 8577, pp. 231–244, 2014.
DOI: 10.1007/978-3-319-08389-6_19, © Springer International Publishing Switzerland 2014

clinical patients with some disorders, while others may be classified as healthy, we can then use this collection of data to train a classifier to allow the diagnosis of new human subjects. Some results in the literature [3,15,21] have already been achieved by other researchers, which shows that this ambitious goal is not out of reach of computational cognitive modeling. This type of computational tools with the ability to capture cognitive phenomena has also the potential to simulate and help study some mental states and processes such as those linked to creativity [12].

We aim to develop of a computational cognitive model with two main very long-term goals. The first is to help understand and simulate cognitive phenomena such as perception, emotion, learning and reasoning, creativity and, ultimately, different kinds of psychological features and personality traits. The second is to provide a tool that can aid in the simulation, diagnosis and, eventually, treatment, of said pathologies.

Challenges: Taking an incremental development and validation approach, in these first steps we focus only on the following features: 1) the model must be able to receive (sensory-like) input information from the outside, 2) it must allow for the representation and simulation of the dynamic time-variant cognitive state (initially set by the input data), 3) it must be able to learn and abstract the patterns in its cognitive state, and 4) it must be able to exhibit creative behavior by creating new concepts and by generating new patterns of cognitive state. While developing mechanisms to represent the input data may not be hard, simulating a dynamic cognitive state and corresponding learning, and creative processes are not trivial. We do so by modeling certain psychological phenomena concerned with the reception and response to the received stimulus; and also two different kinds of learning. Specifically, we aim, at this stage, to model the effect of Priming, the occurrence of False Memories and the Habituation and Sensitization processes — we describe these phenomena in detail in subsection 2.1 – as these will allow for the simulation of a simple cognitive state that changes through time. The learning mechanism we implemented at this stage progressively identifies correlations between elements of the cognitive state. At a later stage, left for now for future work, we will address more complex kinds of learning (including the ones capable of creating new concepts) and reasoning, and emotion representation, elicitation, and processing.

Paper Structure: Section 2 gives an overview of the basic psychological phenomena we model in this early stage of our work. We also describe the cognitive modeling approaches in the literature that are most relevant to our model, as it draws inspiration and features from those approaches. In section 3 we present the overarching principles behind the design our Regulated Activation Networks (RANs) cognitive model, and also discuss the desiderata properties for the future full version of the model, as well as its global characteristics, some of which are already implemented in the preliminary version detailed in section 4.

In section 5 we show our first preliminary results; and in section 6 we provide conclusions and description of future work, which includes the design of a validation plan we will use to assess the reliability regarding the ability to simulate the identified psychological phenomena.

2 Background

First, we review the main psychological phenomena our current simplified RANs model is intended to capture, and why these phenomena matter for the future full version. Then, we show the cognitive modeling approaches in the literature that we found to have features that are important for our purposes – the RANs model includes an innovative combination of these features, plus other new ones we describe later.

2.1 Psychological Phenomena

Priming: Priming is a phenomenon in which a response to a stimulus is influenced by a previous exposure to the same or a similar stimulus. It happens in a non-conscious way, making it an implicit memory effect [11]. Priming effects can be divided in two main types, according to the relation between the stimulus. Perceptual priming stands for stimulus with a similar form and conceptual/semantic priming for stimulus with similar meaning. E.g., if someone is shown a list of words containing the word "mystery", and then the subject is asked to do a word completion task in which there is the incomplete pattern "_ys_e_y", the probability that the person will select the word "mystery" is higher than if the person had not been primed [19]. This is a form of perceptual priming for the stimulus relate in its form. As an example of conceptual priming, if we consider the concept "fruit", that stimulus will have positive effects on the response to semantically related concepts, such as "apple", leading to a faster response to that stimulus [13]. This can occur even when the first concept is consciously forgotten. Other priming types have been suggested (e.g. associative priming) in which stimulus are not related semantically but are frequently associated or have a high probability of occurring together. Another similar effect is context priming, in which context is used to deal in a faster way with stimulus more likely to occur in the context.

The simulation of the priming effect is central to the modeling of psychological phenomena related to implicit memory such as time-varying cognitive and emotional context, bias, predisposition, prejudice, and many others, such as learning associations and recalling related concepts.

Deese-Roediger-McDermott (DRM) Paradigm: The DRM paradigm is a procedure initialized by J. Deese in the 1950s, and extended by H.L. Roediger and K. McDermott in the 1990s with the aim to study false memory and false recall phenomena [20]. The process typically involves reading a list of words to a subject, being all the words semantically related to a non-present word. E.g., the words "bed", "rest", "awake" and the non-present word "sleep". After hearing the list, the subject is asked to recall the words or to select those words from a new list. In both cases, the non-present target word is recalled with the same frequency as the other words. Also, a high percentage of subjects assure remembering hearing the word, suggesting the occurring of a false memory. This phenomenon is quite similar to the conceptual priming described above, suggesting that the underlying mechanisms in both effects are the same. However, the false memory in the DRM paradigm implies that, in the retrieval phase, the target concept reaches an activation level similar to the other concepts making the subject believe to have heard the word and not only facilitating the response to it.

We will use the modeling of the false memory mechanism to simulate thought-drifting, dreams, delirium, and other divergent cognitive processes that may be necessary to capture a variety of healthy mind processes and psychological pathologies.

Habituation: Habituation is a behavioral response decrease, common in humans and animals, that occurs after repeated exposure to the same stimulus. This process is distinct from sensory adaptation, in which sensory receptors change their sensitivity to the stimulus, and that distinction is demonstrated by the inverse process (dishabituation) and also by stimulus discrimination and spontaneous recovery of the habituation process [7]. The following are some of the characteristics of the habituation process we intend to model. Repeated exposure to a stimulus results in a gradual change of the response to an asymptotic level. In most cases this results in an exponential decrease, but linear habituation can also occur. Before the habituation, a response may show facilitation due to a simultaneous process of sensitization (response amplification to the stimulus). The decrease in habituation can be observed in response frequency, magnitude or both. Habituation is a recoverable process. When given enough time without exposure to the stimulus, the response recovers in a partial or total way (spontaneous recovery). After repeated series of habituation and spontaneous recovery, the habituation is potentiated, i.e. the decrease in response occurs progressively faster and more intensely. Potentiation can also occur by increasing the frequency of the stimulus, that will also result in faster and more intense decrease of response. The frequency of stimulation, after the response reaches an habituated level, has been suggested to determine the rate of recovery [17]. Repeated stimulation in this phase may delay the onset of spontaneous recovery. Associated with this process are also the concepts of stimulus generalization and discrimination. Once a response to a stimulus is habituated, a similar novel stimulus will also have a certain degree of response decrease, according to the rate of similarity between the novel and previous stimulus (generalization). Discrimination is observed when a different stimulus does not have its response altered by the habituation of a previous stimulus. Dishabituation occurs when the presentation of another stimulus results in the recovery of the habituated response of the original stimulus.

Modeling these processes is central for simulating learning processes, surprise, response to, and recovery from, traumatic stress, among others.

2.2 Cognitive Modeling Approaches

Spreading Activation: Spreading Activation is a theory of memory [1] based on Collins and Quillian's computer model [4] which has been widely used for the cognitive modeling of human associative memory and in other domains such as information retrieval [5]. It intends to capture both the way information is represented and how it is processed. According to the theory, long-term memory is represented by nodes and associative links between them, forming a semantic network of concepts. The links are characterized by a weight denoting the associative or semantic relation between the concepts. The model assumes activating one concept implies the spreading of activation to related nodes, making those memory areas more available for further cognitive processing. This activation decays over time, and the further it spreads, which can occur through multiple levels [14], the weaker it is. That is usually modeled using a decaying factor for activation. The method of spreading activation has been central in many cognitive models due to its tractability and resemblance of interrelated groups of neurons in the human brain [18]. The connection between spreading activation theories and priming effects is clear: when an activated node propagates activation to a semantically related node semantic priming effects can be observed. I.e., automatic spreading of activation between concepts are the underlying mechanisms for conceptual priming. As for the creation of false memories it has been discussed whether or not the automatic spreading of activation would be sufficient to originate the high rates of false recall observed in the DRM paradigm. Evidence from [22] leads to the conclusion that the target concept can be activated with such mechanisms.

Hopfield Networks: Invented by John Hopfield in 1982, Hopfield Networks are recurrent neural networks. Each node is a binary threshold unit, i.e. it only assumes two possible values, normally 1 and -1, determined by whether the node's input is above or under its threshold. Each pair of nodes has a connection characterized by a weight w, being the connections symmetrical: $w_{ij} = w_{ji}, \forall i, j$, and a node is not allowed to connect to itself. Hence, the input of a node is the sum of other nodes' states multiplied by the weight of the connection between them.

One of the most interesting properties of Hopfield Networks is their ability to store and retrieve patterns, working as an associative memory. This model uses an energy function to determine the current state of the network, and remembering a learned pattern is achieved by descending a gradient of energy toward a local minimum corresponding to the pattern. Learning patterns results from training the network by lowering the energy of the state that the network should remember. A common way to do this is using Hebbian learning, strengthening the weights of connections between simultaneously activated nodes, and reducing the weight of the connection otherwise. This allows the network to recover a "stored pattern" when given an incomplete version of that pattern. With this training process, Hopfield Networks can store, approximately, $0.14 * n$ patterns, n being the number of nodes [23].

Conceptual Spaces: Traditionally there are two main approaches to the problem of modeling representations in artificial intelligence and cognitive science. One is the symbolic approach, in which information is represented by symbols that when combined give rise to expressions that relate to each other in a logical way. Processing information in the symbolic paradigm corresponds to manipulating symbols, not regarding their semantic content. The other is the connectionist approach, from which artificial neural networks are the prime example. In this approach, cognitive processes are represented by the dynamic activity of patterns of several interconnected units. Peter Gärdenfors argues that none of these approaches can model some aspects of cognitive phenomena and proposes a new way to represent information based on the use of geometrical structures, rather than symbols or connections between neurons. However, this approach is not a substitute for previous approaches, but an intermediate level explaining how symbolist representations can arise from connectionist ones [9]. This level of representation is called conceptual for it provides a way to describe concept formation [10]. The thesis focuses on the existence of conceptual spaces as a way to locate concepts in a domain, being the conceptual space formed by a set of quality dimensions which represent object properties and are used to specify relations between them. Examples for quality dimensions can be height, width, depth, temperature, color, etc. These dimensions are endowed with the appropriate geometrical structure for its representation. For instance, Henning's tetrahedron[6] representation for the human gustatory space could be used for the quality dimension "taste". The spatial location of a point in the conceptual space allows for the calculation of distance between points and the measure of similarity between concepts, which would be impossible to do in a symbolic approach. However, the conceptual spaces theory imposes some constraints on what kinds of subspaces can be considered concepts, namely requiring them to be convex, which may compromise its applicability in general. We need a more general geometric notion of concept and that requires a more powerful way of extracting the features, from particular examples, that define the shape of the concept the examples belong to.

Deep Learning: "Deep learning" is a recent family of machine learning methods that attempt to model high-level abstractions in data by using architectures composed of multiple layers [2]. They usually resort to (restricted) Boltzmann Machines in each layer using a feed forward input layer with no lateral connections. Although deep learning techniques are very powerful indeed for extracting features from complex data and creating new representations for those more abstract concepts, they are not fit for representing the direct relationships between same level concepts, vis-à-vis their lack of "lateral" connections between same layer nodes, which is crucial for the simulation of priming and DRM.

3   Principles of Regulated Activation Networks

The RANs model must be capable of representing and simulating the dynamic cognitive state of an agent, its learning and recall processes, the association of ideas, and the creation of new, more abstract, concepts. We assume these can be broken down into, and emerge from, more basic cognitive phenomena simulated by simpler computational processes. Particularly, in order to simulate the dynamic cognitive state, we need at least 1) a notion of a time-variant activation state of a given concept in the agent's mind; and 2) an adaptive notion of relation and influence between two concepts dependent on their respective activations in a given instant. These two constructs are in principle enough to simulate the Priming phenomenon and, along with a sufficiently time-condensed sequence of activation of concepts, enough to simulate as well the False Memory phenomenon – when several concepts (e.g., representing words) are activated by input, the concepts positively related to them should become more active as well, and if their received combined activation is strong enough, the concept should be sufficiently active to be considered "remembered" by the agent. For the Habituation and Sensitization phenomena we need also our model to 3) be able to dynamically change the parameters controlling its behavior. Finally, we also need 4) a learning mechanism that can create new more abstract concepts as patterns of activation are detected among the existing ones.

The first step of the RANs model development is the creation of a connectionist layer of nodes, each one representing a dimension of a concept – for very simple concepts, a single node might suffice to represent the whole concept. A possible extension of the RANs model consists in establishing a correspondence between individual nodes and concepts in a user-defined ontology. Under this setting, a high activation level of a given node may be interpreted as the detection of an instance of the related ontological concept, and the spreading of activation may afford a kind of inference.

Fig. 1. Initialized single-layer RAN model (connections not shown)

In this way it resembles a semantic network, in which each node has an activation state representing its importance at the moment, and related nodes are associated through a link with a weight representing the strength of the relation. This interpretation of the nodes also allows the representation of a conceptual space where each node stands for a dimension of the space and its activation level corresponds to a particular value along that dimension. This layer of nodes receives input information, responding to the stimulus and learning an internal representation of that stimulus through Hebbian-like weight changing.

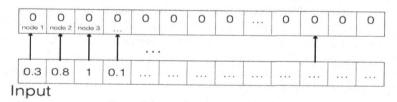

Fig. 2. "Sensory" activation input to nodes in a single-layer RAN model (connections not shown)

Fig. 3. Hebbian learning in a RAN

This Hebbian learning means, from a conceptual spaces perspective, the identification of a correlation between the dimensions of the concept. Also, since each node is connected to every other node in the same layer, it resembles a Hopfield Network, allowing the learning, and the emerging, of certain patterns of nodes' activations which then function as attractors. These attractors, points in the conceptual space, may be seen as prototypical examples of the concept represented by all the points that converge, via the spreading of activation, to the attractor. This representation has the advantage, over

Gärdenfors specification, of not imposing the restriction that all concepts must be convex subspaces. The particular geometric shape of the concept will emerge from the Hebbian learning on the layer.

Within a given cognitive level of abstraction, mechanisms such as spreading activation, and Hebbian correlation identification may be enough for some of our cognitive modeling goals. In particular, a single-layer RANs model (as the one we have implemented and describe herein) only allows us to model the learning of simple correlations between concepts. However, if we wish our RANs model to capture the generalization and abstraction processes involved in higher-level cognitive processes, it must contain some mechanism to progressively build new more abstract concepts into the model. The RANs model will have to combine the "lateral" connectivity typical of models where spreading activation can be applied, with deep learning capabilities for producing new, more abstract layers of concepts. In the future full version of the RANs model a higher learning mechanism will be triggered when, within one layer, the nodes' activation states stabilize: at that moment the RANs model will capture the network state, and represent it by creating a new corresponding node in a new layer of higher abstraction. This capturing can be done via a (restricted) Boltzman Machine, as it is usually the case in deep learning architectures, or any other process that affords the extraction of the relevant features in the pattern. This way, an instance of the RANs model evolves into a deep structured set of layers, each one with a superior level of semantic abstraction, reducing the number of dimensions (a pattern of activation in lower nodes is mapped into a single higher-layer one), potentially drifting apart from connectionism, into a gradually more symbolic representation. With these features, we intend the RANs model to allow for the incremental learning of bottom-up deep representations of progressively more abstract concepts in conceptual spaces. While the semantics of the input "sensory" nodes might be user specified, the new nodes that are eventually created for representing detected patterns, and/or features, of activation may have no immediately obvious semantics. However, it may be easier to recognize complex features/concepts represented in the top level most abstract nodes.

Recalling concepts in the RANs model can be elicited at any desired level of abstraction. The user just has to input activation into any set of nodes and let the activation spread across the RAN until it stabilizes in a fixed-point pattern. When a node N at level n is activated, it spreads its activation, not only to its companion nodes in the same n layer, but also to the nodes in layer $n - 1$ below which correspond to the pattern the node N represents. These in turn repeat the process spreading activation "laterally" to nodes in layer $n - 1$ and also downwards. Whenever the intra-layer spreading of activation causes a stable fixed-point state to emerge, the nodes in the layer above get also activated according to the similarity between the pattern of activation in the layer below they represent and the current pattern active in the layer below. This mechanism thus allows for activation to be spread 1) "upwards" whenever a stable state emerges in a lower layer, 2) "downwards" whenever a node is activated for recall, and 3) "laterally" to nodes in the same layer in all cases. Naturally, all these dynamics depend strongly on the parameters like the decay factor (which controls how much the activation decays inside a node before it is spread), the learning rates (which control how strongly to update the weights when learning), and others.

4 Architecture of the Single-layer RANs model

We now overview the basic characteristics of the single-layer RANs model.

Network Topology: The RANs' topology, at this point, consists of one layer of nodes. We have only tested the model with a full connectivity, but different configurations will be tested in the future, namely: each node linked only to a fixed percentage of the total of nodes; and layers with a small-world connectivity pattern.

Node Properties: Each node has an internal state, represented by its activation level, which ranges from some minimum value to some maximum value and varies continuously in time (the range $[-1, 1]$ was the one implemented, but we are currently experimenting with the range $[0, 1]$). The mean value in the domain is called the rest state. The activation of node at time t is denoted by $A_{ni}(t)$. When a node has a positive activation state (above rest state) we consider it to be active, when it has negative activation (below rest state) we consider it repressed, and if it is equal to the rest state it is considered indifferent. The semantics of the activation values may need to be redefined for the $[0, 1]$ interval.

In the absence of input activation from other nodes or input injected in the network, the activation level of each node gradually converges to the rest state, accordingly with a reposition rate (previously called decay factor), denoted by R_{n_i}, which ranges in $[0, 1]$.

Each node n_i has a threshold variable through time between the minimum and the maximum activation level. Other configurations not including a threshold are currently under consideration.

Activation Propagation: Links between nodes have a weight associated, which represents the importance of the activation from the source node to the next activation level of the target node. The weight of the link from node n_i to node n_j is denoted by w_{ij} and at the beginning all weights are set to zero: $\forall i, j, w_{ij} = 0$. This initialization to zero, drastically different from what happens in traditional feedforward networks where weights are initialized to random values, is inspired by the synaptogenesis process in the human brain where neurons have initially very few connections, and grow new ones as the child grows up.

At each processing step, each node will propagate activation and update its threshold if

$$|A_{n_i}(t)| > |T_{n_i}(t)| \tag{1}$$

In that case, the threshold will be updated according to the formula

$$T_{n_i}(t+1) = T_{n_i}(t) - \delta * \Delta TA \tag{2}$$

where $0 \le \delta \le 1$ is the threshold's learning rate, and $\Delta TA = |A_{n_i}(t) - T_{n_i}(t)|$.

The activation propagated to each node n_j linked with n_i is

$$A_{n_i}(t) * w_{ij} \tag{3}$$

When some node receives activation from its neighbors it is combined with the activation of the node itself. The activation level of node at time $t + 1$ is:

$$A_{n_i}(t + 1) = \lambda_{n_i} * ((1 - R_{n_i}) * A_{n_i}(t)) + (1 - \lambda_{n_i}) * f(\sum_j A_{n_j}(t) * w_{ij}) \tag{4}$$

where $0 \le \lambda_{n_i} \le 1$ is the relative importance the node gives to its own activation versus the activation received from other nodes, and it is called the *solipsism factor*, and $f : \Re \to$ [minimum activation level, maximum activation level] is a sigmoid function, e.g. $f(x) = \frac{1}{1+e^{-\beta x}} * 2 - 1$.

In a similar way, when an input is injected in the network, that component is pondered with the node activation level, therefore being the activation level of node n_i at time $t + 1$ given by the formula:

$$A_{n_i}(t + 1) = \lambda_{n_i} * ((1 - R_{n_i}) * A_{n_i}(t)) + (1 - \lambda_{n_i}) * I(i) \tag{5}$$

where $I(i)$ is the input to node n_i.

Learning: Learning in the RANs model consists of two different processes. The first one is the activation correlation process, in which weights are updated accordingly to the correlation between the nodes' activation level, in a Hebbian influenced learning. The goal is to strengthen the links between nodes with similar levels of activation and to weaken the links otherwise. This way, each time an input is received, for each node n_i, connections with other nodes are updated according to the level of similarity between its activation states pondered by a learning rate. In this case, as activations vary between -1 and 1, the similarity rate is transformed to that same interval. The variation of weights is thus

$$\Delta W_{ij} = \mu_{ac} * (2 * similarityRate - 1) \tag{6}$$

where $0 \le similarityRate = 1 - \dfrac{|A_{ni}(t) - A_{nj}(t)|}{maxDif} \le 1$; $0 \le \mu_{ac} \le 1$ is the learning rate for activation correlation; maxDif = actMax - actMin; actMax and actMin are, respectively, the maximum and minimum level of activation. After that process, the network is given one time instant to spread activation. The inclusion of a second error-driven learning process, similar to a backpropagation, is currently under consideration.

Simulation Procedure: The general procedure for running a simulation with the RANs model comprehends the following steps: Considering N the number of nodes, initialize NxN weight matrix and 1xN activation and threshold vectors to the same value as the rest state. Initialize 1xN reposition rate and solipsism factor vectors and threshold learning rate, activation correlation learning rate and simulation time with its respective values (these are currently being subject to grid-search experimentation). Schedule a set of 1xN input patterns and the time steps for their injection. Initialize time variable to 1.

Until time reaches the established simulation time, run the following execution cycle: In case of a pattern scheduled to that time instant, apply equation 5 to calculate the new level of activation for the nodes and equation 6 to update the weights according to the current activation level. Otherwise, apply equation 4 to update the nodes' activation level according to the output generated in the previous time step by the same nodes. After that, for each node to which the condition 1 applies, calculate its activation output using equation 3 and update its threshold using equation 2. In the future, these outputs may be subject to some transformation (e.g. by regulatory mechanisms) before being used as inputs. Finally, collect any desired data (current activation levels, weights, etc.) and update the time variable.

The execution cycle is summarized in the following pseudo code:

Algorithm 1. RANs cycle

```
while time < simulation time do
    if isPatternScheduled(time) then
        inject input pattern (5)
        do activation correlation learning process (6)
    else
        activation input (4)
    end if
    for each node do
        if node activation above threshold then (1)
            calculate node output (3)
            update threshold (2)
        end if
    end for
    collect data
    time ← time + 1
end while
```

---

5 Preliminary Results

The results herein shown regard a specific parametrization of the model. However, the architecture of the RANs model will be subject to detailed exploration and experimentation concerning its topological properties and the influence of different types of connectivity, learning, regulation processes and parameters on the network's behavior and utility to the modeling of the intended phenomena. Still, these preliminary results serve as an appetizer for the model's capabilities while providing some insight on how we are dealing with the input patterns.

The simulations used the following parameters: Num. of nodes: 50; Connectivity: Total; Activations in $[-1, 1]$; Weights in $[-\infty, \infty]$; $\forall i, R_{ni} = 0.05$; $\forall i, \lambda_{ni} =$ Random value (uniform distribution over $[0, 1]$); $\forall i, A_{ni}(0) = 0$; $\forall i, \delta_{ni} = 0.1$.

The simulation process involves generating a random pattern, and injecting it periodically (in this case, each 100 time steps) in the network. Fig. 4 shows how nodes' activation evolve through time (each line represents a single node activation state). From

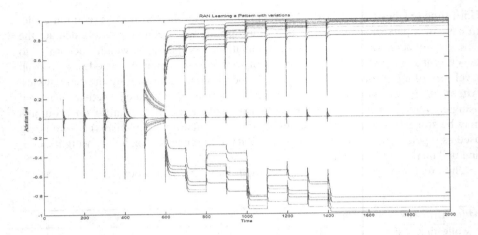

Fig. 4. RANs preliminary experiments with learning a pattern

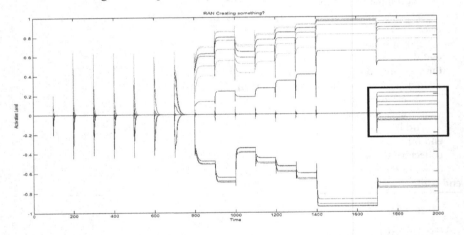

Fig. 5. RANs preliminary experiments with creative processes – "new zone" boxed

a global perspective, we observe that when an input is injected the network has an immediate response (for that input is directly injected in the nodes) and the activation is reposed to the rest state value (in this case, 0). Each time the pattern is injected, a round of Hebbian updating of weights takes place, and as a consequence of that learning, the reposition phase becomes progressively longer, i.e, the network takes longer to return to a neutral cognitive state. When an input is given in a non-neutral state ($t = 600$) the network alters its behavior and activations start to converge to a stable fixed-point state. Subsequent pattern injections only reveal minor changes in the value for each node activation, and can be considered as slight adjustments to the representation of the pattern previously learned.

A first attempt at simulating creativity was experimented as well. The process was very similar to the previous one, with the exception that at a time step where the network has already converged to a stable state ($t = 1400$, fig. 5) we stopped injecting the input patterns, and at $t = 1700$ a new random pattern was inserted. The interest in fig. 5 comes from the difference between the activations before and after that moment. Besides some minor changes to some nodes' activations, we observe a totally new "zone of activation" that was previously neutral. It is premature to assume that the production of the new RANs' state can be described as creative, but the fact that our model can integrate two different states in a new representation can be a good starting point for modeling creative processes.

6   Conclusions and Future Work

Prevalence of psychological and psychiatric diseases, such as depression, schizophrenia and others, is a growing concern in the industrialized world. Computational cognitive models can help in understanding and simulating mental processes, both healthy and pathological ones, hopefully contributing to the diagnosis and treatment of the latter. Also, there is a recent growing interest in understanding and potentiating creative processes, both in humans and in computers. For these reasons, we put forward the desiderata, and first simplified version, of our Regulated Activation Networks model, a new computational cognitive model capable of simulating a variety of psychological phenomena including, among others, Priming, False Memory, Habituation, different kinds of Learning, and Memory. The current preliminary version of the RANs model has only one fully connected layer of nodes representing concepts, uses a Hebbian learning rule, and resorts to spreading activation as means for inference and recall.

We are also currently experimenting with a probabilistic approach at the Hebbian learning of weights and the spreading of activation. Future work includes implementing the full version of the model with a deep learning mechanism that will afford the dynamic creation of new nodes representing more abstract concepts, as well as the implementation of regulatory parameters and mechanisms. We will develop and execute a validation plan in collaboration with psychologists, in order to assess how realistic and reliable are the simulations with the RANs model.

Acknowledgements. This work is part of our contribution to the project ConCreTe. The project ConCreTe acknowledges the financial support of the Future and Emerging Technologies (FET) programme within the Seventh Framework Programme for Research of the European Commission, under FET grant number 611733.

References

1. Anderson, J.R.: A spreading activation theory of memory. J. of Verbal Learning and Verbal Behavior 22(3), 261–295 (1983)
2. Bengio, Y., Courville, A.C., Vincent, P.: Representation learning: A review and new perspectives. IEEE Trans. Pattern Anal. Mach. Intell. 35(8), 1798–1828 (2013)
3. Braver, T.S., Barch, D.M., Cohen, J.D.: Cognition and control in schizophrenia: A computational model of dopamine and prefrontal function. Bio. Psychiatry 46, 312–328 (1999)

4. Collins, A.M., Quillian, M.R.: Retrieval time from semantic memory. J. of Verbal Learning and Verbal Behavior 8(2), 240–247 (1969)
5. Crestani, F.: Application of spreading activation techniques in information retrieval. Artificial Intelligence Review 11, 453–482 (1997)
6. Erickson, R.P., Ohrwall, H., von Skramlik, E., Henning, H.: Ohrwall, Henning and von Skramlik; the foundations of the four primary positions in taste. Neurosci. Biobehav. Rev. 8(1), 105–127 (1984)
7. Rankin, C.H., et al.: Habituation revisited: an updated and revised description of the behavioral characteristics of habituation. Neurobiol. Learn. Mem. 92(2), 135–138 (2009)
8. Wittchen, H.U., et al.: The size and burden of mental disorders and other disorders of the brain in Europe 2010. European Neuropsychopharmacology 21(9), 655–679 (2011)
9. Gärdenfors, P.: Symbolic, conceptual and subconceptual representations. In: Human and Machine Perception: Information Fusion, pp. 255–270 (1997)
10. Gärdenfors, P.: Conceptual spaces as a framework for knowledge representation. Mind and Matter 2, 9–27 (2004)
11. Jacoby, L.L.: Perceptual enhancement: persistent effects of an experience. J. Exp. Psychol. Learn. Mem. Cogn. 9(1), 21–38 (1983)
12. Kyaga, S., Landén, M., Boman, M., Hultman, C.M., Långström, N., Lichtenstein, P.: Mental illness, suicide and creativity: 40-year prospective total population study. J. of Psychiatric Research 47(1), 83–90 (2013)
13. Matsukawa, J., Snodgrass, J.G., Doniger, G.M.: Conceptual versus perceptual priming in incomplete picture identification. J. Psycholinguist. Res. 34(6), 515–540 (2005)
14. McNamara, T.P., Altarriba, J.: Depth of spreading activation revisited: Semantic mediated priming occurs in lexical decisions. J. of Memory and Language 27(5), 545–559 (1988)
15. O'Reilly, R.C.: Biologically based computational models of high-level cognition. Science 314(5796), 91–94 (2006)
16. World Health Organization. Global burden of mental disorders and the need for a comprehensive, coordinated response from health and social sectors at the country level (December 2011)
17. Rankin, C.H., Broster, B.S.: Factors affecting habituation and recovery from habituation in the nematode Caenorhabditis elegans. Behav. Neurosci. 106(2), 239–249 (1992)
18. Roediger, H.L., Balota, D.A., Watson, J.M.: Spreading activation and arousal of false memories. In: The Nature of Remembering: Essays in Honor of Robert G. Crowder, pp. 95–115 (2001)
19. Roediger, H.L., Blaxton, T.A.: Effects of varying modality, surface features, and retention interval on priming in word-fragment completion. Mem. & Cognition 15(5), 379–388 (1987)
20. Roediger, H.L., Mcdermott, K.B.: Creating false memories: Remembering words not presented in lists. Journal of Experimental Psychology: Learning, Memory, and Cognition 21(4), 803–814 (1995)
21. Rolls, E., Loh, M., Deco, G., Winterer, G.: Computational models of schizophrenia and dopamine modulation in the prefrontal cortex. Nat. Rev. Neurosci. 9(9), 696–709 (2008)
22. Seamon, J.G., Luo, C.R., Gallo, D.A.: Creating false memories of words with or without recognition of list items: Evidence for nonconscious processes. Psychological Science 9(1), 20–26 (1998)
23. Wei, G., Yu, Z.: Storage capacity of letter recognition in hopfield networks

Computing Concept Lattices
from Very Sparse Large-Scale Formal Contexts

Lenka Pisková and Tomáš Horváth

University of Pavol Jozef Šafárik, Košice, Slovakia
{lenka.piskova,tomas.horvath}@upjs.sk

Abstract. This paper introduces a new algorithm for computing concept lattices from very sparse large-scale formal contexts (input data) where the number of attributes per object is small. The algorithm consists of two steps: generate a diagram of a formal context and compute the concept lattice of the formal context using the diagram built in the previous step. The algorithm is experimentally evaluated and compared with algorithms AddExtent and CHARM-L.

1 Introduction

A well-known issue in the area of Formal Concept Analysis (FCA) is its computational complexity, i.e. the number of concepts grows exponentially with the amount of data, which are in the form of a binary relation. In our research, we were motivated by large-scale data sets where the maximum number of attributes per object is small. Examples of such data are drug prescriptions databases, jewelery shopping transactions, book store transactions, etc. All these relations (tables) share similar characteristics, namely that there are hundreds of thousands or millions of objects (rows) and thousands of attributes/items (columns) but the data is very sparse, i.e. despite a very large number of items, a drug prescription contains usually only a few drugs, people do not buy dozens of jewels at once as well as they buy only a few books during a visit of the bookstore, etc. We present an algorithm for computing formal concepts from very sparse large-scale data in this paper and show some of its characteristics.

1.1 Formal Concept Analysis [8]

A *formal context* is a triple (X, Y, R) consisting of a set X of objects, a set Y of attributes and an incidence relation $R \subseteq X \times Y$ between them. We write $(x, y) \in R$ to express that the object x has the attribute y. For a set $A \subseteq X$ of objects and a set $B \subseteq Y$ of attributes we define $A^{\uparrow R} = \{y \in Y : (\forall x \in A)(x, y) \in R\}$ and $B^{\downarrow R} = \{x \in X : (\forall y \in B)(x, y) \in R\}$. $A^{\uparrow R}$ is the set of attributes common to the objects in A and $B^{\downarrow R}$ is the set of objects which have all the attributes in B. A *formal concept* of (X, Y, R) is a pair (A, B) where $A \subseteq X, B \subseteq Y, A^{\uparrow R} = B$ and $B^{\downarrow R} = A$. A and B are called the *extent* and the *intent* of (A, B), respectively. The set of all concepts of (X, Y, R) is denoted by $\mathcal{B}(X, Y, R)$. $A \subseteq X$ $(B \subseteq Y)$

N. Hernandez et al. (Eds.): ICCS 2014, LNAI 8577, pp. 245–259, 2014.
DOI: 10.1007/978-3-319-08389-6_20, © Springer International Publishing Switzerland 2014

is an extent (intent) if and only if $A^{\uparrow_R \downarrow_R} = A$ ($B^{\downarrow_R \uparrow_R} = B$). We define a partial order $\leq$ on $\mathcal{B}(X, Y, R)$ by $(A_1, B_1) \leq (A_2, B_2)) \Leftrightarrow A_1 \subseteq A_2$ (equivalently, $B_1 \supseteq B_2$). The set of all concepts of (X, Y, R) ordered by $\leq$ constitutes the *concept lattice* $(\mathcal{B}(X, Y, R), \leq)$ of (X, Y, R).

Example 1: Table 1 shows a formal context, which induces 11 formal concepts.

Table 1. The formal context (left) and all formal concepts of the context (right)

	J	K	L	M	N
1	×		×	×	
2	×			×	
3		×	×	×	
4		×			×
5	×	×			

Formal concepts		
$(\{1, 2, 3, 4, 5\}, \emptyset)$	$(\{5\}, \{J, K\})$	$(\{1\}, \{J, L, M\})$
$(\{1, 2, 5\}, \{J\})$	$(\{1, 2\}, \{J, M\})$	$(\{3\}, \{K, L, M\})$
$(\{3, 4, 5\}, \{K\})$	$(\{4\}, \{K, N\})$	$(\emptyset, \{J, K, L, M, N\})$
$(\{1, 2, 3\}, \{M\})$	$(\{1, 3\}, \{L, M\})$	

1.2 Related Work

The major problem in computing formal concepts is that some concepts are computed multiple times which significantly slows down the computation. One possible solution is to generate concepts in a specific order, e.g. Ganter's NextClosure algorithm [7] is based on the lexicographical ordering of formal concepts.

Recently, attention has been paid to the CbO algorithm [11] leading to various modifications of it [2, 3, 10]. The order of attributes in a formal context (input table) can yield different CbO trees, and the number of concepts that are computed multiple times is different. The algorithms based on CbO significantly outperform other algorithms. In the competition[1] between algorithms for mining formal concepts, FCbO [10] took first place and the runner-up was In-Close [2]. Next, we focus on algorithms that generate all formal concepts and Hasse diagrams of concept lattices [4–6, 9, 13, 14, 16, 17, 19].

Lindig's algorithm [13] utilizes the idea that if (A, B) is a concept, then for each $x \in X \setminus A$ it holds that $((A \cup \{x\})^{\uparrow_R \downarrow_R}, (A \cup \{x\})^{\uparrow_R})$ is a concept which is greater than (A, B) and deals with the fact that $((A \cup \{x\})^{\uparrow_R \downarrow_R}, (A \cup \{x\})^{\uparrow_R})$ is not necessary an upper neighboor of (A, B).

The algorithm proposed in [16] constructs a lexicographic tree of concepts used to build the diagram graph of a concept lattice. It has the smallest computational complexity, but experiments have shown that other algorithms outperform it [12]. The incremental version of the algorithm was proposed in [17].

Another approach to solve the problem that some concepts are generated multiple times is to search for a new concept in the space of already computed concepts, but this is time consuming. One way to speed up the computation is to reduce the search space by dividing the set of concepts into disjoint sets.

The incremental algorithm proposed by Godin [9] divides the set of concepts into buckets according to the cardinality of their intents. The main idea of the

---

[1] http://www.upriss.org.uk/fca/fcaalgorithms.html

algorithm is that if $(C, D) = \inf \{(A, B) \in \mathcal{B}_{(X,Y,R)} : D = B \cap \{x\}^{\uparrow_R}\}$ and there is no concept of the form $(E, D) \in \mathcal{B}_{(X,Y,R)}$, then (A, B) is the generator of a new concept $(C, D) = (A \cup \{x\}, B \cap \{x\}^{\uparrow_R})$.

The drawback of the Godin algorithm is that it iterates through almost all concepts from $\mathcal{B}_{(X,Y,R)}$ so it would be appropriate to restrict the set of concepts. AddIntent [14] deals with this problem outperforming other algorithms for building lattices ([4, 7, 9, 15, 16]) except for sparse contexts where Bordat algorithm [4] is the fastest one and AddIntent [14] a close second.

FCA and Frequent Itemset Mining are two interconnected areas. The problem of computing (intents of) all formal concepts is equivalent to the problem of mining all closed frequent itemsets using a minimum support equal to zero [18]. The Titanic algorithm [19] for building lattices was inspired by the Apriori algorithm [1] for mining frequent itemsets.

Finally, CHARM (and CHARM-L) [20] explores the itemset and tidset space over an IT-tree and exploits some properties of itemset-tidset pairs for fast identification of frequent closed itemsets. The algorithm uses a vertical data representation called diffsets. To our knowledge, CHARM-L is the only algorithm for generating frequent closed itemsets along with their lattice structure.

2   The Algorithm

For the purposes of describing the algorithm it is useful to view each set of objects $A \in 2^X$ as a word in the alphabet X. We can define a lexicographic order on the set 2^X. This allows us to write $\{x_1, \ldots, x_k\} \in 2^X$ as a word $x_1 \ldots x_k$ where $x_1 < x_2 < \cdots < x_k$. Similarly for sets of attributes, we can write $y_1 \ldots y_l$ where $y_1 < y_2 < \cdots < y_l$ instead of $\{y_1, \ldots, y_l\} \in 2^Y$.

The basic structure the algorithm constructs is a diagram $\mathcal{D}_{(X,Y,R)}$, an example of which is shown in Fig. 1. Each node of the diagram (element of $\mathcal{N}_{(X,Y,R)}$) is a pair $(B^{\downarrow_R}, B)$ consisting of a set of objects $B^{\downarrow_R} \subseteq X$ and a set of attributes $B \subseteq Y$ such that $B \subseteq \{x\}^{\uparrow_R}$ for some $x \in X$. Observe that all nodes in $\mathcal{D}_{(X,Y,R)}$ are different from each other, but there are nodes having the same object parts in $\mathcal{D}_{(X,Y,R)}$. For each node $(B^{\downarrow_R}, B)$ it holds that $B^{\downarrow_R} \neq \emptyset$ except for the so-called bottom node $(\emptyset^{\uparrow_R \downarrow_R}, \emptyset^{\uparrow_R})$ (the node $(\emptyset, JKLMN)$ in Fig. 1) which is added to the diagram at the end of the construction of $\mathcal{D}_{(X,Y,R)}$ in case $\emptyset^{\uparrow_R \downarrow_R} = \emptyset$ (otherwise, the node $(\emptyset^{\uparrow_R \downarrow_R}, \emptyset^{\uparrow_R})$ is in the diagram already).

After adding the bottom node $(\emptyset^{\uparrow_R \downarrow_R}, \emptyset^{\uparrow_R})$ to the diagram of a formal context, this diagram contains all formal concepts of the formal context. Obviously, the diagram can still contain nodes that are not formal concepts, e.g. $(1, JL)$.

The set of nodes $\mathcal{N}_{(X,Y,R)}$ is partially ordered by the $\leq$ relation. The partial order relation $\leq$ on $\mathcal{N}_{(X,Y,R)}$ is defined as follows: $(A_1, B_1) \leq (A_2, B_2)$ if and only if $B_1 \supseteq B_2$. Next, we define the cover relation on $\mathcal{N}_{(X,Y,R)}$ for $\leq$. For two nodes $(A_1, B_1), (A_2, B_2) \in \mathcal{N}_{(X,Y,R)}$, if $(A_1, B_1) \leq (A_2, B_2)$ and there is no node $(A_3, B_3) \in \mathcal{N}_{(X,Y,R)}$ distinct from both (A_1, B_1) and (A_2, B_2) such that $(A_1, B_1) \leq (A_3, B_3) \leq (A_2, B_2)$, then (A_1, B_1) is called a lower cover (or a child) of (A_2, B_2) and (A_2, B_2) is called an upper cover (or a parent) of (A_1, B_1).

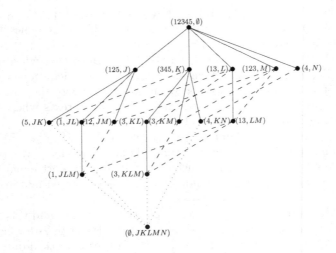

Fig. 1. The diagram $\mathcal{D}_{(X,Y,R)}$ of the formal context in Table 1. Edges (elements of $\mathcal{E}_{(X,Y,R)}$) represent the cover relation between nodes (elements of $\mathcal{N}_{(X,Y,R)}$).

Each (oriented) edge (element of $\mathcal{E}_{(X,Y,R)}$) of the diagram represents the cover relation, i.e. there is an edge between the nodes $(A_1, B_1), (A_2, B_2)$ in the diagram $\mathcal{D}_{(X,Y,R)}$ if and only if (A_1, B_1) is a lower cover of (A_2, B_2). After the node $(\emptyset^{\uparrow_R\downarrow_R}, \emptyset^{\uparrow_R})$ is added to $\mathcal{N}_{(X,Y,R)}$ (provided that $\emptyset^{\uparrow_R\downarrow_R} = \emptyset$), it is connected by an edge to all nodes that do not have any child in the diagram, what is indicated by dotted lines.

Considering all nodes $(B^{\downarrow_R}, B) \in \mathcal{N}_{(X,Y,R)}$ such that $B^{\downarrow_R} \neq \emptyset$ and only the solid lines between the nodes, the diagram in Fig. 1 becomes a tree. All nodes at a certain level (depth) of the tree have the same cardinality of attribute parts. This tree is in fact a lexicographic tree similar to the one used by CHARM[2] [20].

The algorithm we present uses a two step approach:

1. Generate a diagram $\mathcal{D}_{(X,Y,R)}$ from the formal context (X, Y, R)
2. Construct the concept lattice $(\mathcal{B}(X, Y, R), \leq)$ of (X, Y, R) using $\mathcal{D}_{(X,Y,R)}$

2.1   Algorithm to Generate $\mathcal{D}_{(X,Y,R)}$

We first present how to modify the diagram (the set of nodes and the set of edges) of a formal context when introducing a new object into the context.

Let us suppose we are adding to a formal context (X, Y, R) a new object $x \notin X$ having attributes $\{y_1, \ldots, y_n\}$. We do not assume any overlap of Y and $\{y_1, \ldots, y_n\}$, i.e. $\{y_1, \ldots, y_n\}$ can contain attributes not present in Y. Denote the

---

[2] The itemset-tidset search tree (IT-tree) of CHARM contains also nodes $(B^{\downarrow_R}, B)$ where $B^{\downarrow_R} = \emptyset$, e.g. $JKLM$ and KMN, however, by using some properties of IT-pairs, CHARM can quickly enumerate all closed frequent itemsets.

incidence relation between $\{x\}$ and $\{y_1, \ldots, y_n\}$ by $R_x \subseteq \{x\} \times \{y_1, \ldots, y_n\}$ and the new formal context with the object $\{x\}$ added by the triplet (X', Y', R'), where $X' = X \cup \{x\}$, $Y' = Y \cup \{y_1, \ldots, y_n\}$, and $R' \subseteq X' \times Y'$ such that $R' \cap (X \times Y) = R$, $R' \cap (\{x\} \times \{y_1, \ldots, y_n\}) \subseteq R_x$ and $R' \cap (X \times (\{y_1, \ldots, y_n\} \setminus Y)) = R' \cap (\{x\} \times (Y \setminus \{y_1, \ldots, y_n\})) = \emptyset$.

Observe that if $B \subseteq Y \setminus \{y_1, \ldots, y_n\} \neq \emptyset$, then $B^{\downarrow R'} = B^{\downarrow R}$, and if $B' \subseteq \{y_1, \ldots, y_n\} \setminus Y \neq \emptyset$, then $B'^{\downarrow R'} = B'^{\downarrow R_x} = \{x\}$. Otherwise, for any $B' \subseteq Y \cap \{y_1, \ldots, y_n\}$ it holds that $B'^{\downarrow R'} = B'^{\downarrow R} \cup \{x\}$. Further, if $A \subseteq X$, then $A^{\uparrow R'} = A^{\uparrow R}$. Obviously, $\{x\}^{\uparrow R'} = \{x\}^{\uparrow R_x} = \{y_1, \ldots, y_n\}$.

From the above presented ideas, for any $B' \subseteq \{y_1, \ldots, y_n\}$, if there is a node $(A, B) \in \mathcal{N}_{(X,Y,R)}$ such that $B' = B$, then the node $(A', B') \in \mathcal{N}_{(X',Y',R')}$ with the same attribute part $B' = B$ as the node (A, B) is a modified node enlarging the object part of (A, B) by the object x. Otherwise, if $B' \supset B$, then the node $(A', B') = (\{x\}, B') \in \mathcal{N}_{(X',Y',R')}$ is a new node. Finally, if $B \subseteq Y \setminus \{y_1, \ldots, y_n\} \neq \emptyset$, then the node (A, B) is old.

In addition, the edges of the diagram $\mathcal{D}_{(X',Y',R')}$ have to be updated. Each new node in $\mathcal{N}_{(X',Y',R')}$ can be connected by an edge only to modified or new nodes in $\mathcal{N}_{(X',Y',R')}$ (never to old nodes). If a node (A, B) is new in $\mathcal{N}_{(X',Y',R')}$, then (A', B') will be a child of a node (C', D') in the diagram $\mathcal{D}_{(X',Y',R')}$ if and only if $B' \supset D'$ and $|B'| = |D'| + 1$.

Remark 1. The new object x "adds" to the diagram $\mathcal{D}_{(X,Y,R)}$ a Boolean algebra, i.e. modified and new nodes in $\mathcal{D}_{(X',Y',R')}$ form a Boolean algebra.

Lemma 1. *Let (X, Y, R) be a formal context and (X', Y', R') be the formal context originated from (X, Y, R) by adding a new object $x \notin X$ having attributes $\{y_1, \ldots, y_n\}$. If $\mathcal{B}(X, Y, R) \subseteq \mathcal{N}_{(X,Y,R)} \cup \{(\emptyset^{\uparrow R \downarrow R}, \emptyset^{\uparrow R})\}$, then $\mathcal{B}(X', Y', R') \subseteq \mathcal{N}_{(X',Y',R')} \cup \{(\emptyset^{\uparrow R' \downarrow R'}, \emptyset^{\uparrow R'})\}$.*

Proof. $\mathcal{B}(X', Y', R')$ can be obtained from $\mathcal{B}(X, Y, R)$ by taking all formal concepts in $\mathcal{B}(X, Y, R)$ and modifying the extent of the formal concepts $(A, B) \in \mathcal{B}(X, Y, R)$ for which $B \subseteq \{y_1, \ldots, y_n\}$ by adding x. The concepts that remain intact are called old and the others, modified concepts in $\mathcal{B}(X', Y', R')$. In addition, new concepts are created. These new concepts are always in the form $(A \cup \{x\}, B \cap \{y_1, \ldots, y_n\})$ for some formal concept $(A, B) \in \mathcal{B}(X, Y, R)$. [9] Note that, obviously, $B \cap \{y_1, \ldots, y_n\} \subseteq \{y_1, \ldots, y_n\}$.

For any $B' \subseteq \{y_1, \ldots, y_n\}$ it holds that $(A', B') \in \mathcal{N}_{(X',Y',R')}$, i.e. (A', B') is modified or new node in $\mathcal{N}_{(X',Y',R')}$, and thus $x \in A'$ (see above).

From the assumption $\mathcal{B}(X, Y, R) \subseteq \mathcal{N}_{(X,Y,R)} \cup \{(\emptyset^{\uparrow R \downarrow R}, \emptyset^{\uparrow R})\}$ and the previous considerations it follows that $\mathcal{B}(X', Y', R') \subseteq \mathcal{N}_{(X',Y',R')} \cup \{(\emptyset^{\uparrow R' \downarrow R'}, \emptyset^{\uparrow R'})\}$. □

The procedure *updateDiagram* depicted in Algorithm 1 computes the diagram $\mathcal{D}_{(X',Y',R')}$ of the formal context (X', Y', R') when adding a new object x having attributes $\{y_1, \ldots, y_n\}$ into the diagram $\mathcal{D}_{(X,Y,R)}$. The procedure accepts as its arguments a diagram $\mathcal{D} = (\mathcal{N}, \mathcal{E})$ of a formal context (X, Y, R), a new object x and attributes $\{y_1, \ldots, y_n\}$ of the object x. The procedure outputs the updated diagram $\mathcal{D} = (\mathcal{N}, \mathcal{E})$.

Algorithm 1. Procedure $updateDiagram(\mathcal{D}, x, \{y_1, \ldots, y_n\})$

---

1. $i = 1$
2. $array[0] \leftarrow \emptyset$
3. **for all** $y \in \{y_1, \ldots, y_n\}$ **do**
4.   **for** j **from** 0 **to** $i - 1$ **do**
5. $array[i + j] \leftarrow array[j] \cup \{y\}$
6. $B \leftarrow array[i + j]$
7.   **if** there is A such that $(A, B) \in \mathcal{N}$ **then**
8. $A \leftarrow A \cup \{x\}$
9.   **else**
10. $A \leftarrow \{x\}$
11.   add node (A, B) to $\mathcal{N}$
12.   **for all** $(C, D) \in \mathcal{N}, B \supseteq D, |B| = |D| + 1$ **do**
13.   add edge $(A, B) \rightarrow (C, D)$ to $\mathcal{E}$
14. $i \leftarrow i * 2$

The key step when processing the object x having attributes $\{y_1, \ldots, y_n\}$ is the generation of all subsets of $\{y_1, \ldots, y_n\}$ by using local variables $array$ as a temporary storage for generated subsets of the set of attributes $\{y_1, \ldots, y_n\}$.

After a new subset of $\{y_1, \ldots, y_n\}$ is generated (line 5), the content of B is updated (line 6). Observe that B is the copy of this subset. The procedure checks if there is a node in the diagram $\mathcal{D}$ having B as its attribute part (line 7). The node is denoted by (A, B). If the test succeed, the node (A, B) is modified by augmenting A by the object x (line 8). If the test in line 7 fails, a new node $(\{x\}, B)$ is added to the diagram $\mathcal{D}$ (lines 10, 11) and it is connected to the corresponding nodes (lines 12-13).

Observe the order in which the subsets of $\{y_1, \ldots, y_n\}$ are generated. If D is a subset of B, then D has been generated before B. Therefore, whenever a new node (A, B) is created, it can be connected to all nodes (C, D) such that $B \supseteq D$ and $|B| = |D| + 1$ (because all these nodes were created a step before).

Example 2: We illustrate Algorithm 1 on the following example: Consider the first two objects (rows) of the formal context in the table 1. The diagram $\mathcal{D}$ of the correspoding formal context contains the following 8 nodes: $N_1 = (12, \emptyset)$, $N_2 = (12, J)$, $N_3 = (1, L)$, $N_4 = (1, JL)$, $N_5 = (12, M)$, $N_6 = (12, JM)$, $N_7 = (1, LM)$, $N_8 = (1, JLM)$, and, the following 12 edges: $N_2 \rightarrow N_1$, $N_3 \rightarrow N_1$, $N_4 \rightarrow N_2$, $N_4 \rightarrow N_3$, $N_5 \rightarrow N_1$, $N_6 \rightarrow N_2$, $N_6 \rightarrow N_5$, $N_7 \rightarrow N_3$, $N_7 \rightarrow N_5$, $N_8 \rightarrow N_4$, $N_8 \rightarrow N_6$, $N_8 \rightarrow N_7$. When processing the third object (row), the following 4 new nodes are added: $N_9 = (3, K)$, $N_{10} = (3, KL)$, $N_{11} = (3, KM)$, $N_{12} = (3, KLM)$, and, the following 8 new edges are created: $N_9 \rightarrow N_1$, $N_{10} \rightarrow N_9$, $N_{10} \rightarrow N_3$, $N_{11} \rightarrow N_9$, $N_{11} \rightarrow N_5$, $N_{12} \rightarrow N_9$, $N_{12} \rightarrow N_3$, $N_{12} \rightarrow N_5$. In this step, the node N_1 is modified to $N_1 = (123, \emptyset)$, N_3 to $N_3 = (13, L)$, N_5 to $N_5 = (123, M)$ and N_7 to $N_7 = (13, LM)$.

Algorithm 1 can be easily used to generate the diagram $\mathcal{D}_{(X,Y,R)}$ of a formal context (X, Y, R) incrementally, i.e. processing objects of the formal context one by one. The function $generateDiagram$ in Algorithm 2 accepts as its argument a formal context (X, Y, R) and outputs the diagram $\mathcal{D} = (\mathcal{N}, \mathcal{E})$ of (X, Y, R).

Algorithm 2. Function $generateDiagram((X, Y, R))$

---

1. $\mathcal{N} \leftarrow \emptyset$
2. $\mathcal{E} \leftarrow \emptyset$
3. **for all** $x \in X$ **do**
4. $updateDiagram(\mathcal{D}, x, \{x\}^{\uparrow R})$
5. **if** $(\emptyset^{\uparrow R \downarrow R}, \emptyset^{\uparrow R}) \notin \mathcal{N}$ **then**
6. $\mathcal{N} \leftarrow \mathcal{N} \cup \{(\emptyset^{\uparrow R \downarrow R}, \emptyset^{\uparrow R})\}$
7.   **for all** $(A, B) \in \mathcal{N}$ **do**
8.   **if** there is no $(C, D) \rightarrow (A, B)$ in $\mathcal{E}$ **then**
9.   add edge $(\emptyset^{\uparrow R \downarrow R}, \emptyset^{\uparrow R}) \rightarrow (A, B)$ to $\mathcal{E}$
10. **return** $\mathcal{D}$

Initially, the diagram $\mathcal{D}$ is empty, i.e. contains no nodes nor edges (lines 1, 2). Then, $updateDiagram$ is invoked repeatedly for each object $x \in X$ (lines 3 - 4). Finally, if there is the node $(\emptyset^{\uparrow R \downarrow R}, \emptyset^{\uparrow R})$ in $\mathcal{D}$, i.e. the test in line 5 fails, the diagram $\mathcal{D}$ is the diagram of the formal context (X, Y, R). Otherwise, i.e. the test in line 5 pass, the node $(\emptyset^{\uparrow R \downarrow R}, \emptyset^{\uparrow R})$ is added to the diagram $\mathcal{D}$ (line 6) and it is connected to all nodes in $\mathcal{D}$ that do not have any child in $\mathcal{D}$ (lines 7 - 9).

Remark 2. The outcome of $generateDiagram$ function in Algorithm 2 does not depend on the order in which the objects of the formal context are acquired.

Theorem 1. *The function generateDiagram in Algorithm 2 is correct, i.e. given a formal context (X, Y, R), it stops after finitely many steps and returns the diagram $\mathcal{D}_{(X,Y,R)}$ that satisfies the following conditions:*

1. *$\mathcal{D}_{(X,Y,R)}$ contains all formal concepts of the formal context (X, Y, R), i.e. $\mathcal{B}(X, Y, R) \subseteq \mathcal{N}_{(X,Y,R)}$.*
2. *The diagram $\mathcal{D}_{(X,Y,R)}$ is the Hasse diagram of a complete lattice.*

Proof. 1. Obviously, $(\emptyset^{\uparrow R \downarrow R}, \emptyset^{\uparrow R}) \in \mathcal{N}_{(X,Y,R)}$. The theorem is therefore a direct consequence of Lemma 1.

2. We prove that for each $(A, B), (C, D) \in \mathcal{N}_{(X,Y,R)}$ there exists a greatest lower bound (a) and a least upper bound (b).

a First, we show that for each $(A, B), (C, D) \in \mathcal{N}_{(X,Y,R)}$ there exists $(E, F) \in \mathcal{N}_{(X,Y,R)}$ such that $(E, F) \leq (A, B)$ and $(E, F) \leq (C, D)$.
Let $F = B \cup D$. If $(E, F) \in \mathcal{N}_{(X,Y,R)}$, then $(E, F) \leq (A, B)$ and $(E, F) \leq (C, D)$ (because $F \supseteq B$ and $F \supseteq D$). Moreover, for each $(A, B), (C, D) \in \mathcal{N}_{(X,Y,R)}$ it holds if $(G, H) \leq (A, B)$ and $(G, H) \leq (C, D)$, then $(G, H) \leq (E, F)$ (if $H \supseteq B$ and $H \supseteq D$, then $H \supseteq B \cup D = F$). Otherwise, if $(E, F) \notin \mathcal{N}_{(X,Y,R)}$, then the greatest lower bound of (A, B) and (C, D) is $(\emptyset^{\uparrow R \downarrow R}, \emptyset^{\uparrow R}) \in \mathcal{N}_{(X,Y,R)}$.

b Next, we show that for each $(A, B), (C, D) \in \mathcal{N}_{(X,Y,R)}$ there exists $(E, F) \in \mathcal{N}_{(X,Y,R)}$ such that $(A, B) \leq (E, F)$ and $(C, D) \leq (E, F)$.
Let $F = B \cap D$. Then, $(A, B) \leq (E, F)$ and $(C, D) \leq (E, F)$ ($B \supseteq F$ and $D \supseteq F$). Moreover, from the fact that $(A, B) \in \mathcal{N}_{(X,Y,R)}$ and $F \subseteq B$ it follows that $(E, F) \in \mathcal{N}_{(X,Y,R)}$. Further, for each $(A, B), (C, D) \in \mathcal{N}_{(X,Y,R)}$ it holds that if $(A, B) \leq (G, H)$ and $(C, D) \leq (G, H)$, then $(E, F) \leq (G, H)$ (if $B \supseteq H$ and $D \supseteq H$, then $F = B \cap D \supseteq H$).

The diagram $\mathcal{D}_{(X,Y,R)}$ is finite and, clearly, every finite lattice is complete. Therefore, $\mathcal{D}_{(X,Y,R)}$ is the Hasse diagram of a complete lattice.   □

2.2   Algorithm to Construct $(\mathcal{B}(X, Y, R), \leq)$ using $\mathcal{D}_{(X,Y,R)}$

The diagram $\mathcal{D}_{(X,Y,R)}$ contains all formal concepts of the formal context (X, Y, R) and possibly some other nodes (see Theorem 1). In this section we show that if we (properly) remove all non-concept nodes from $\mathcal{D}_{(X,Y,R)}$, then $\mathcal{D}_{(X,Y,R)}$ will be the Hasse diagram of the concept lattice $(\mathcal{B}(X, Y, R), \leq)$.

The identification of formal concepts in the diagram $\mathcal{D}_{(X,Y,R)}$ is simple. Naturally, formal concepts are these nodes (A, B) of $\mathcal{D}_{(X,Y,R)}$ for which $A^{\uparrow R} = B$. The other way to identify whether a node (A, B) is the formal concept or not is by using the cover relation on the nodes of $\mathcal{D}_{(X,Y,R)}$ what we will see later.

First, we define an equivalence relation θ on the nodes of the diagram $\mathcal{D}_{(X,Y,R)}$.

Definition 1. *We define an equivalence relation θ on $\mathcal{N}_{(X,Y,R)}$ as follows: for nodes $(A, B), (C, D) \in \mathcal{N}_{(X,Y,R)}$ it holds $(A, B)\theta(C, D)$ if and only if $A = C$. The equivalence class of (A, B) is given by $[(A, B)]_\theta = \{(C, D) \in \mathcal{N}_{(X,Y,R)} : (A, B)\theta(C, D)\}$.*

All nodes from $\mathcal{N}_{(X,Y,R)}$ having the same object parts are in the same equivalence class, and, only one node in an equivalence class $[(A, B)]_\theta$, namely $(A, A^{\uparrow R})$, i.e. $(A^{\uparrow R \downarrow R}, A^{\uparrow R})$, is the formal concept of the formal context (X, Y, R).

Theorem 2. *Let (X, Y, R) be a formal context and let $(A, B) \in \mathcal{N}_{(X,Y,R)}$. Then, $(A, A^{\uparrow R})$ is the smallest element of the equivalence class $[(A, B)]_\theta$ w.r.t. $\leq$.*

Proof. Obviously, $(A, A^{\uparrow R}) \in \mathcal{N}_{(X,Y,R)}$. Next, we have to show that if $(C, D) \in [(A, B)]_\theta$, then $(A, A^{\uparrow R}) \leq (C, D)$.
    Let $(C, D) \in [(A, B)]_\theta$. Then, it holds that $A = C$. Moreover, from the fact that $(C, D) \in \mathcal{N}_{(X,Y,R)}$ it follows that $C = D^{\downarrow R}$, and thus $A \subseteq D^{\downarrow R}$. Since $\uparrow R$, $\downarrow R$ form a Galois connection [8], it holds $A^{\uparrow R} \supseteq D$ which implies $(A, A^{\uparrow R}) \leq (C, D)$.
    □

The previous theorem allows us to identify non-concept nodes of the diagram $\mathcal{D}_{(X,Y,R)}$ using the cover relation on $\mathcal{N}_{(X,Y,R)}$ as follows: $(A, B) \in \mathcal{N}_{(X,Y,R)}$ is not the formal concept of (X, Y, R) if and only if (A, B) has a child having the same object part in the diagram $\mathcal{D}_{(X,Y,R)}$. Actually, it suffices to consider only the cardinality of object parts of the nodes (see Corollary 1).

Corollary 1. *Let* (X, Y, R) *be a formal context and let* $(A, B) \in \mathcal{N}_{(X,Y,R)}$. $(A, B) \notin \mathcal{B}(X, Y, R)$ *if and only if there exists* $(C, D) \in \mathcal{N}_{(X,Y,R)}$ *such that* $|C| = |A|$ *and* (C, D) *is a child of* (A, B) *in* $\mathcal{D}_{(X,Y,R)}$.

Proof. The claim follows from the fact that if $C \subseteq A$ and $|C| = |A|$, then $C = A$ (provided that C and A are finite sets).   □

Now we are able to determine whether a node of the diagram is the formal concept or not. In the following, we describe how to remove a non-concept node from the diagram. More precisely, we show how to modify the set of edges of the diagram after a non-concept node is removed.

Let $\mathcal{D}_{(X,Y,R)}$ be the diagram of a formal context (X, Y, R) and let us suppose we are removing a node $(A, B) \notin \mathcal{B}(X, Y, R)$ from the diagram $\mathcal{D}_{(X,Y,R)}$. Denote by $\mathcal{D}'_{(X,Y,R)}$ the diagram constructed from the diagram $\mathcal{D}_{(X,Y,R)}$ by removing the node (A, B) and by updating the set of edges.

$\mathcal{D}'_{(X,Y,R)}$ is constructed from $\mathcal{D}_{(X,Y,R)}$ as follows: First, the node (A, B) and all edges connecting this node to all its children (let S be a set of children of (A, B)) as well as its parents (let T be a set of parents of (A, B)) are removed from the diagram $\mathcal{D}_{(X,Y,R)}$. Clearly, for each $(S_1, S_2) \in S$ and $(T_1, T_2) \in T$ it holds that $(S_1, S_2) \leq (T_1, T_2)$. Possibly, (S_1, S_2) is the child of (T_1, T_2) (or (T_1, T_2) is the parent of (S_1, S_2)) in the diagram $\mathcal{D}'_{(X,Y,R)}$. If there exists a path that goes upward from (S_1, S_2) to (T_1, T_2) and does not pass through the node (A, B) in the diagram $\mathcal{D}_{(X,Y,R)}$, then (S_1, S_2) is not the child of (T_1, T_2) in the diagram $\mathcal{D}'_{(X,Y,R)}$. Otherwise, (S_1, S_2) is the child of (T_1, T_2) in $\mathcal{D}'_{(X,Y,R)}$, and thus the edge from (S_1, S_2) to (T_1, T_2) is added to the diagram $\mathcal{D}_{(X,Y,R)}$.

Lemma 2. *Let* $\mathcal{D}_{(X,Y,R)}$ *be the diagram of a formal context* (X, Y, R), $(A, B) \in \mathcal{N}_{(X,Y,R)}$ *and* $(A, B) \notin \mathcal{B}(X, Y, R)$. *Let* $\mathcal{D}'_{(X,Y,R)}$ *be the diagram constructed from* $\mathcal{D}_{(X,Y,R)}$ *by removing* (A, B) *as described above. There is a path that goes upward from a node* (C, D) *to a node* (E, F) *in* $\mathcal{D}'_{(X,Y,R)}$ *if and only if* $(C, D) \leq (E, F)$.

Proof. Let $((C, D) = P_1, \ldots, P_i = (E, F))$ be a path that goes upward from (C, D) to (E, F) in the diagram $\mathcal{D}'_{(X,Y,R)}$. We differentiate the two cases:
1. The path $(P_1, \ldots, P_i)$ does not pass through the node (A, B) in the diagram $\mathcal{D}_{(X,Y,R)}$, i.e. $P_j \neq (A, B)$ for each $j \in \{1, \ldots, i\}$. In this case the proposition follows from the assumption.
2. The path passes through the node (A, B) in $\mathcal{D}_{(X,Y,R)}$ (which is not contained in the diagram $\mathcal{D}'_{(X,Y,R)}$), i.e. there exists $j \in \{1, \ldots, i\}$ such that $P_j = (A, B)$. Then, the path also goes through a child (S_1, S_2) and a parent (T_1, T_2) of (A, B). It holds that either there exists a path that goes upward from (S_1, S_2) to (T_1, T_2) in $\mathcal{D}'_{(X,Y,R)}$, or the edge from (S_1, S_2) to (T_1, T_2) is added to the diagram when the diagram $\mathcal{D}'_{(X,Y,R)}$ is constructed, and thus the assertion holds.   □

The procedure in Algorithm 3 removes a node (A, B) from the diagram $\mathcal{D}_{(X,Y,R)}$ of (X, Y, R) with adjusting the set of edges of the diagram. The procedure *removeNode* accepts as its arguments a diagram $\mathcal{D} = (\mathcal{N}, \mathcal{E})$ of a formal

---

Algorithm 3. Procedure $removeNode(\mathcal{D}, (A, B))$

1. $S \leftarrow$ children of (A, B) in $\mathcal{D}$
2. $T \leftarrow$ parents of (A, B) in $\mathcal{D}$
3. $\mathcal{N} \leftarrow \mathcal{N} \setminus \{(A, B)\}$
4. **for all** $(S_1, S_2) \in S$ **do**
5.   remove edge $(S_1, S_2) \rightarrow (A, B)$ from $\mathcal{E}$
6. **for all** $(T_1, T_2) \in T$ **do**
7.   remove edge $(A, B) \rightarrow (T_1, T_2)$ from $\mathcal{E}$
8. **for all** $(S_1, S_2) \in S$ **do**
9.   **for all** $(T_1, T_2) \in T$ **do**
10.   **if** there is no path from (S_1, S_2) to (T_1, T_2) in $\mathcal{D}$ **then**
11.   add edge $(S_1, S_2) \rightarrow (T_1, T_2)$ to $\mathcal{E}$

---

context (X, Y, R) and a node (A, B) of the diagram. The procedure outputs the modified diagram $\mathcal{D} = (\mathcal{N}, \mathcal{E})$.

First, the node (A, B) and all edges connecting this node with its children as well as its parents are removed from the diagram $\mathcal{D}$ (lines 3 - 7). Then, if there is no path that goes upward from a child (S_1, S_2) to a parent (T_1, T_2) of (A, B) in $\mathcal{D}$, the egde from (S_1, S_2) to (T_1, T_2) is added to $\mathcal{D}$ (lines 8 - 11).

Algorithm 3 is used to construct the Hasse diagram of the concept lattice $(\mathcal{B}(X, Y, R), \leq)$ from the diagram $\mathcal{D}_{(X,Y,R)}$. The procedure $computeLattice$ in Algorithm 4 accepts as its argument a diagram $\mathcal{D} = (\mathcal{N}, \mathcal{E})$ of a formal context (X, Y, R) and computes the Hasse diagram of $(\mathcal{B}(X, Y, R), \leq)$.

Algorithm 4. Procedure $computeLattice(\mathcal{D})$

1. **for all** $(A, B) \in \mathcal{N}$ **do**
2.   **if** (A, B) has a child (C, D) in $\mathcal{D}$ such that $|C| = |A|$ **then**
3. $removeNode(\mathcal{D}, (A, B))$

The idea of $computeLattice$ procedure is the following: Nodes of the diagram $\mathcal{D}$ are processed one by one (line 1). If the test in line 2 passes, i.e. a node (A, B) of the diagram $\mathcal{D}$ is not the formal concept of $\mathcal{B}(X, Y, R)$, then the node (A, B) is removed from $\mathcal{D}$ using $removeNode$ procedure described above (line 3).

Theorem 3. *The procedure computeLattice in Algorithm 4 is correct, i.e. for a given diagram $\mathcal{D}_{(X,Y,R)}$, it stops after finitely many steps and computes the Hasse diagram of the concept lattice $(\mathcal{B}(X, Y, R), \leq)$.*

Proof. The diagram $\mathcal{D}_{(X,Y,R)}$ contains all formal concepts of the formal context according to Theorem 1. The statement follows from Corollary 1 (only non-concept nodes of $\mathcal{D}_{(X,Y,R)}$ are removed) and Lemma 2. □

2.3   Complexity of the Algorithm

In this section we analyze the worst-case time complexity of the algorithm.

Let n be the maximum number of attributes per object in a formal context (X, Y, R), i.e. $n = \max\{|\{x\}^{\uparrow_R}| : x \in X\}$. Observe that each node of $\mathcal{D}_{(X,Y,R)}$ has at most n parents and n is the depth of the diagram $\mathcal{D}_{(X,Y,R)}$.

The outer loops of Algorithm 1 (lines 3 and 4), i.e. generating the power set of $\{x\}^{\uparrow_R}$, require $O(2^n)$ time. Assuming that the test to check whether each generated set of attributes B is the attribute part of a node already present in the diagram (line 7) is done using a binary search, this part takes $O(n \log |Y|)$. Each new node (A, B) is connected to the corresponding nodes (lines 12 - 13) in time $O(n)$. Thus, the time complexity of Algorithm 1 is $O(2^n n^2 \log |Y|)$.

The complexity of the algorithm 2 is $O(|X| 2^n (n^2 \log |Y| + 1))$.

In Algorithm 3, all edges connecting the node (A, B) with its parents as well as its children are removed from the diagram $\mathcal{D}_{(X,Y,R)}$ (lines 4 - 7) in time $O(n + |Y|)$. The test whether there is a path (consisting of egdes of $\mathcal{D}_{(X,Y,R)}$) connecting a child (S_1, S_2) and a parent (T_1, T_2) of (A, B) in $\mathcal{D}_{(X,Y,R)}$ (lines 10 - 11) takes $O(n!)$. Thus, the time complexity of Algorithm 3 is $O(n! (n + |Y|))$.

The complexity of the algorithm 4 is $O(|X| 2^n n! (n + |Y|))$.

The maximum number of attributes per object and the total number of nodes of the diagram $\mathcal{D}_{(X,Y,R)}$ have significant impact on the efficiency of the algorithm.

3   Experimental Evaluation

We have first evaluated the proposed algorithm by comparing the number of formal concepts of a given context and the number of nodes of the diagram constructed by the algorithm. Definitely, the higher the average number of attributes per object in a formal context, the higher the number of nodes of the diagram generated by the algorithm that are not formal concepts (but due to the lack of space we omit the details of this experiment).

Next, we have experimentally compared the performance of the proposed algorithm against AddExtent (an attribute-incremental implementation of AddIntent [14]) that has been provided to us by one of its authors[3], and CHARM-L [20] that is publicly available. The experiments were conducted on the computing node with 16 cores equipped with 24 GB RAM memory running GNU/Linux openSUSE 12.1. We have run the experiments on randomly generated data and measured CPU times with outputs of algorithms turned off.

The results of experiments are depicted in Fig. 2 and Fig. 3. The names of the graphs in the figures indicate the total number of attributes, the average and maximum number of attributes per object in data sets, respectively.

Let us first compare how the algorithms perform on data sets in which the average number of attributes per object is 2 (top of Fig. 2 and Fig. 3). On these data sets, our algorithm is the best choice. The number of objects does not affect the performance of our algorithm much unlike AddExtent and CHARM-L.

Next, let us analyze the results of experiments on data sets with the average number of attributes per object set to 3 (middle of Fig. 2 and Fig. 3). If the total number of attributes is low (1000, 5000), our algorithm has similar performance

---

[3] We thank Sergei O. Obiedkov for providing the source code for AddExtent.

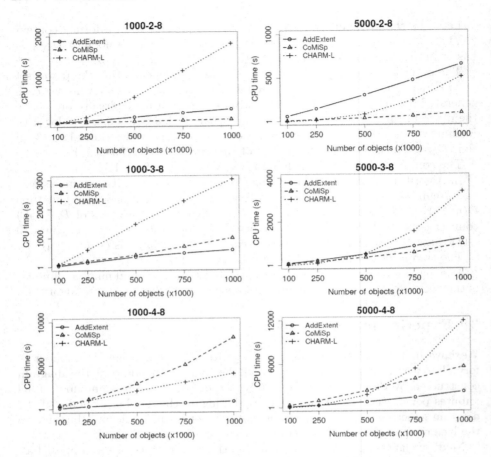

Fig. 2. The performance of algorithms (CPU times in seconds) on random data sets having 1000 attributes (left) and 5000 attributes (right). The average number of attributes per object is 2 (top), 3 (middle) and 4 (bottom), and the maximum number of attributes per object is set to 8.

as AddExtent. On data sets having the total number of attributes 10000 and 50000 (middle of Fig. 3), the behaviour of our algorithm is curious. Our algorithm outperforms both AddExtent and CHARM-L on all data sets except for those where the number of objects is 1000000.

Finally, let us consider the performance of algorithms on data sets where the average number of attributes per object is 4 (bottom of Fig. 2 and Fig. 3). AddExtent outperforms both other algorithms on data sets where the total number of attributes is 1000. If the total number of attributes is increased to 5000, our algorithm performs better than CHARM-L, but AddExtent still outperforms our algorithm. If the total number of attributes is 10000 or 50000 and the maximum number of attributes per object is lowered to 6 (bottom of Fig. 3), our algorithm performs better than AddExtent and have similar performance as CHARM-L.

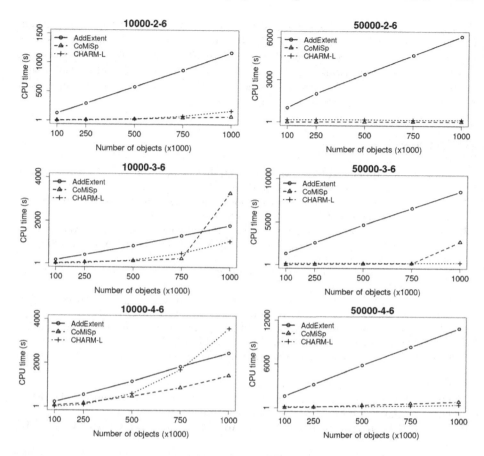

Fig. 3. The performance of algorithms (CPU times in seconds) on random data sets having 10000 attributes (left) and 50000 attributes (right). The average number of attributes per object is 2 (top), 3 (middle) and 4 (bottom), and the maximum number of attributes per object is set to 6.

4   Conclusion

We presented an algorithm for mining formal concepts from very sparse large-scale data. We identified some issues for our future work, the most important of which are (i) to identify some speed-up possibilities of our algorithm both on theoretical as well as implementation side, (ii) provide more exhaustive experiments on larger – and possibly real – datasets and (iii) investigate the way of parallelization of our algorithm. Even if some work remains to fine-tune our algorithm, we believe that it is worth of further investigation.

Acknowledgements. This publication is the result of the Project implementation: University Science Park *TECHNICOM* for Innovation Applications Supported by Knowledge Technology, *ITMS: 26220220182*, supported by the Research & Development Operational Programme funded by the ERDF and partially supported by the projects Center of knowledge and information systems in Košice *CEZIS, ITMS: 26220220158* and *VEGA 1/0832/12*.

References

1. Agrawal, R., Imielinski, T., Swami, A.: Mining association rules between sets of items in large databases. In: SIGMOD, pp. 207–216 (1993)
2. Andrews, S.: In-close, a fast algorithm for computing formal concepts. In: Supplementary Proceedings of ICCS. CEUR WS, vol. 483 (2009)
3. Andrews, S.: In-close2, a high performance formal concept miner. In: Andrews, S., Polovina, S., Hill, R., Akhgar, B. (eds.) ICCS 2011. LNCS (LNAI), vol. 6828, pp. 50–62. Springer, Heidelberg (2011)
4. Bordat, J.P.: Calcul pratique du trelis de galois dune correspondance. Math. Sci. Hum. 96 (1986)
5. Carpineto, C., Romano, G.: Galois: An order-theoretic approach to conceptual clustering. In: Proc. of the 10th Conf. on Machine Learning (1993)
6. Chein, M.: Algorithme de recherche des sous-matrices premiéres dune matrice. Bull. Math. Soc. Sci. Math. R. S. Roumanie 13 (1969)
7. Ganter, B.: Two basic algorithms in concept analysis. Tech. Rep. FB4 No. 831 (1984)
8. Ganter, B., Wille, R.: Formal concept analysis. Springer (1999)
9. Godin, R., Missaoui, R., Alaoui, H.: Incremental concept formation algorithms based on galois (concept) lattice. Computational Intelligence 11 (1995)
10. Krajča, P., Outrata, J., Vychodil, V.: Advances in alg. based on CbO. In: CLA 2010 (2010)
11. Kuznetsov, S.O.: A fast algorithm for computing all intersections of objects in a finite semi-lattice. Aut. Docum. and Math. Linguistics 27(5), 11–21 (1993)
12. Kuznetsov, S.O., Obiedkov, S.A.: Comparing performance of algorithms for generating concept lattices. Jour. of Exp. and Theor. Art. Intelligence 14(23) (2002)
13. Lindig, C.: Fast concept analysis. In: Working with Conceptual Structures - Contributions to ICCS 2000 (2000)
14. van der Merwe, D., Obiedkov, S., Kourie, D.: AddIntent: A new incremental algorithm for constructing concept lattices. In: Eklund, P. (ed.) ICFCA 2004. LNCS (LNAI), vol. 2961, pp. 372–385. Springer, Heidelberg (2004)
15. Norris, E.M.: An algorithm for computing the maximal rectangles in a binary relation. Revue Roumaine de Mathmatiques Pures et Appliqués 23(2), 243–250 (1978)
16. Nourine, L., Raynaud, O.: A fast algorithm for building lattices. Information Processing Letters 71, 199–204 (1999)
17. Nourine, L., Raynaud, O.: A fast incremental algorithm for building lattices. Jour. of Exp. and Theor. Art. Intelligence 14, 217–227 (2002)

18. Pasquier, N., Bastide, Y., Taouil, R., Lakhal, L.: Pruning Closed Itemset Lattices for Association Rules. In: Actes Bases De Données Avancées Bda 1998, Hammamet, Tunisie, pp. 177–196 (1998)
19. Stumme, G., Taouil, R., Bastide, Y., Pasquier, N., Lakhal, L.: Computing iceberg concept lattices with titanic. Knowl. and Data Eng. 42(2), 189–222 (2002)
20. Zaki, M.J., Hsiao, C.: Efficient algorithms for mining closed itemsets and their lattice structure. IEEE TKDE 17(4), 462–478 (2005)

Swip: A Natural Language to SPARQL Interface Implemented with SPARQL

Camille Pradel, Ollivier Haemmerlé, and Nathalie Hernandez

IRIT, University of Toulouse, France
{Camille.Pradel,Ollivier.Haemmerle,Nathalie.Hernandez}@irit.fr

Abstract. The *Swip* approach aims at translating into SPARQL queries expressed in natural language exploiting query patterns. In this article, we present the main module of the prototype implementing this approach which entirely relies on SPARQL. All steps of the interpretation process which are carried out in this module are indeed completely performed on RDF triple stores through SPARQL updates. Thus, the implementation gets benefit from SPARQL engine capabilities, which prevent us from worrying about graph manipulation and matching.

The purpose of the approach we propose is to provide end-users with a means to query ontology based knowledge bases using natural language (NL) queries and thus hide the complexity of formulating a query expressed in a graph query language such as SPARQL. The process we proposed to interpret NL queries as well as its implementation making an extensive use of SPARQL contribute to the originality of our approach. Section 1 summarizes works and systems sharing our objectives in order to situate our overall approach named *Swip* (for *Semantic Web Interface using Patterns*). Section 2 introduces the framework in which the *Swip* system takes place. Section 3 details first steps of the interpretation process in order to give a precise idea of the implementation. Section 4 discusses the benefits and drawbacks of this approach.

1 Positioning Our Work

We make in Subsection 1.1 a quick overview of approaches from the literature aiming at helping users to query graph based KBs (note that a complete survey on this topic can be found in [1]), and then position our work with regard to others in Subsection 1.2.

1.1 Related Work

We stick here to the classification proposed in [2], and present the main approaches following a formality continuum, from most formal to most permissive, beginning with the query languages themselves.

N. Hernandez et al. (Eds.): ICCS 2014, LNAI 8577, pp. 260–274, 2014.
DOI: 10.1007/978-3-319-08389-6_21, © Springer International Publishing Switzerland 2014

On an extreme side of this continuum are the formal graph languages, such as SPARQL. They are the targets of the systems we are presenting in this section. Such languages present obvious usability constraints, which make them unsuited to end-users. Expressing formal queries implies knowing and respecting the language syntax used, understanding a graph model and, most constraining, knowing the data schema of the queried KB. The work presented in [3] aims at extending the SPARQL language and its querying mechanism in order to take into account keywords and wildcards when the user does not know exactly the schema he/she wants to query on. Here again, such an approach requires that the user knows the SPARQL language.

Similar are approaches assisting the user during the formulation of queries in such languages. Very light interfaces such as *Flint*[1] and *SparQLed*[2] implement simple features such as syntactical coloration and autocompletion. Other approaches rely on graphical interfaces such as [4,5] for SPARQL queries. Even if these graphical interfaces are useful and make the query formulation work less tedious, we believe they are not well suited to end-users since they do not overcome the previously mentioned usability limits of formal graph query languages.

Sewelis [6] introduces *Query-based Faceted Search*, a new paradigm combining faceted search and querying, the most popular paradigm in semantic data search, while other approaches such as *squall2sparql* [7] define a controlled natural language whose translation into SPARQL is straightforward.

Other works aim at generating formal queries directly from user queries expressed in terms of keywords or NL. Our work is situated in this family of approaches [8,9,10,11]. The user expresses his/her information need in an intuitive way, without having to know the query language or the knowledge representation formalism. In these systems, the generation of the query requires the following steps: (i) matching the keywords to semantic entities defined in the KB, (ii) building query graphs linking the entities previously detected by exploring the KB, (iii) ranking the built queries, (iv) making the user select the right one. The existing approaches focus on several main issues: optimizing the first step by using external resources (such as WordNet or Wikipedia)[8,12], optimizing the knowledge exploration mechanism for building the query graphs [9,10], enhancing the query ranking score [12], and improving the identification of relations using textual patterns [11]. *Autosparql* [13] extends the previous category: after a basic interpretation of the user NL query, the system interacts with the user, asking for positive and negative examples, in order to refine the initial interpretation by performing a learning algorithm. In [14] the interpretation process of this same system is improved by determining a SPARQL template from the syntactic structure of the natural language question.

1.2   The *Swip* Approach

One of the originalities of our approach lies in the use of pre-built query patterns in order to lead the interpretation process. The main postulate underlying our

---

[1] http://openuplabs.tso.co.uk/demos/sparqleditor
[2] http://sindicetech.com/sindice-suite/sparqled/

work, justified in [15], states that, in real applications, the submitted queries are variations of a few typical query families. Each query pattern represents one of these query families. Query patterns are made of a graph using the vocabulary of the targeted ontology and representing the information need, and of a descriptive sentence template.

The query interpretation process is detailed [16]; it consists of two main steps, with an intermediate result which is the *pivot query*. Pivot query is a structure half-way between the NL query and the targeted formal query, and aims at storing results from the first interpretation step. It is made of keywords and allows expressing relations between these keywords. For instance, the pivot query ?"person": "produce"= "In Utero". "In Utero": "album" is obtained while interpreting the NL query "Who produced the album In Utero?" (extracted from the QALD-3 challenge[3] dataset). This organization brings two advantages: the pivot query explicitly represents important information extracted from the NL query syntactical analysis, and it facilitates the implementation of multilingualism by means of a common intermediate format: to adapt the approach to a new language, it is only required to implement a new module of translation from NL to pivot; the pivot query translation step remains unchanged.

The first step, the natural language query interpretation, roughly consists in the identification of named entities, a dependency analysis and the translation of the obtained dependency graph into a pivot query. The web application implementing this module follows traditional principles: web services performing basic subtasks of the interpretation process are exploited by a workflow which is itself accessible under the form of a web service. Each service exploits as an input the output of one or more previous services. The first module and the set of services it is made of were presented in [17,15].

In the second step, detailed in [16], query patterns are mapped to the pivot query; we thus obtain a list of potential interpretations of the user query, which are then ranked according to their estimated relevance and proposed to the user in the form of NL queries formulated using the NL sentence templates of each pattern. The implementation of this module represents for us a real novelty: the interpretation of pivot queries is performed on an RDF triple repository by emitting SPARQL updates. This approach prevents us from implementing functions to manipulate and match graphs and allows exploiting SPARQL engine capabilities. Indeed, the SPARQL core feature consists in graph mapping, which is exactly the purpose of this interpretation step. This task is thus entirely carried out through SPARQL queries. The result of each step is systematically committed into the KB, using SPARQL updates, which makes it available for subsequent steps. To our knowledge, no approach relies on such an exploitation of SPARQL. This second module is presented in the following.

---

[3] http://greententacle.techfak.uni-bielefeld.de
/~cunger/qald/index.php?x=task1&q=3

2   Framework

We describe here the framework that is required for the deployment of our approach. We present in 2.1 the ontologies we built in order to define a logical framework for the implementation and in 2.2 the few needed logistic facilities.

2.1   Ontologies for Patterns and Queries

RDF triples which are manipulated and generated during the interpretation process exploit the vocabulary of two ontologies which are accessible from the *Swip* home page[4]. This section introduces these ontologies.

In the rest of this paper, we express URIs with prefixes without making them explicit; they are common prefixes and must be considered with their usual value. We also define new prefixes, specific to entities defined in our ontologies: to the prefix name `patterns` is associated the URI `http://swip.univ-tlse2.fr/ontologies/patterns`, to `queries` is associated the URI `http://swip.univ- tlse2.fr/ontologies/queries`.

Our ontologies are built according to the principles of *normalization* design pattern, introduced in [18]. This method enables the development of modular and reusable ontologies, defining classes by the properties their instances should fulfill. For this reason we present the *patterns* ontology by first introducing the main properties then the main classes composing it.

Figure 1 presents the object property hierarchy of the *patterns* ontology; properties are characterized by their domains and ranges. To each property corresponds its inverse property.

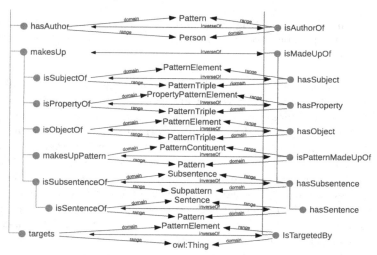

Fig. 1. Object properties of the *patterns* ontology

---

[4] `http://swip.univ-tlse2.fr/SwipWebClient/welcome.html`

Property `patterns:makesUp` is the generic relation of meronymy; it is specialized by different subproperties which are chosen according to the type of elements building the pattern, such as `patterns:isSubjectOf`, `patterns:isPropertyOf`, `patterns:isObjectOf`, `patterns:isSentenceOf` and `patterns:isSubsentenceOf`. The property `patterns:targets` expresses the relation between a pattern element and the resource (class, property or datatype) of the target ontology this element is referring to; it thus specifies mapping possibilities.

Data properties `patterns:hasCardinalityMin` and `patterns:hasCardinalityMax` specify cardinalities of subpatterns.

Figure 2 shows the hierarchy of the *patterns* ontology main classes, as well as their meronymy relations.

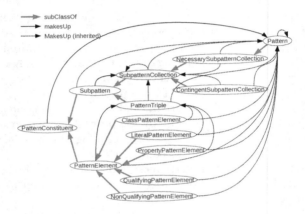

Fig. 2. The main classes describing query patterns, the taxonomic relations ordering them, as well as the possible meronymy relations between instances of these classes

Subpatterns are embodied by the class `patterns:Subpattern` which subsumes distinct classes `patterns: PatternTriple` and `patterns:SubpatternCollection`.

An instance of the class `patterns:PatternTriple` is a triple of the pattern graph. A triple is characterized by its subject (`patterns:hasSubject` property), its property (`patterns:hasProperty` property) and its object (`patterns:hasObject` property). Values of these properties are instances of the class `patterns:PatternElement`. A pattern element references a literal type (`patterns:LiteralPatternElement`), a class (`patterns:ClassPatternElement`) or a property (`patterns:PropertyPatternElement`) of the target ontology. For example, an instance of `patterns:ClassPatternElement` involved in a pattern will be linked by the relation `patterns:targetKBElement` to a class of the target ontology; this class will then, in this context, be considered as an instance. This method, called *punning*, is possible in OWL2. However, its usage cannot be modeled at the ontology level as this would require the usage of a part of reserved vocabulary (`owl:Class`, `owl:Property`, `owl:Literal`). `owl:Thing` is used instead of each of these forbidden entities, which does not differentiate between the three subclasses of `patterns:PatternElement` other than by declaring them distinct.

The *queries* ontology used to represent pivot queries and the results of the interpretation process is not detailed here as it was designed on the same principles as *patterns* and is much simpler. As explained in [15], a pivot query is a set of subqueries, which are 1/2/3-tuples of query elements. The classes and properties of the *queries* ontology logically reflect this structure, their names are self-explaining, and an example of a pivot query and part of its interpretation results expressed in RDF and based on this ontology is used in the following subsections.

2.2 Infrastructure

The interpretation process is entirely based on SPARQL queries. Thus, to set the approach up, the main needed infrastructure is a SPARQL server which will receive update queries emitted by the user interface. Triples which are generated by each query exploit the vocabulary of the *Queries* ontology and are added to the triples already contained in the engine.

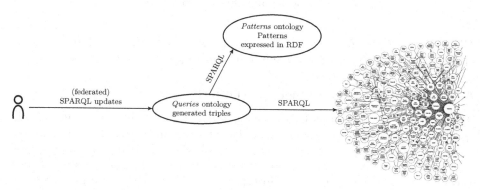

Fig. 3. Example deployment of the *Swip* system

To successfully complete the interpretation process, we obviously need to access the dataset targeted by the user query and to the query patterns (previously translated in RDF triples according to the *Patterns* ontology vocabulary). These elements can be directly included in the triple base of the SPARQL server, or simply distributed on the web of data and accessed via SPARQL 1.1 federation. Figure 3 illustrates a possible scenario of deployment.

3 Description of the First Steps

Subsections 3.1, 3.2 and 3.3 detail the first steps of the pivot query interpretation process. Then following steps are shortly presented in Subsection 3.4. Figure 4 shows all implementation steps, each one corresponding to a SPARQL

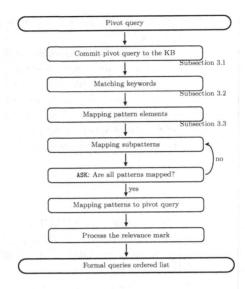

Fig. 4. Details of the pivot query formalization steps

UPDATE or ASK (for emulating loops, as explained later), and these corresponding SPARQL queries are given on the Swip presentation web page[5].

3.1 Committing Pivot Query to the KB

The system first processes a URI which is unique for each set of equivalent pivot queries. If this URI already exists in the KB, this query has been previously processed and the saved results can be exploited as they are. If not, the system generates an RDF graph representing the pivot query and commits it into the KB before starting the query interpretation. The RDF graph produced for the query ?"person": "produce"= "In Utero". "In Utero": "album" and based on the *Queries* ontology is shown on the left side of Figure 5. This figure shows the initial RDF data and also the results of the SPARQL updates which are processed during the matching step (cf. Subsection 3.2). RDF resources are represented in rounded nodes, literals in rectangle nodes, and properties are logically materialized by labeled edges. For the sake of readability, literal types are not shown, and resource classes are shown only when relevant.

3.2 Matching Query Elements to KB Resources

The first step of the pivot query interpretation consists in matching each element of the pivot query to KB entities (classes, properties and instances) or to literal types associating with each match a trust mark which represents the supposed quality of the matching and its likelihood.

---

[5] http://swip.univ-tlse2.fr/SwipWebClient/welcome.html

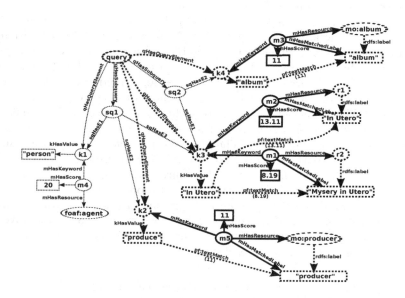

Fig. 5. The matching step on the example pivot query

```
INSERT
{
    ?matchUri a queries:Matching;
              queries:matchingHasKeyword ?keyword;
              queries:matchingHasResource ?r;
              queries:matchingHasScore ?score;
              queries:matchingHasMatchedLabel ?l.
    ?keyword queries:keywordAlreadyMatched "true"^^xsd:boolean.
}
WHERE
{
    <[queryUri]> queries:queryHasQueryElement ?keyword.
    ?keyword a queries:KeywordQueryElement;
             queries:queryElementHasValue ?keywordValue.
    FILTER NOT EXISTS {
        ?keyword queries:keywordAlreadyMatched "true"^^xsd:boolean. }
    (?l ?score) pf:textMatch (?keywordValue 6.0 5).
    ?r rdfs:label ?l.
    BIND (UUID() AS ?matchUri)
}
```

Fig. 6. SPARQL update used for matching query elements

Keyword matching is performed by processing a similarity measure between pivot query keywords and KB resources labels. To carry out this task, we use LARQ[6], an extension of SPARQL proposed by the ARQ[7] SPARQL engine, which exploits search functionalities of the Lucene[8] query engine. This extension introduces a new syntax, which determines both the literals matching a given string and a value representing the likeliness between them, called Lucene score: the "triple" (?lit ?score) pf:textMatch '+text' binds to ?lit all literals which are similar to the text string and to ?score the corresponding Lucene score. The SPARQL query used to perform this task is shown in Figure 6.

Figure 5 shows a subset of the bindings obtained by this query execution and the generated matching instances. The graph as it is before the update is shown in dotted lines; the part of this graph that is matched by the WHERE clause is highlighted in bold, and the committed resources and triples are shown in full lines. The following figures use the same presentation rules.

Note that although this step is performed using a (nonstandard) extension of the SPARQL language which makes it not very portable, an alternative can be implemented using standard features such as REGEX and CONTAINS string functions, or a simple string comparison.

3.3   Mapping Pattern Elements to Query Elements

Before being able to map the entire patterns to the user query, the first task consists in figuring out for each pattern element all conceivable mappings to user query elements – called *element mappings* – and their respective trust marks. This step consists in creating a mapping between a query element and a pattern element when this query element was matched to a resource which is linked in some way to the pattern element target. We define several cases where a link is determined between a matched resource r and a pattern element target t:

1. r is a subclass of t (this case includes r is the class t itself),
2. r is a subproperty of t (this case includes r is the property t itself),
3. r is an instance of t,
4. r refers to the same literal type as t.

To each element mapping is associated a trust note whose value is the same as the involved matching score. Figure 7 shows the instantiation of some element mappings through a SPARQL update. Pattern element cd_info_element5 which targets class foaf:agent is mapped to keyword k1 ("person") which matched the same class. Pattern element cd_info_element1 which targets class mo:Record is mapped twice to keyword k3 ("In Utero") which matched two instances of this same class. Pattern element cd_info_element4 which targets property mo:producer is mapped to keyword k2 ("produce") which matched the same property. The following paragraphs present extensions of previous cases added to better match real life requirements.

---

[6] http://jena.apache.org/documentation/larq/
[7] http://jena.apache.org/documentation/query/
[8] http://lucene.apache.org/

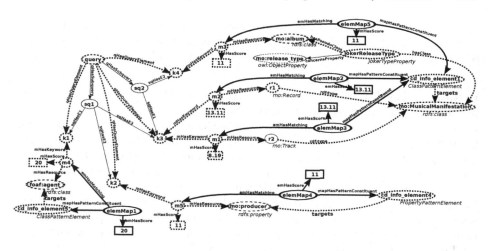

Fig. 7. The element mapping step on the example pivot query

"*Type" properties considered harmful The instantiation of the element mapping `elemMap5` is issued from an extension of the first case stated above. This extension is due to the observation of a recurrent modelling choice made by some ontology developers, which consists in classifying instances of a class c by defining an object property with the class c as domain, an (enumerated) class c' as range and a set of predifined instances of c' which are the different ways to classify instances of c. For instance, in the *music ontology* [19], instances of class `mo:Record` can be involved as a subject in a triple with predicate `mo:releaseType` and with object an instance of class `mo:ReleaseType` (`mo:Album`, `mo:Single`, `mo:Live`, `mo:Soundtrack`...). It seems to us that this choice, although it must have been guided by some requirements, is not relevant, as it ignores the classification mechanism proposed by RDFS and OWL, i.e. typing instances using (sub)classes, to express a piece of knowledge that is indeed a classification. We support our thesis by pointing out two features we consider as symptomatic of this modelling flaw and that are present in our example:

- in NL, the terms relative to the instances of c' are used in the same way as terms relative to classes; for instance, in the NL query (extracted from the QALD-3 challenge) "Give me all soundtracks composed by John Williams", the term "soundtracks" has the same place and function as the term "songs" in "Give me all songs by Aretha Franklin."
- the way the ontology developers themselves named the object property betrays the true nature of this property; indeed a property whose name ends with "Type" will very probably be used for typing instances and as such should be a subproperty of `rdf:type`.

Our approach overcomes this nongeneric modelling by explicitly identifying these cases. In Figure 7, the instance `wildcardReleaseType`, of type `WildcardTypeProperty` states that the resource `mo:album` should be considered during the mapping process as a subclass of `mo:Record`, which maps `cd_info_element1` to keyword `k4`; it also specifies that the typing property (used while generating the final SPARQL query) is in this case `mo:release_type` instead of `rdf:type`.

Instance-Class Element Mappings. A last type of element mapping can be produced by the previously generated ones. These mappings, called *instance-class element mappings*, are issued from observing how users, when expressing themselves in NL, often specify a term referring to an instance by another term referring to the class of this instance. Some examples from the QALD-3 challenge are "the band Dover", "the album In Utero", "the song Hardcore Kids".

According to [15], such NL formulations are translated in the pivot language into a two element subquery, composed of the keyword referring to the instance qualified by the keyword referring to the class; for instance the part of the NL query "the album In Utero" becomes "In Utero: album". We defined a particular kind of element mapping to handle this case. Such a mapping maps one pattern element to two keywords. Its score is the sum of trust notes of the two originally involved element mappings.

The implementation we propose handle this special (but recurrent) case in a simple way. Figure 8 uses the same example and illustrates the instantiation of an instance-class element mapping. `elemMap1` and `elemMap4` map `cd_info_element1` to respectively `k3` ("In Utero") and `k4` ("album"), and the resource matched by `k3` is an instance (taking into account the previous remark on "*Type" properties) of the resource matched by `k4`; this provides a new element mapping `elemMap6` mapping the considered pattern element to both query elements.

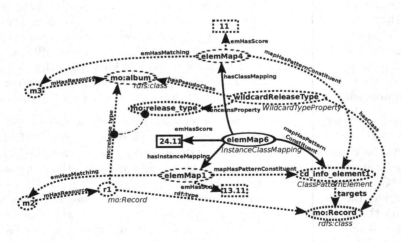

Fig. 8. An instance-class element mapping on the example pivot query

3.4   Following Steps

Following steps are not detailed here for the sake of brevity. Again, constructed SPARQL queries and the ontology organizing the generated triples are available on the home page of the *Swip* system.

As specified in [16], the subpattern mapping step implies complex operations like combinations of sets of elements or Cartesian products between sets whose number cannot be determined in advance. Moreover, a patterns association implies to first associate most simple subpatterns, and then subpatterns directly containing them, and so on until having associated the pattern itself. These needs are traditionally overcome in procedural programming by using loop control structures, which are themselves built on the conditional jump. A single succession of SPARQL updates cannot fill this need. We thus added the possibility to make a conditional jump in our implementation. This conditional jump is performed in a simple way through a SPARQL ASK: there are two possible branches following the execution of such a query and the one which is actually taken depends on the result of this execution. As showed on Figure 4, this method is used at the end of the subpattern mapping step, at the (ASK) query *"Are all patterns mapped?"* If the answer is (FALSE), the taken branch returns to the previous query execution and the test will be repeated later. If the answer is (TRUE), then the next query in the process is executed. Thus, we create a loop ensuring that all subpatterns have been associated before going on.

4   Discussions on the Implementation

The architecture described above presents several benefits. One of the most obvious is the ease of use of a cache system. As the result of each processing step is committed into the RDF repository, it is straightforward to reexploit previously generated data. For instance, as explained in Subsection 3.1, the Swip system realises that an incoming pivot query has previously been processed when its URI is already present in the repository and the result of its interpretation can directly be returned. Also, the matching of a given keyword is performed only once for all queries containing this same keyword.

Moreover, with this architecture, there is a total and natural asynchronism between client and server sides: once the query URI is processed, it is returned to the client which can then update results of the interpretation progressively and independantly from the server, by directly querying the RDF repository.

This approach also ensures a total homogeneity and consistency as input, intermediary and output data is stored and manipulated through standard technologies. Even the configuration and tuning of the system can be made that way; for instance, the identified "*TypeProperties" (cf. Subsection 3.3) are directly expressed in RDF and exploited as such during the interpretation. It is also worth noticing that, although the first part of the approach was implemented in a more "traditional" way, using web services, this could be changed exploiting the *NLP2RDF* recent initiative [20] which aims at providing results of popular NLP tools in a common format, called *NIF* (NLP Interchange Format) and integrated in the semantic web framework.

Finally, this architecture allows to deploy the *Swip* system in a distributed way, which would facilitates scaling the approach. Indeed, as explained earlier, *Swip* simply needs to access queried data and query patterns via SPARQL. Thanks to recent federation features of SPARQL[9], all this data can be aggregated and located on a single SPARQL server or distributed on several servers. Patterns can also be grouped or distributed, brought together with the knowledge base they target or isolated. This standards based architecture is thus very flexible and permits multiple variations. Moreover, processing is performed by SPARQL servers, which makes the initial program (the one which emits the SPARQL queries) very light. We can thus conceive of this program being executed on the client side setting in motion several servers distributed on the web via SPARQL federation and orchestrating them to perform a query interpretation.

Two major drawbacks mar these qualities. The first one is the lack of control on the SPARQL queries processing and thus on the overall system performance. The SPARQL server is used as a blackbox and its efficiency directly impacts that of the interpretation process. Experience showed that ARQ server is not good with big queries (more than twenty triples in several subqueries) and that splitting these complex queries into an equivalent set of successive queries was much more efficient. Moreover, SPARQL is still a relatively young recommendation and, despite the novelties brought by SPARQL 1.1, it proposes a limited range of features. For instance, there are very few arithmetic functions; only the basic ones are supported by the recommendation, which is not the case of the power function. As a consequence, a degraded solution had to be found to process the final relevance mark which initially implies this function.

5   Perspectives

We now want to explore new leads opened by the implementation described in this paper. We believe that the approach we used can be generalized and exploited in other use cases: we could thus implement any type of algorithm, particularly algorithms manipulating graphs. Moreover, for reasons exposed in 4, this approach seems particularly well suited for web applications and could easily be adapted to some recently proposed frameworks aiming at integrating web APIs to the web of data, such as *RDF-REST* [21].

We want to formalize this new way of programming and compare it to close paradigms, such as data-driven programming, or more pragmatic approaches, such as PL-SQL. This work would probably lead us to propose an extension to SPARQL allowing the conditional branch in order to obtain a programming language particularly suited for implementing graph algorithms.

---

[9] http://www.w3.org/TR/sparql11-federated-query/

References

1. Lopez, V., Uren, V., Sabou, M., Motta, E.: Is question answering fit for the semantic web?: a survey. Semantic Web 2(2), 125–155 (2011)
2. Kaufmann, E., Bernstein, A.: Evaluating the usability of natural language query languages and interfaces to semantic web knowledge bases. Web Semantics: Science. Services and Agents on the World Wide Web 8(4), 377–393 (2010)
3. Elbassuoni, S., Ramanath, M., Schenkel, R., Weikum, G.: Searching rdf graphs with sparql and keywords. IEEE Data Eng. Bull. 33(1), 16–24 (2010)
4. Russell, A., Smart, P.R.: Nitelight: A graphical editor for sparql queries. In: International Semantic Web Conference (Posters & Demos) (2008)
5. Clemmer, A., Davies, S.: Smeagol: A "Specific-to-general" semantic web query interface paradigm for novices. In: Hameurlain, A., Liddle, S.W., Schewe, K.-D., Zhou, X. (eds.) DEXA 2011, Part I. LNCS, vol. 6860, pp. 288–302. Springer, Heidelberg (2011)
6. Ferré, S., Hermann, A.: Reconciling faceted search and query languages for the semantic web. International Journal of Metadata, Semantics and Ontologies 7(1), 37–54 (2012)
7. Ferré, S.: SQUALL: A controlled natural language for querying and updating RDF graphs. In: Kuhn, T., Fuchs, N.E. (eds.) CNL 2012. LNCS, vol. 7427, pp. 11–25. Springer, Heidelberg (2012)
8. Lei, Y., Uren, V.S., Motta, E.: SemSearch: A search engine for the semantic web. In: Staab, S., Svátek, V. (eds.) EKAW 2006. LNCS (LNAI), vol. 4248, pp. 238–245. Springer, Heidelberg (2006)
9. Zhou, Q., Wang, C., Xiong, M., Wang, H., Yu, Y.: SPARK: Adapting keyword query to semantic search. In: Aberer, K., et al. (eds.) ISWC/ASWC 2007. LNCS, vol. 4825, pp. 694–707. Springer, Heidelberg (2007)
10. Tran, T., Wang, H., Rudolph, S., Cimiano, P.: Top-k exploration of query candidates for efficient keyword search on graph-shaped (rdf) data. In: ICDE, pp. 405–416 (2009)
11. Cabrio, E., Cojan, J., Aprosio, A., Magnini, B., Lavelli, A., Gandon, F.: Qakis: an open domain qa system based on relational patterns. In: International Semantic Web Conference (Posters & Demos) (2012)
12. Wang, H., Zhang, K., Liu, Q., Tran, T., Yu, Y.: Q2Semantic: A lightweight keyword interface to semantic search. In: Bechhofer, S., Hauswirth, M., Hoffmann, J., Koubarakis, M. (eds.) ESWC 2008. LNCS, vol. 5021, pp. 584–598. Springer, Heidelberg (2008)
13. Lehmann, J., Bühmann, L.: Autosparql: Let users query your knowledge base. In: Antoniou, G., Grobelnik, M., Simperl, E., Parsia, B., Plexousakis, D., De Leenheer, P., Pan, J. (eds.) ESWC 2011, Part I. LNCS, vol. 6643, pp. 63–79. Springer, Heidelberg (2011)
14. Unger, C., Bühmann, L., Lehmann, J., Ngonga Ngomo, A., Gerber, D., Cimiano, P.: Template-based question answering over rdf data. In: Proceedings of the 21st International Conference on World Wide Web, pp. 639–648. ACM (2012)
15. Pradel, C., Haemmerlé, O., Hernandez, N.: Natural language query interpretation into SPARQL using patterns. In: COLD@ISWC 2013, Sydney, Australia (October 2013)
16. Pradel, C., Haemmerlé, O., Hernandez, N.: A semantic web interface using patterns: The SWIP system. In: Croitoru, M., Rudolph, S., Wilson, N., Howse, J., Corby, O. (eds.) GKR 2011. LNCS (LNAI), vol. 7205, pp. 172–187. Springer, Heidelberg (2012), http://www.springerlink.com

17. Pradel, C., Peyet, G., Haemmerlé, O., Hernandez, N.: Swip at qald-3: results, criticisms and lesson learned (working notes). In: CLEF 2013, Valencia, Spain, September 23-26 (2013)
18. Rector, A.: Modularisation of domain ontologies implemented in description logics and related formalisms including owl. In: Proceedings of the 2nd International Conference on Knowledge Capture, pp. 121–128. ACM (2003)
19. Raimond, Y., Abdallah, S., Sandler, M., Giasson, F.: The music ontology (2007)
20. Hellmann, S., Lehmann, J., Auer, S., Brümmer, M.: Integrating nlp using linked data
21. Champin, P.A.: RDF-REST: a unifying framework for web APIs and linked data (2013)

Environmental Scanning and Knowledge Representation for the Detection of Organised Crime Threats

Ben Brewster, Simon Andrews, Simon Polovina,
Laurence Hirsch, and Babak Akhgar

CENTRIC, Sheffield Hallam University, UK
{B.Brewster,S.Andrews,S.Polovina,L.Hirsch,
B.Akhgar}@SHU.ac.uk

Abstract. ePOOLICE aims at developing an efficient and effective strategic early warning system that utilises environmental scanning for the early warning and detection of current, emergent and future organised crime threats. Central to this concept is the use of environmental scanning to detect 'weak signals' in the external environment to monitor and identify emergent and future threats prior to their materialization into tangible criminal activity. This paper gives a brief overview of the application of textual concept extraction and categorization, and the Semantic Web technologies Formal Concept Analysis and Conceptual Graphs as part of the systems technological architecture, describing their benefits in aiding effective early warning.

Keywords: ePOOLICE, Environmental Scanning, Open-Source Intelligence (OSINT), Formal Concept Analysis, Conceptual Graphs, Ontological Knowledge Representation.

1   Introduction

As a result of the popularity and near ubiquitous nature of the Internet, organised crime has become ever more diverse in nature [1]. Criminal groups increasingly benefit from increased levels of collaboration, mobility around the EU and access to dynamic infrastructure, enhancing the capacity and capability of their criminal practices [2]. By their very nature, criminal groups are constantly seeking to exploit new avenues in order to sustain their illicit practices. These include, but are not limited to; drug crimes such as dealing, trafficking, cultivation and trafficking in human beings [3]. Numerous reports have studied and discussed the factors that facilitate and enable organised crime [2], [4], with others researching the mining of open-source data in order to detect communities [5] and the utility of information fusion in the detection of weak signals and the provision of strategic early warning [6].

The requirement to integrate the concepts cited has been identified in order to develop an approach to provide strategic early warning of organised crime. Although the system proposed here does not provide a comprehensive solution, it does demonstrate as a prototype how environmental scanning and semantic web technologies can be applied in order to enhance Law enforcement agencies capability in the early, strategic identification

N. Hernandez et al. (Eds.): ICCS 2014, LNAI 8577, pp. 275–280, 2014.
DOI: 10.1007/978-3-319-08389-6_22, © Springer International Publishing Switzerland 2014

combatting such threats [7], [8]. ePOOLICE proposes the development of a prototype environmental scanning system, applying a variety of state-of-the-art technological solutions, including Formal Concept Analysis and ontological knowledge representation through the use of Conceptual Graphs. At this stage it is important to note that although this paper focuses on the application of these technologies specifically, it does not form an accurate, holistic representation of the ePOOLICE project in its entirety. When considering the application of open-source scanning in this way, public privacy and surveillance fears are a key concern that must be accounted for [9]. In order to preserve the privacy of citizens, ePOOLICE refrains from the identification of specific individuals and instead focuses on the identification of patterns and observations ensuring that a 'Privacy by Design' approach to systems development is followed. For a more in depth discussion of the privacy and ethical considerations related to ePOOLICE, please see [10]. ePOOLICE integrates environmental crawling via the application of open-source scanning alongside a semantic knowledge repository, for the storage and retrieval of new and existing domain knowledge. Although not strictly part of the systems overall architecture, open-sources underpin the analytical capability of ePOOLICE through providing access to indicators, sentiment, location data and other potentially relevant concepts that may provide information that aids in increasing threat awareness. A variety of text mining and analytical approaches will be used to extract information and meaning from a number of disparate, un-structured and structured, open-source repositories. Utilising techniques that process and parse textual data in real time, the system aims to inform decision makers to assist in combatting not only current and emergent organised crime threats but also in assessing the potential for future threats.

Central to ePOOLICE, is the input from, and collaboration with end user partners from the law enforcement domain. These partners consist of law enforcement agencies themselves, criminologists and other domain experts. The knowledge provided by the end-user partners will guide the multidisciplinary, pan-European research team in the identification of poignant, current and future organised crime issues.

2   Objectives

At its core, the ePOOLICE system aims to address two main requirements;

1. The detection of organised crime: This is the early detection of existing/current and emerging organised criminal threats and criminal organisations.
2. The prediction of future organised criminal threats: Using environmental scanning it is possible to strategically assess the potential for future threats based upon historical data, patterns, and indicators derived from open-sources.

These objectives guide the principal design of ePOOLICE and are achieved through the use of what CISC (Criminal Investigation Service Canada) [11] define as 'temporal proximity'. In this definition 'primary' indicators refer to information that may be directly related to, or that occur as a result of, an organised crime (OC) threat. 'Secondary' indicators however consist of information that enables or promotes the potential for an OC threat to occur. Secondary indicators are likely to consist of information that may be identified during a PESTLE (political, economic, sociological, legal and environmental) analysis such as the identification of features including economic or political instability and legislation changes; factors that could enhance the potential for organised crime and/or increase criminal entities capacity to commit it.

3 System Architecture

Three of the key components of ePOOLICE's technical architecture (as shown in figure 1) are the environmental knowledge repository (EKR) and data fusion and analysis tools (further analysis). These components are then integrated and represented using a number of visualisation tools displaying threat indicators geographically along with subsidiary information such as trends and meta-data detailing the source and validity of indicators; information that can be applied by decision makers.

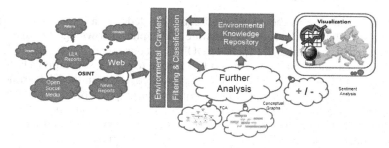

Fig. 1. ePOOLICE technical architecture overview

3.1 Data Acquisition, Content Categorization and Concept Extraction

The data acquisition aspect of ePOOLICE is concerned with the crawling and extraction of information from disparate, structured and unstructured, open sources. During extraction, data is parsed and normalised into a structured, unified format (such as XML or JSON), removing sensitive data that may contribute to victims being directly identifiable thus enabling all data to be processed and analysed via a single document pipeline, which is then marked up in preparation for further textual mining, classification and visualisation.

Post extraction, data is categorized, and key concepts identified to discern the data's relevance to specific subject areas, events and geographic locations. In addition, potential indicators or characteristics of illicit activities (both 'primary' and 'secondary" in nature) are extracted. This processing is conducted using statistical and rule based textual analysis techniques that utilize technologies such as natural language processing (NLP) to discern meaning from disparate content through the assessment of structure, linguistic patterns and concept references in the source data. For this purpose, a combination of linguistic rules, regular expressions and Boolean syntax will be applied to extract relevant concepts and identify content categories from within crawled data. For example, a newly published report such as EUROPOL's serious organised crime threat assessment [12] may contain pertinent information such as new transit routes, transportation methods and types of exploitation that can be identified using regular expressions and Boolean syntax.

Relevant indicators, terminology, known locations and routes, alongside other domain relevant knowledge is derived from the expertise of end users and semantically translated to form the taxonomy and ontology components of the system. Figure 2 gives an example of named entities that may be extracted in this way using an extract from a EUROPOL case study [13].

<DATE>On 26 March 2012</DATE>, a highly organised <CRIME>drug trafficking</CRIME> network was brought to trial in <LOCATION>Sweden</LOCATION>. Eight members of the group faced criminal charges for trafficking multi-tonne shipments of high-quality <DRUG>cocaine</DRUG> from <ORIGIN>South America</ORIGIN> to <DESTINATION>Europe</DESTINATION>. Another trial on the <CRIME>money laundering</CRIME> .....

Fig. 2. Concept Extraction

3.2   Environmental Knowledge Repository: Conceptual Graphs

Conceptual Graphs enable the representation of knowledge in a format that is discernable not only by humans but also by software and capture knowledge through the use of an ontological vocabulary [14]. In ePOOLICE, conceptual graphs are utilised to tangibly represent environmental domain knowledge using the semantic web. Formats such as RDF and OWL are used to house this knowledge within the EKR triple-store. The domain knowledge represented using conceptual graphs, corresponds directly to the taxonomy used to extract and categorize data during its acquisition. The domain knowledge embedded in the EKR provides a model upon which the system can use to effectively 'understand' crawled data sources, defining the relationships between valuable concepts such as locations and indicators.

Conceptual Graphs add value through the identification of patterns in the underlying concepts that can be identified in order to relate data to other data in the conceptual graph's vocabulary. In ePOOLICE, these value-adding features enable insights to be derived through the identification of relationships between weak indicators, which, in isolation may not seem in any way related to organised crime activity. However, by modelling existing domain knowledge using conceptual graphs, it may be possible to project and therefore discern that several weak indicators together constitute a valid indicator of illicit organised crime activity.

3.3   Situation Assessment: Formal Concept Analysis

One aspect of ePOOLICE's analytical armoury is Formal Concept Analysis (FCA), a semantic data analysis method that captures, categorizes and delivers data meaning in real time to influence decision makers through ontologically modelling the relationships between objects and their attributes [15]. FCA presents the sources of organised crime indicators as formal objects. These sources may be made up of data from open-sources and police reports. The indicators themselves are presented as formal attributes, with attributes forming the characteristics of objects. Attributes are made up of named entities such as the identification of location, time and information such as drug references, thus enabling situation assessment. As a result, formal concepts represent frequent groups of indicators along with situation assessment information. The more frequent the group, the more weight can be given to the evidence. Situation assessment allows information to be appropriately visualised depending on the requirements of the analyst. For example, a map-based system to represent threat indicators geographically may form the basis of one such approach to visualisation. This can also include a measure displaying the frequency level itself, giving the analyst control over the levels of support that they are interested in.

In FCA, the scaling of continuous attributes, such as geospatial (see Figure 3) and temporal values gives the analyst control over situation assessment. Applying a 'zoom' like functionality allows the analyst to see a more aggregated, strategic perspective when 'zoomed out', while 'zooming in' enables the focus to be concentrated on the scenarios that make up events or sources, enabling the manual assessment of extreme or erroneous data. In temporal terms, 'zooming out' gives a more strategic, historical view of events, with 'zooming in' focusing solely on the current situation.

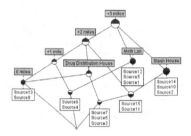

Fig. 3. FCA scaling of geospatial information

4   Concluding Remarks

ePOOLICE applies a variety of novel, state-of-the-art technologies in order to inform decision makers of current emergent, and potential future organised crime threats. Central to this capability are the semantic web technologies; FCA and Conceptual Graphs. In this overview we have presented a brief introduction to ePOOLICE, identified the rationale behind the project and described the key technological components that underpin the systems architecture.

Disclaimer
In this document, terms indicating origin or ethnicity are not being used to imply anything general or stereotypical of that origin or ethnic group. Such terms are only used as instances of factual reporting and are not be taken as a reference to any race or ethnic group as a whole. Nevertheless, for a system to monitor organized crime to operate effectively, the identification of certain elements such as gender, nationality and ethnicity, in addition to the explicit identification of crime gangs, victim groups and modus operandi are often important. For instance, Vietnamese victims are trafficked into Europe (supported by several reliable sources) and, therefore, the reference to the Vietnamese origin of criminal groups is crucial to investigate such cases as it forms a key characteristic of the phenomena described. Furthermore, when sensitive or personal data is being handled, it will be done so in accordance with laws protecting the privacy and human rights of individuals, including data protection laws.

Acknowledgement. This project has received funding from European Union Seventh Framework Programme FP7/2007 - 2013 under grant agreement n° FP7-SEC-2012-312651.

References

1. Wright, A.: Organised crime. Routledge (2013)
2. Europol. Organised Crime Threat Assessment, https://www.europol.europa.eu/sites/default/files/publications/octa2011.pdf
3. Crown Office & Procurator Fiscal Service. Solicitor General launches Serious and Organised Crime Division, http://www.copfs.gov.uk/media-site/media-releases/243-solicitor-general-launches-serious-and-organised-crime-division-socd
4. Abadinsky, H.: Organized crime. Cengage Learning (2009)
5. Tang, L., Liu, H.: Community detection and mining in social media. Synthesis Lectures on Data Mining and Knowledge Discovery 2(1), 1–137 (2010)
6. Ng, G.S., Quek, C., Jiang, H.: FCMAC-EWS: A bank failure early warning system based on a novel localized pattern learning and semantically associative fuzzy neural network. Expert Syst. Appl. 34(2), 989–1003 (2008)
7. Zenko, M., Friedman, R.R.: UN early warning for preventing conflict. Int. Peacekeeping 18(1), 21–37 (2011)
8. ePOOLICE. ePOOLICE - About, https://www.epoolice.eu/EPOOLICE/about.jsp
9. Omand, D., Bartlett, J., Miller, C.: Introducing social media intelligence (SOCMINT). Intelligence and National Security 27(6), 801–823 (2012)
10. Gerdes, A., Larsen, H.L., Rouces, J.: Issues of Security and Informational Privacy in Relation to an Environmental Scanning System for Fighting Organized Crime. In: Larsen, H.L., Martin-Bautista, M.J., Vila, M.A., Andreasen, T., Christiansen, H. (eds.) FQAS 2013. LNCS, vol. 8132, pp. 155–163. Springer, Heidelberg (2013)
11. Criminal Intelligence Service Canada (CISC). Strategic Early Warning for Criminal Intelligence, http://www.cisc.gc.ca/products_services/sentinel/document/early_warning_methodology_e.pdf
12. EUROPOL. EU Serious and Organised Crime Threat Assessment, https://www.europol.europa.eu/sites/default/files/publications/socta2013.pdf
13. EUROPOL. EU Drug Markets Report: A Strategic Analysis, https://www.europol.europa.eu/content/eu-drug-markets-report-strategic-analysis
14. Sowa, J.F.: Conceptual structures: information processing in mind and machine. Addison-Wesley Longman Publishing Co., Inc. (1984)
15. Wille, R.: Formal Concept Analysis as Mathematical Theory of Concepts and Concept Hierarchies. In: Ganter, B., Stumme, G., Wille, R. (eds.) Formal Concept Analysis. LNCS (LNAI), vol. 3626, pp. 1–33. Springer, Heidelberg (2005)

Combining Business Intelligence
with Semantic Technologies: The CUBIST Project

Frithjof Dau[1] and Simon Andrews[2]

[1] SAP AG, Germany
Frithjof.dau@sap.com
[2] Sheffield Hallam University, UK
S.Andrews@shu.ac.uk

Abstract. This paper describes the European Framework Seven CUBIST project, which ran from October 2010 to September 2013. The project aimed to combine the best elements of traditional BI with the newer, semantic, technologies of the Sematic Web, in the form of RDF and FCA. CUBIST's purpose was to provide end-users with "conceptually relevant and user friendly visual analytics" to allow them to explore their data in new ways, discovering hidden meaning and solving hitherto difficult problems. To this end, three of the partners in CUBIST were use-cases: recruitment consultancy, computational biology and the space industry. Each use-case provided their own requirements and evaluated how well the CUBIST outcomes addressed them.

1   Introduction

CUBIST - Combining and Uniting Business Intelligence and Semantic Technologies – is an EU-funded research project which ran from Oct. 2010 until Sept. 2013. This paper summarizes key achievements and results of CUBIST.

CUBIST is the joint effort of seven partners, namely SAP AG/SAP (Germany), Ontotext/ONTO (Bulgaria), Sheffield Hallam University/SHU (UK), Centrale Recherche S.A./CRSA (France), Heriot-Watt Universit/HWU (UK), Space Applications Services NV/SAS (Belgium), and Innovantage/INN (UK). SAP, ONTO, SHU and CRSA haven acted as technical partners, developing the CUBIST prototype. HWU, SAS and INN in turn have served as use-case partners.

The CUBIST project developed methodologies and a platform that combines essential features of Semantic Technologies and BI. The most-prominent deviations from traditional BI-platforms are:

- The data persistency layer in the CUBIST-prototype based on a BI enabled triple store, thus CUBIST enables a user to perform BI operations over semantic data.

- In addition to some traditional charts, CUBIST provides novel graph-based visualizations to analyse the data. Formal Concept Analysis is used as the mathematical foundation for meaningfully clustering the data.

From a user's perspective, CUBIST provides three different means to access the data:

N. Hernandez et al. (Eds.): ICCS 2014, LNAI 8577, pp. 281–286, 2014.
DOI: 10.1007/978-3-319-08389-6_23, © Springer International Publishing Switzerland 2014

- Factual search: A semantic search that allows the user to query the data in order to retrieve entities which satisfy user-defined constraints.

- Explorative search: A graph-based view that allows the user to interactively explore the data.

- Visual analytics: Clusters and aggregations of data can be visually analyzed using traditional charts or novel visualizations. The selection of the visualized data as well as the visualizations are highly interactive, thus CUBIST provides 'BI as a self-service'.

2   Architecture and Software Components

In the project, a reference architecture for a semantic BI-system has been defined. Figure 1 depicts this reference architecture. It consists of five layers:

- The Data Layer includes all structured and unstructured data sources relevant to the CUBIST system. Examples for such data are: structured relational databases and Excel files, unstructured web documents, semi-structured XML documents.

- The Semantic ETL Layer comprises different software products and components (TalenD open Studio, Revelytix Spyder, D2R Server, Gate Data Extraction) that help accessing, extracting and transforming data into a unified RDF data model so that the legacy data can be stored in the RDF data warehouse.

- The Semantic Data Warehouse is a persistence layer containing a high-performance RDF database, responsible for storing and querying semantically enriched data and its related schema information (ontology). This is provided by OWLIM: a highly scalable triple store from ONTO. In addition, this layer contains a Data Management API which enabled the semantic ETL tools to store and update data in the RDF data warehouse; as well as a SPARQL endpoint providing a data query and access interface (1).

- The CUBIST Services Layer includes the Search Service component (which offers application logic to navigate and search the semantic data warehouse) provided by NowaSearch frontend and Search Service from SAP and the FCA Service component (which manages the formal context lifecycle) provided by SHU and is a new development within CUBIST. FCAService provides features of the standalone tools FcaBedrock (2) and In-Close2 (3) as web-services.

- Finally, the CUBIST Frontend Layer is comprised of all GUI tools used by the end users including CUBIX (4) by CRSA which is a standalone FCA visualization and analysis tool newly developed for CUBIST and which serves as the main visual analytics frontend.

OWLIM is a commercial product but with a free version. FCAService and CUBIX have been published on GitHub under the under the Apache 2.0 license. NowaSearch is currently being reviewed in an SAP-internal open source approval process.

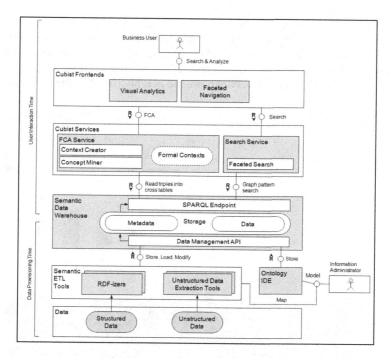

Fig. 1. FMC[1] diagram of the CUBIST reference architecture

3   User Workflow

The data schema of each use-case was modelled in RDFS. The start screen of the prototype is the "search and select view", which follows the faceted search paradigm. The facets shown there correspond to the RDFS-classes in the backend. A facet of interest (FoI) is selected (later on, entities of this class will serve as formal objects in the FCA-based visual analytics). Attributes (datatype properties) can be chosen to be visualized and used for filtering. If the attribute is a nominal attribute, checkboxes of the attribute values are used to include only objects carrying the chosen value(s). If the attribute is ordinal, the user can select intervals of values for filtering. For example, a user can select a FoI and use attributes of other facets to filter down the result set, being a subset of his FoI. In the backend, the CUBIST query generator finds the first-found smallest subgraph where all needed types are included, and uses this subgraph in its search engine. For the result-set, there is a listview where the found FoI-elements with their FoI-specific attributes are shown. A tabular view shows only the names of the FoI elements, but attributes of possibly different facets and can be visually analyzed with traditional charts or graph- and FCA-based visualizations.

If the resultset consists of only a few entities, they can serve as starting point for exploring their neighborhood in the in graph-based data. The "graph-exploration

---

[1] Fundamental Modelling Concepts (FMC).

view" provides a node-edge visualization, where the nodes depict entities (instances) in the data and labelled edges depict the selected relations between the entities. Via different user interactions, nodes and edges can be added to or removed from the view. Zooming and panning, as well as restricting the displayed nodes to regions of interest allow the user to interactively explore the neighborhood of the result set.

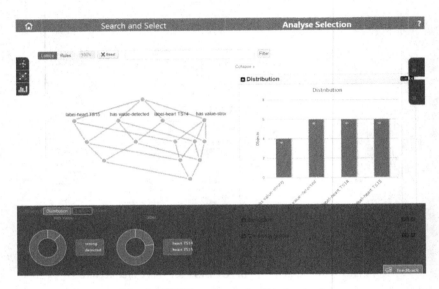

Fig. 2. CUBIX prototype for VA-frontend

A result set defined in the search and select view can be visually analyzed with FCA methods. The result set consists of entities with selected attributes, which can be strings, numbers, or date-time-values. This result set has first to be converted into a formal context. This is done in the "conceptual scaling view", where the use can scale many-valued attributes into binary formal attributes using a number of FCA techniques (5, 6). Conceptual scaling in CUBIST is the counterpart of data aggregation methods in traditional BI systems.

Finally, the scaled result set can be analyzed with FCA methods in the "analyze view". In this view, several kinds of diagrams are provided. First of all, the well-known Hasse-diagrams are utilized, as well as Sankey diagrams. Via duplication of formal concepts, the concept lattice can be turned into a tree, and different tree visualizations like sunburst-diagrams and a standard tree visualization can be used. Apart from the user-chosen main diagram, standard charts like bar charts or a graph visualization are connected to the main diagram with linking-and-brushing show specific details. The analyze view provides functionalities to interactively filter and drill down the presented data. Finally, apart from visualizing the concept lattice, association rules can be displayed, filtered and explored.

Figures 2 and 3 show some FCA-based visualizations produced for the HWU EMAGE use-case.

4   Use Cases

The three use case partners HWU, SAS and INN provided three use cases with different needs to analyze the data.

HWU provided a biomedical informatics use case. In this use case, gene expression data of mouse embryos is analyzed (7) to explore the co-occurrence of genes in different tissues (body parts of the mouse), and the change of level of expressiveness of a gene during the development of a mouse. Traditional BI tools fall short for these kind of questions, and in fact, before CUBIST there were no tools to analyze the gene expression data in this use case. The dataset of this use case contained six RDFS classes and ~1.400.000 triples.

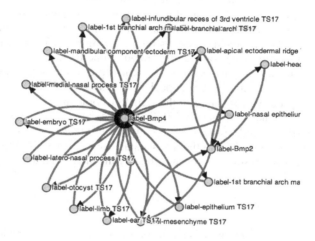

Fig. 3. Attribute implication

SAS provided a Control Center operations use case: In mission control rooms in space Control Centers very large volumes of data are obtained from heterogeneous sources. The CUBIST prototype was used to analyze telemetry data of solar equipment. This equipment logs ~200 attributes per second, and CUBIST was used to detect patterns of attributes and attribute values in this data which appear during failures of the equipment. Before CUBIST, SAS had no proper tools to conduct this task. The SAS dataset contains only one RDFS-class with ~200 properties and contained ~500.000.000 triples.

INN provided a market and competitive intelligence use case, in which information from job advertisements crawled by CUBIST and an existing firmographic database had been combined. CUBIST elicited market intelligence (insights about who is recruiting, and where and when and how they recruit) and competitive intelligence to help employers track and better understand the recruitment activity of their competitors.. The INN dataset comprised ~57.000.000 triples and seven types.

5   Evaluation and Conclusion

A detailed evaluation was carried out by end-users for each use-case, involving questionnaires, walk-throughs and problem solving tests. Overall, CUBIST was more highly rated by the HWU and SAS users than the INN users. In other words: CUBIST is better suited for the novel ways to analyse information required by the HWU and SAS use-cases and less suited for the traditional BI approach required by INN (8). The CUBIST approach is not improving, but *complementing* traditional BI-means.

A main feature of CUBIST was the integration factual search, explorative search and visual analytics data (9). The evaluation showed that supporting these different types of information need was appreciated by the users, and each of the corresponding components was rated useful for specific information needs.

The visual analytics in CUBIST are dominated by novel visualizations which show clusters and dependencies, instead of numerical results, linked to some traditional visualizations. The visual analytic features target the more novel information needs. Having different types of visualizations integrated in one BI-tool was positively rated, even if some visualizations (such as an FCA lattice) are in the beginning hard to understand. The ease of use and visual appeal were less well-rated, however, perhaps reflecting the still prototypical nature of the CUBIST components.

Finally, the users concluded that CUBIST was an expert tool. The novel approaches taken in CUBIST need some learning effort in the beginning. Nevertheless, CUBIST provided new insight into the users' data. The effort required to use the tools and read the FCA-based results was outweighed by the answers that they provided.

References

1. Dau, F.: Towards scalingless generation of formal contexts from an ontology in a triple store. International Journal of Conceptual Structures and Smart Applications 1(1), 18–37 (2013)
2. Andrews, S., Orphanides, C.: FcaBedrock, a formal context creator. In: Croitoru, M., Ferré, S., Lukose, D. (eds.) ICCS 2010. LNCS, vol. 6208, pp. 181–184. Springer, Heidelberg (2010)
3. Andrews, S.: In-Close2, a high performance formal concept miner (2011)
4. Melo, C., Mikheev, A., Le-Grand, B., Aufaure, M.-A.: Cubix: A visual analytics tool for conceptual and semantic data. In: Proceedings of the 12th International Conference on Data Mining Workshops. IEEE (2012)
5. Wolff, K.E.: A first course in formal concept analysis: How to understand line diagrams. Advances in Statistical Software 4, 429–438 (1993)
6. Andrews, S.: Data conversion and interoperability for FCA (2009)
7. Richardson, L., Venkataraman, S., Stevenson, P., Yang Y., Burton, N., Rao, J., Fisher, M., et al.: EMAGE mouse embryo spatial gene expression database: 2010 update. Nucleic Acids Res. 38(Database issue), D703–D709 (2010)
8. Orphanides, C.: Exploring the applicability of formal concept analysis on market intelligence data (2011)
9. Taylor, A., McLeod, K., Burger, A.: Semantic visualisation of gene expression information. In: Proceedings of the 3rd CUBIST Workshop. CEUR (2013)

FCA-Based Recommender Models and Data Analysis for Crowdsourcing Platform Witology

Dmitry I. Ignatov[1], Alexandra Yu. Kaminskaya[1], Natalia Konstantinova[4], Alexander Malioukov[2], and Jonas Poelmans[1,3]

[1] National Research University Higher School of Economics, Moscow, Russia
dignatov@hse.ru
http://www.hse.ru
[2] Witology, Moscow, Russia
http://www.witology.com/en
[3] KU Leuven, Belgium
[4] University of Wolverhampton, UK

Abstract. This paper considers a recommender part of the data analysis system for the collaborative platform Witology. It was developed by the joint research team of the National Research University Higher School of Economics and the Witology company. This recommender system is able to recommend ideas, like-minded users and antagonists at the respective phases of a crowdsourcing project. All the recommender methods were tested in the experiments with real datasets of the Witology company.

Keywords: collaborative and crowdsourcing platforms, data mining, Formal Concept Analysis, biclustering, recommender systems.

1 Introduction

The success of modern collaborative technologies is marked by the appearance of many novel platforms for distributed brainstorming or carrying out so called "public examination". There are a lot of such crowdsourcing companies in the USA (Spigit[1], BrightIdea[2], InnoCentive[3] etc.) and Europe (Imaginatik[4]). There is also a Kaggle platform[5] which is very beneficial for data practitioners and companies that want to select the best solutions for their data mining problems. In 2011 Russian companies decided to launch business in data mining area as well. Witology[6] and Wikivote[7] are the two most representative examples of such

---

[1] http://spigit.com/
[2] www.brightidea.com/
[3] http://www.innocentive.com/
[4] http://www.imaginatik.com/
[5] http://www.kaggle.com
[6] http://witology.com/
[7] http://www.wikivote.ru/

N. Hernandez et al. (Eds.): ICCS 2014, LNAI 8577, pp. 287–292, 2014.
DOI: 10.1007/978-3-319-08389-6_24, © Springer International Publishing Switzerland 2014

Russian companies. Several all-Russian projects have already been finished successfully (for example, Sberbank-21[8], National Entrepreneurial Initiative-2012[9] etc.). Socio-semantic networks constitute the core of such crowdsourcing systems [1,2] and new approaches are needed to analyze data underlying these networks.

The term "crowdsourcing" is a portmanteau of "crowd" and "outsourcing", coined by Jeff Howe in 2006 [3]. There is no general definition of crowdsourcing, however it has a set of specific features. Crowdsourcing is a process, both online and offline, that includes a task solving by a distributed large group of people who usually come from different organisations, and are not necessarily paid for their work. As a rule, while participating in a project, users of crowdsourcing platforms discuss and solve one common problem, propose possible solutions and evaluate ideas of each other as experts [3]. This process usually results in a reliable ranking of ideas and users who generated them. However, special means are needed in order to develop a deeper understanding of users's behavior and to perform complex dynamic and statistic analyses. Traditional methods of clustering, community detection and text mining should be adapted or even fully redesigned. Moreover, these methods require ingenuity for their effective and efficient use that will allow to find non-trivial results. We try to bridge this gap and propose new methodology that can be helpful for these crowdsourcing platforms.

This paper extends on our previously published paper [4] and is devoted to the modeling of a recommender system for crowdsourcing platforms. In the previously published paper we have already described the general methodology, however, this paper provides further details as well as through experiments. The paper has the following structure: Section 2 provides more details about Witology platform, Section 3 briefly describes FCA-based data representations and methods, Section 4 discusses the selected results of the experiments, Section 5 concludes the paper.

2   Witology Platform

For the better understanding of the problem, we provide more details about the Witology crowdsourcing project. The crowdsourcing process at this platform features eight main stages: "Solution's generation", "Selection of similar solutions", "Generation of counter-solutions", "Total voting", "Solution's improvements", "Solution's stock", "Final improvements" and finally "Solution's review".

The first stage "Solution's generation" is performed individually by each user. A key difference between the traditional brainstorming and the "Solution's generation" stage is that the participant are not aware of the ideas of other participants. The main similarity is the absence of criticism which was postponed till the later stages. In the "Selection of similar solutions" phase participants are selecting similar ideas (solutions) and their aggregated opinions are transformed into clusters of similar ideas.

---

[8] http://sberbank21.ru/
[9] http://www.asi.ru/en/

Counter-solutions generation includes criticism (pros and cons) and evaluation of the proposed ideas by means of communication between an author and experts. During this stage author of the idea can invite other experts to his team taking into account their contribution to the discussion and criticism. "Total voting" is performed by evaluating each proposed idea by all the users where votes indicate users' attitude as well as quality of the proposed solution (marks are integers between -3 and 3). Two stages, i.e. "Solution's improvements" and "Final improvements", involve active collaboration of experts and authors who work to improve their solutions together.

"Solution's stock" is one of the most interesting game stages of the project. At this stage all the participants who were able to accumulate positive reputation at the previous stages receive money in internal currency "wito" and can take part in stock trade.

The way crowdsourcing platforms work is very different from the principles underlying online-shops or specialized music/films recommender sites. Crowdsourcing projects consist of several stages where the results of each stage greatly depend on the results of a previous one. This is the reason why existing recommender models should be adapted considerably. We have developed new methods for making recommendations based on well-known mathematical approaches and propose methods for idea recommendation (for voting), like-minded persons recommendation (for collaborating) and antagonists recommendation (for contridea generation stage). To the best of our knowledge, it is the first time the last recommendation type is proposed.

3   FCA-Based Models for Crowdsourcing Data

At the initial stage of analysis of the data acquired from the collaborative platform, two data types were identified: data without keywords (links, evaluations, user actions) and data with keywords (all user-generated content).

For the analysis of the 1st type of data (without keywords) we applied Social Network Analysis (SNA) methods, clustering (biclustering and triclustering [5,6], spectral clustering), Formal Concept Analysis (FCA) [7] (concept lattices, implications, association rules) and its extensions for multimodal data, triadic, for instance [8]; recommender systems [9,10] and statistical methods of data analysis [11] (the analysis of distributions and average values).

Basic definitions of FCA and OA-biclustering can be found in [4].

It is easy to show that all key crowdsourcing platform data can be described in FCA terms by means of formal contexts (single-valued, multi-valued or triadic).

1. The data below is described by a single-valued formal context $\mathbb{K} = (G, M, J)$.

Let $\mathbb{K}_P = (U, I, P)$ be a formal context, where U is a set of users, I is a set of ideas, and $P \subseteq U \times I$ shows which user proposed which idea. Two other contexts, $\mathbb{K}_C = (U, I, C)$ and $\mathbb{K}_E = (U, I, E)$, describe binary relations of idea commenting and idea evaluation respectively.

The user-to-user relationships can also be represented by means of a single-valued formal context $\mathbb{K} = (U, U, J \subseteq U \times U)$, where $u_1 J u_2$ can designate, for example, that user u_1 commented some idea proposed by u_2. Relationships between content items can be modelled in the same way, e.g. $\mathbb{K} = (T, T, J \subseteq T \times T)$, where $t_1 J t_2$ shows that t_1 and t_2 occurred together is some text (idea or comment).

2. A multi-valued context $\mathbb{K}^W = (G, M, W, J)$ can be useful for representing data with numeric attributes.

Let $\mathbb{K}^F = (U, K, F, J)$ be a multi-valued context, where U is a set of users, K is a set of keywords, F is a set of keyword frequency values, $J \subseteq U \times K \times F$ shows how many times a particular user u applied a keyword k in an idea description or while discussing some ideas. The context $\mathbb{K}^F$ can be reduced to a plain context by means of (plain) scaling.

The commenting and evaluation relations can be described through multi-valued contexts in case we count each comment or evaluation for a certain topic. E.g, the multi-valued context $\mathbb{K}^V = (U, I, V = \{-3, -2, -1, 0, 1, 2, 3\}, J)$ describes which mark a particular user u assigns to an idea i, where V contains values of possible marks; it can be written as $u(i) = w$, where $v \in V$.

For more information about triadic models see [4].

3.1   FCA-Based Recommender Model

Two kinds of recommendations seem to be potentially useful for crowdsourcing. The first one is a recommendation of like-minded people to a particular user, and the second one is the ability to find antagonists, users which discussed the same topics as a target one, but with opposite marks.

1. Recommendations of Like-Minded Persons and Interesting Ideas.
Let $\mathbb{K}_P = (U, I, P)$ be a context which describes idea proposals. Consider a target user $u_0 \in U$, then every formal concept $(A, B) \in \mathfrak{B}_P(U, I, P)$ containing u_0 in its extent provides potentially interesting ideas to the target user in its intent and prospective like-minded persons in $A \setminus \{u_0\}$.

Consider the set $\mathfrak{R}(u_0) = \{(A, B) | (A, B) \in \mathfrak{B}_P(U, I, P)$ and $u_0 \in A\}$ of all concepts containing a target user u_0. Then the score of each idea or user to recommend to u_0 can be calculated as follows $score(i, u_0) = \frac{|\{u | u \in A, (A,B) \in \mathfrak{R}(u_o) \text{ and } i \in u'\}|}{|\{u | u \in A \text{ and } (A,B) \in \mathfrak{R}(u_o)\}|}$ or $score(u, u_0) = \frac{|\{A | u \in A \text{ and } (A,B) \in \mathfrak{R}(u_o)\}|}{|\mathfrak{R}(u_o)|}$ respectively. As a result we have a set of ranked recommendations $R(u_0) = \{(i, score(i)) | i \in B \text{ and } (A, B) \in \mathfrak{R}\}$. One can select the topmost N of recommendations from R ordered by their score.

2. Recommendations of Antagonists
Consider two evaluation contexts: the multi-valued context $\mathbb{K}^W = (U, I, W = \{-3, -2, -1, 0, 1, 2, 3\}, J)$ and binary context $\mathbb{K}_E = (U, I, E)$. Then consider (X, Y) from $\mathfrak{R}(u_0) = \{(A, B) | (A, B) \in \mathfrak{B}_P(U, I, P) \text{ and } u_0 \in A\}$. Set X contains people that evaluated the same set of topics Y, but we cannot say that all of them are like-minded people w.r.t relation E. However, we can introduce a distance measure, which shows for every pair of users from X how distant they are in marks of ideas evaluation:

$$d_{(X,Y)}(u_1, u_2) = \sum_{\substack{u_1, u_2 \in X \\ i \in Y}} |i(u_1) - i(u_2)|. \tag{1}$$

As a result we again have a set of ranked recommendations $R_{(X,Y)}(u_0) = \{(u, d(i)) | u \in U \text{ and } (A, B) \in \mathfrak{R}\}$. The topmost pairs from $R_d(u_0)$ with the highest distance contain antagonists, that is persons with the opposite views on most of the topics which u_0 evaluated. To aggregate $R_{(X,Y)}(u_0)$ for different (X, Y) from $\mathfrak{R}(u_0)$ into a final ranking we can calculate

$$d_{(u_0, u)} = \max\{d_{(X,Y)}(u_0, u) | (X, Y) \in \mathfrak{R}(u_0) \text{ and } u_0, u \in X\}. \tag{2}$$

The proposed models were tuned and validated, and they also had several variations such as using biclusters instead of formal concepts and other ways of final distance calculation.

4   Results of the Experiments

We proposed and implemented several methods for antagonists recommendation task: bicluster-based, cosine (or correlation) based, simple and bicluster-based Hamming methods. In Figure 1 we can see that for small Top-N output the best result is achieved by biclustering with Hamming metric, but for large Top-N output both methods, biclustering with correlation and simple Hamming distance, show almost equally good results. Taking into account that our goal was to reduce user's time spending on crowdsourcing tasks, we need small Top-N, and the best choice is biclustering with Hamming distance (almost 0.5 precision).

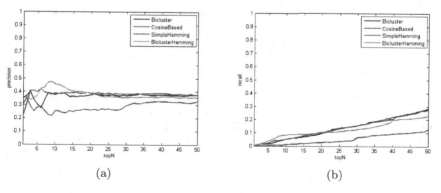

Fig. 1. Antagonists recommendation precision (a) and recall (b)

5 Conclusion

The paper presented a new methodology that can be applied to the data acquired from crowdsourcing platforms. The results of our experiments suggest that the developed methods are useful for making recommendations in the Witology crowdsourcing system and able to support user's activity on the platform.

Acknowledgments. The main part of this work was performed by the project and educational group "Algorithms of Data Mining for Innovative Projects Internet Forum". Further work was supported by the Basic Research Program at the National Research University Higher School of Economics in 2012-2014 and performed in the Laboratory of Intelligent Systems and Structural Analysis. First author was also supported by Russian Foundation for Basic Research (grant #13-07-00504)

References

1. Roth, C., Cointet, J.P.: Social and semantic coevolution in knowledge networks. Social Networks 32, 16–29 (2010)
2. Yavorsky, R.: Research Challenges of Dynamic Socio-Semantic Networks. In: Ignatov, D., Poelmans, J., Kuznetsov, S. (eds.) CDUD 2011 - Concept Discovery in Unstructured Data. CEUR Workshop Proceedings, vol. 757, pp. 119–122 (2011)
3. Howe, J.: The rise of crowdsourcing. Wired (2006)
4. Ignatov, D.I., Kaminskaya, A.Y., Bezzubtseva, A.A., Konstantinov, A.V., Poelmans, J.: FCA-based models and a prototype data analysis system for crowdsourcing platforms. In: Pfeiffer, H.D., Ignatov, D.I., Poelmans, J., Gadiraju, N. (eds.) ICCS 2013. LNCS (LNAI), vol. 7735, pp. 173–192. Springer, Heidelberg (2013)
5. Barkow, S., Bleuler, S., Prelic, A., Zimmermann, P., Zitzler, E.: Bicat: a biclustering analysis toolbox. Bioinformatics 22(10), 1282–1283 (2006)
6. Ignatov, D.I., Kuznetsov, S.O., Poelmans, J., Zhukov, L.E.: Can triconcepts become triclusters? International Journal of General Systems 42(6), 572–593 (2013)
7. Ganter, B., Wille, R.: Formal Concept Analysis: Mathematical Foundations, 1st edn. Springer-Verlag New York, Inc., Secaucus (1999)
8. Jäschke, R., Hotho, A., Schmitz, C., Ganter, B., Stumme, G.: TRIAS–An Algorithm for Mining Iceberg Tri-Lattices. In: Proceedings of the Sixth International Conference on Data Mining, ICDM 2006, pp. 907–911. IEEE Computer Society, Washington, DC (2006)
9. Ignatov, D.I., Kuznetsov, S.O.: Concept-based Recommendations for Internet Advertisement. In: Belohlavek, R., Kuznetsov, S.O. (eds.) Proc. CLA 2008. CEUR WS, vol. 433, pp. 157–166. Palacky University, Olomouc (2008)
10. Ignatov, D.I., Poelmans, J., Dedene, G., Viaene, S.: A New Cross-Validation Technique to Evaluate Quality of Recommender Systems. In: Kundu, M.K., Mitra, S., Mazumdar, D., Pal, S.K. (eds.) PerMIn 2012. LNCS, vol. 7143, pp. 195–202. Springer, Heidelberg (2012)
11. Clauset, A., Shalizi, C.R., Newman, M.E.J.: Power-law distributions in empirical data. SIAM Rev. 51(4), 661–703 (2009)

Conceptual Structures
in LEADing and Best Enterprise Practices

Simon Polovina[1], Mark von Rosing[2], and Wim Laurier[3]

[1] Conceptual Structures Research Group
Sheffield Hallam University, UK
S.Polovina@shu.ac.uk
www.shu.ac.uk/research/c3ri
[2] LEADing Practice ApS, Denmark
mvr@leadingpractice.com
www.leadingpractice.com
[3] Université Saint-Louis – Bruxelles, Belgium
Universiteit Gent, Belgium
www.usaintlouis.be
www.ugent.be

Abstract. Conceptual Structures, namely Conceptual Graphs (CGs) and Formal Concept Analysis (FCA) are beginning to make an impact in Industry. This is evidenced in LEAD as it seeks to provide its 3100+ industry practitioners in many Fortune 500 and public organisations with capabilities that can handle ontology and semantics. The existing ontology and semantics work in LEAD, supported by the Global University Alliance, is described and how CGs, FCA and their tools (e.g. CoGui, CG-FCA) enhance this endeavour.

1 Introduction

Despite its long incubation period, Conceptual Graphs (CGs) and Formal Concept Analysis (FCA) are beginning to make an impact in Industry. Like the Semantic Web and the emerging use of Ontology tools such as Protégé in Enterprise tools such as Essential Project, Conceptual Structures (that CGs and FCA epitomise) reflect industry's need for enriched knowledge capture and reasoning tools that extend beyond the existing provision [1, 2, 3, 4, 5]. The LEADing Practice Community (LEAD) is one such recognition, as it seeks to provide its 3100+ industry practitioners in many Fortune 500 and public organisations with tools that can handle ontology and semantics. In LEAD, Conceptual Structures are becoming pivotal in the way of thinking, working and modelling around LEAD's enterprise and industry standards.

2 LEAD

Founded in 2004, LEAD was originally an acronym for *Layered Enterprise Architecture Development* but now simply refers to itself as the LEADing Practice Community that develops Enterprise Standards [6]. This is because it now includes

N. Hernandez et al. (Eds.): ICCS 2014, LNAI 8577, pp. 293–298, 2014.
DOI: 10.1007/978-3-319-08389-6_25, © Springer International Publishing Switzerland 2014

Enterprise Modelling (including Business Process Management) and Enterprise Engineering as well as Enterprise Architecture. LEADing Practice's CEO (Chief Executive Officer) Henrik von Scheel describes it as:

"LEADing Practice represents a new breed of Enterprise Standards and is recognized as a paradigm shift by the global business and IT community to empower through its Reference Content, enabling organizations to innovate, transform and deliver value". [6]

LEAD is based on an open source community concept that is adopted by most of the Fortune 500 and public organizations, and is integrated into software solutions such as SAP (ASAP Methodology), IBM Rational, IBM System Architect, iGrafx and Software AG (ARIS). Today, LEADing Practice is the fastest growing open source standard development community, supported by the 2nd largest certified community of 3100+ industry practitioners. LEAD has developed 90 different Enterprise Standards with detailed reference content as well as over 51 different Industry User Group Committees that provides a global platform for industry executives, experts, academics, thought-leaders, practitioners and researchers to develop, use and apply [6]. The LEADing Practice reference content connects to all the major existing enterprise architecture and other frameworks, methods and approaches (such as TOGAF, Zachman, FEAF, ITIL, Prince2, COBIT, DNEAF, and others). This is tabulated by LEAD [7].

To mention some examples, the LEGO Group, famous for its Lego bricks, provides a detailed exposition of the use of LEADing Practices and how to combine business model, process model, performance management, and information aspects. This led to that the LEGO Group receiving the prestigious Gartner Group award of being the Best BPM organization [8]. There are many other success stories that also include non-commercial organisations, and government bodies such as the Government of Germany, US Government, the Government of Canada and many others [6].

To assist its industry practitioners and the development of the discipline as a whole, LEAD provides a multitude of reference content with meta-object, descriptions, templates and Hands-On Modelling rules and tools. In order to enable a repeatable and structured way, LEAD is structured as a "Way of Thinking, Way of Working, Way of Modelling, Way of Implementation and Way of Governance", with each way setting the context of the next one in this list. It reflects the architectural principle of capturing the very purpose of an enterprise (its vision and mission) right down to the individual assets (e.g. purchases, sales, employees, IT support systems) that fulfil that purpose.

LEADing Practice's Enterprise Standards are developed by a) Researching and analysing industry best practice & leading practices, b) Identifying common and repeatable patterns (the basis of LEAD's standards), c) Developing the Enterprise Standards that increase the level of re-usability and replication, and d) Build industry accelerators within the standards, enabling to adopt and reproduce the best & leading practices. LEAD is therefore practically oriented, but based on a strong theoretical base that it gathers from its research partner, the Global University Alliance.

Global University Alliance (GUA). Also founded in 2004 by one of us (von Rosing) and in support of developing Best and Leading Practices, the Global University Alliance (GUA) is stated as a non-profit organisation and international consortium of university tutors and researchers whose aim it is to provide a collaborative platform for academic research, analysis and development and to explore leading practices, best practices as well as to identify missing gaps in those practices [9]. Academic research is therefore combined with industry practice, providing industry practitioners with a strong theoretical basis whilst providing academics with industry experiences (such as the companies referred to earlier) to test that base. Conceptual Structures has a natural fit with this approach, as enterprises need to draw on enriched models to capture their way of thinking in their way of modelling, and ensure they are optimising the intellectual and physical assets they have own to achieve to fulfil their vision and mission. Enterprises are creative knowledge-based endeavors whereas computers prefer simpler structures such as data that they can easily process. Conceptual Structures harmonise the creativity of humans with the productivity of computers. Put simply, Conceptual Structures recognise that organisations work with concepts; machines like structures. It connects the user's conceptual approach to problem solving with the formal structures that computer applications need to bring their productivity to bear. This will be discussed later, once the role of ontology and semantics in LEAD through the GUA is understood.

3   Ontology and Semantics in LEAD

To support LEAD and the various enterprise practitioners, the GUA provides a number of resources that encompass the study of Enterprise DNA, Enterprise Philosophy, Enterprise Ontology, Enterprise Semantics, Information Management, BPM, Enterprise Architecture, Sustainability, and Industry Standards [10]. Two of us (Laurier and Polovina) are responsible for Ontology and Semantics respectively in the GUA and can be seen in the foundation of the Figure on LEAD's home page [6]; accordingly these are described in more detail as follows.

The Enterprise Ontology Reference Framework. Ontology formally represents knowledge as a set of concepts within a domain, and the relationships between those concepts. It is the theory of being. It can be used to model a domain and support reasoning about concepts. The Global University Alliance (GUA) supports LEAD and other contemporary enterprise practices through its Enterprise Ontology Reference Framework (EORF) [11]. It describes ontology as a shared vocabulary and the definition of its objects or concepts, and their properties and relations within and across key business domains as illustrated by the Framework. Along the way, EORF's modelling and architecture principles have attracted software vendors. These include SAP AG, IBM, and iGrafx and have used or adapted the EORF's meta-model. The last mentioned, iGrafx, currently incorporates EORF's modelling aspects into their extended process methods and meta-models [12]. Another is Essential Project, which through its productive use of Protégé in its Essential Manager and its own Visualisation tool (Essential Viewer) requires ontology as its building blocks to help organisations analyse and manage the knowledge needed to make decisions that impact or are impacted by their own enterprise architecture [3, 2]. The GUA describes EORF in further detail including its rapid adoption by industry, taking advantage of academic expertise [11].

The Enterprise Semantics Reference Framework. The GUA also provides the Enterprise Semantics Reference Framework (ESRF) [13]. The ESRF aligns with the EORF above. Together they underpin LEAD's Reference Content. The term Semantics arises from Ancient Greek: σημαντικός - sēmantikós - being the study of meaning. It denotes the relation between signifiers, symbols and objects. Semantics (or Semantic) is about making meaning from the objects using the best possible signifiers and symbols. It includes those signifiers and symbols used to describe the relationship between objects as this comparison further enhances their meaning. Enterprise Semantics is therefore the study of objects and symbols used to describe the enterprise, what they stand for, their underlying formal logics and their relationship and correlation. It supports Enterprise Pragmatics - the sharing of meanings gathered from Enterprise Semantics across its diverse interpretations in practice, leading towards a universal truth whilst maintaining these wide-ranging interpretations and beliefs in the real world. Through this way of thinking and working it sets the way of modelling – the impact of modelling and engineering as it most usefully aligns the relevant parts of the enterprise. The ESRF includes the objects, their properties and relations found throughout Enterprise Architecture, Modelling and Engineering. The ESRF at its core consists of a number of holistic meta-models and fully integrated templates (e.g. maps, matrices and models). The scope of the ESRF is detailed further by the GUA [13]. Like the EORF, the ESRF has attracted software vendors as already described above under the previous EORF heading.

4   Conceptual Structures in LEAD

Conceptual Graphs. As indicated earlier, LEAD consists of a number of meta-models. Like other enterprise meta-models, The LEAD meta-models define a formal structure for the concepts or objects and their relations. These structures ensure self-consistency and to provide a computational basis for enterprise modelling tools (e.g. iGraphx), especially because simple drawing tools alone do not capture the ontology and semantics that enterprise tools require. That requirement enables the tool to assist an organisation in capturing its conceptual structure thus aligning its IT systems with its corporate vision and mission, as described earlier. The GUA therefore uses CGs as they characterise conceptual structures using the same concept-relation fundamental as the meta-models. This is illustrated by Figure 1 that shows the use of the CGs tool CoGui [14]. In CoGui the GUA are then able to explore the meta-models (in this case the Performance meta-model) to identify, for example, novel indirect relations that provide higher-level abstractions, or patterns, in that model or the actual organisation's population of the meta-model.

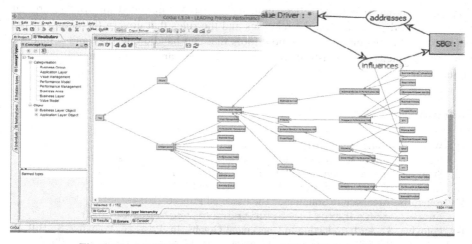

Fig. 1. The LEAD Performance Meta-model in CGs, using CoGui

Formal Concept Analysis. Using CGs the cognitive-level models (drawn by hand from the minds of experts) can be taken further and explored at the logical level, where for example apply logical operators can be applied (e.g. CGs maximal join or projection operations) [4]. Using FCA the GUA has also been able to explore the concepts (objects) and relations (attributes) at the lower but even more rigorous mathematical level to discover e.g. hidden relations or un-named concepts (pointing possibly to a new object). In particular, CG-FCA is used that once visualised in the FCA tool Concept Explorer (ConExp) produces a lattice as illustrated by the extract shown in Figure 2 [15].

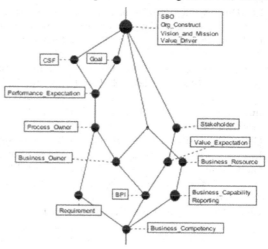

Fig. 2. The LEAD Performance Meta-model in ConExp, after conversion by CG-FCA

5   Concluding Remarks

We have demonstrated how Conceptual Graphs (CGs) and Formal Concept Analysis (FCA) are making an impact in Industry through the GUA into LEAD. As stated, these Conceptual Structures reflect industry's need for enriched knowledge around meta-objects, their attributes, the relations and the rules that govern the nature of the objects and the models. It reveals how tools that can handle ontology and semantics can be applied to industrial-strength enterprise architecture, modelling and engineering.

References

[1] W3C, "Semantic Web" (2013),
http://www.w3.org/standards/semanticweb/ (accessed 2014)

[2] Stanford Center for Biomedical Informatics Research, "Protégé" (2014),
http://protege.stanford.edu/ (accessed 2014)

[3] Enterprise Architecture Solutions, "The Essential Project" (2014),
http://www.enterprise-architecture.org/ (accessed 2014)

[4] Polovina, S.: An Introduction to Conceptual Graphs. In: Priss, U., Polovina, S., Hill, R. (eds.) ICCS 2007. LNCS (LNAI), vol. 4604, pp. 1–14. Springer, Heidelberg (2007)

[5] Andrews, S., Orphanides, C., Polovina, S.: Visualising Computational Intelligence through Converting Data into Formal Concepts. In: Bessis, N., Xhafa, F. (eds.) Next Generation Data Technologies for CCI. SCI, vol. 352, pp. 139–165. Springer, Heidelberg (2011)

[6] LEADing Practice, "Welcome to LEADing Practice",
http://www.leadingpractice.com/ (accessed 2014)

[7] LEADing Practice, "LEADing Practice Interconnects with Main Existing Frameworks" (2014), http://www.leadingpractice.com/about-us/
interconnects-with-main-existing-frameworks/ (accessed 2014)

[8] von Rosing, M., von Scheel, H., Falk Bøgebjerg, A.: The LEGO LEADing BPM Practice Case Story (2013), http://www.leadingpractice.com/wp-content/
uploads/2013/10/LEGO-LEADing-BPM-Practice-Case-Story.pdf
(accessed 2014)

[9] Global University Alliance, "Home - Global University Alliance" (2014),
http://www.globaluniversityalliance.net/ (accessed 2014)

[10] Global University Alliance, "Research Areas" (2014),
http://www.globaluniversityalliance.net/research-areas/
(accessed 2014)

[11] Global University Alliance, "The Enterprise Ontology Reference Framework" (2014),
http://www.globaluniversityalliance.net/
research-areas/enterprise-ontology/ (accessed 2014)

[12] iGrafx, "Enabling Process Excellence" (2014),
http://www.igrafx.com/ (accessed 2014)

[13] Global University Alliance, "The Enterprise Semantics Reference Framework" (2014),
http://www.globaluniversityalliance.net/
research-areas/enterprise-semantics/ (accessed 2014)

[14] LIRMM, "GoGui - A Conceptual Graph Editor" (2014),
http://www.lirmm.fr/cogui/ (accessed 2014)

[15] Andrews, S., Polovina, S.: A Mapping from Conceptual Graphs to Formal Concept Analysis. In: Andrews, S., Polovina, S., Hill, R., Akhgar, B. (eds.) ICCS 2011. LNCS (LNAI), vol. 6828, pp. 63–76. Springer, Heidelberg (2011)

Investigating Oncological Databases
Using Conceptual Landscapes

Christian Săcărea

Babeş-Bolyai University, Department of Computer Science, Cluj-Napoca, Romania
csacarea@math.ubbcluj.ro

Abstract. This paper presents an application of the conceptual land-scapes paradigm in the representation of the knowledge content of on-cological databases. Even if the method is not new, to the best of our knowledge it is the first time when applied in the study of oncologi-cal data. Moreover, building knowledge management systems for medi-cal databases might be of interest for large scale health-care industrial applications of Formal Concept Analysis. Conceptual Landscapes is a paradigm of Knowledge Representation which is grounded on Concep-tual Knowledge Processing. Using the mathematical apparatus of For-mal Concept Analysis and the knowledge management suite ToscanaJ, as well as a triadic extension called Toscana2Trias, we present several issues related to the study of adverse drug reactions in oncology using concep-tual landscapes, as well as building a knowledge management system of a cancer registry database according to the principles of Conceptual Knowledge Processing.

1 Introduction

Formal concept analysis is, by now, an established method for mining knowledge in a variety of fields and it has a number of applications in medical knowledge discovery as well, including adverse drug reaction (ADR) discovery [1]. However, ADR analysis involves more than just discovering new, previously unknown re-actions. For highly toxic or dangerous treatments, such as cancer treatments, understanding the factors that make the appearance of a certain adverse reac-tion more likely or more intense, or, on the contrary, less likely or less intense is of great importance for the well being of the patient. This type of data that involves not only patients and reactions, but also risk factors is called triadic data and is rather difficult to analyze in a coherent manner. Triadic concepts were developed [2] in order to provide a tool for this kind of analysis and some applications have already been developed, mainly dealing with the analysis of folksonomies [3]. More recent contributions deal with triadic FCA and Factor Analysis [4], [5].

This paper presents an application of triconcepts for adverse drug reactions analysis using the TRIAS algorithm [6] applied to an oncology medical database. The application can be used to discover frequent triconcepts that link specific

N. Hernandez et al. (Eds.): ICCS 2014, LNAI 8577, pp. 299–304, 2014.
DOI: 10.1007/978-3-319-08389-6_26, © Springer International Publishing Switzerland 2014

factors in patient treatment, demographics or history to combinations of adverse reactions, providing a useful tool for analyzing known risk factors, as well as discovering new ones, such as drug interactions that increase or decrease the frequency of certain adverse reactions. We also give an insight of using the conceptual landscapes paradigm for knowledge representation of oncological data sets [7]. More precisely, this paper gives a brief overview of a health-care project conducted at the *Dr. I. Chiricuta* Oncological Institute of Cluj-Napoca, Romania aiming to apply the methods of Conceptual Knowledge Processing to oncological data. Two different medical datasets have been used, the cancer registry and a small breast cancer adverse drug reactions dataset.

Even if the methods are classical in the FCA community, the results have a high potential for health-care industrial applications of FCA and its extensions (triadic, relational, temporal FCA), only a few of them being discussed in this article, due to space limitations.

2   Data Sets and Methods

The cancer registry is comprising data about several thousands patients the most relevant attributes being the patient's age, gender, survival status, details about the tumor (topography, morphology, behavior, differentiation, stage), and details about the type of treatment received (surgery, radiotherapy, chemotherapy, or hormonal therapy). In its original form, it contains 25 attributes for each patient.

The main objective of the cancer registry is to collect and classify information on all cases of the cancer in a defined population and to provide analysis on these data. The registry receives records relating to an individual patient from multiple sources. The usual analysis are represented by a) trends in cancer incidence and mortality, b) evaluation of mass screening, c) clinical care evaluation.

The adverse drug reactions dataset consists of 60 entries with a number of 73 attributes. The database contains information pertaining to patient demographics, type and location of malignant cells, treatments and their outcome, and adverse drug reactions (nausea, vomiting, neutropenia, peripheral nephropathy, hair loss, skin-hair-nails, edema).

We have used the knowledge management suite ToscanaJ ([8], [9]) and Toscana2Trias, a triadic extension developed at the Department of Computer Science of the Babes-Bolyai University of Cluj-Napoca, Romania. Toscana2Trias uses the TRIAS algorithm developed by R. Jaeschke et al. [6]. Basically, Toscana2Trias allows to connect to a database, and then displays the table names (or attribute names). The user may define, according to his own view, which are the objects, the attributes and the conditions. The ternary incidence relation is then read from the database. Moreover, if a conceptual schema has been built upon the data set, i.e., the data has been preprocessed for ToscanaJ, then the user has even more control over the selection of the objects, attributes and conditions. From the conceptual schema, a part of the scaled attributes can be considered as conditions, the rest being considered as attributes in the tricontext. Triadic concepts are then computed, using the Trias algorithm. These triconcepts are then displayed in a variety of formats, their connections can be visualized by any graph visualization software. If the data set is larger, this visualization becomes easily obscure because of the number of triconcepts. In this case, we can compute only the frequent triconcepts by setting a treshold. For more about triadic FCA, please refer to [2].

3   Building Conceptual Landscapes

The scaling effort was focused on attributes describing treatment options and their results, the type and location of cancerous cells and the adverse drug reactions. Although the adverse reactions present in the database refer only to the Docetaxel (DTX) treatment, all treatment attributes were scaled and represented as diagrams. For both datasets (Cancer Registry and ADR database), the results illustrate perfectly how conceptual landscapes are useful in the medical domain. For instance, mining all side effects of the Docetaxel (DTX) treatment (dosage and number of cycles) together with the cancer data (localization of the primary tumor, metastasis, cancerous cell types), patient data and the intensity and combination of the adverse reactions, offers a new way of perceiving all implications that may arise.

1. The Patient Landscape: this landscape includes all conceptual diagrams built from patient related data, such as age group, that the medical professionals considered as relevant. Figure 1 illustrates a small part of this conceptual landscape.

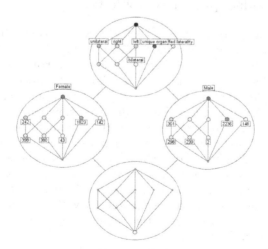

Fig. 1. Nested line diagram

2. The Disease Landscape: this includes all conceptual diagrams built from data related to the disease, including localization of the primary tumor, presence and localization of metastasis or type of cancerous cells. Figure 2 represents the carcinoma map of the conceptual system we have built.

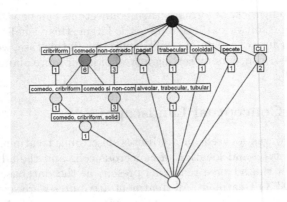

Fig. 2. Carcinoma map

3. The Treatment Landscape: this includes all conceptual diagrams built from data pertaining to the treatment or treatments the patient has undergone (type of drugs, number of cycles, dosage), the evolution of the patient and the adverse drug reactions (intensity and combinations).

Combining these three landscapes, different browsing scenarios can be formed using only a small subset of the diagrams (the same diagram can even be included more than once): Age - Survival - Topography, Treatment - Survival - Vitality - Cause of death, Age - Topography - Vitality - Survival - Cause of death.

4   Triadic Landscapes of Knowledge

A triadic context K is a quadruple $(K1, K2, K3, Y)$ consisting of three sets $K1, K2, K3$ and a ternary relation Y between them, i.e., $Y \subseteq K1 \times K2 \times K3$. The elements of $K1, K2,$ and $K3$ are called objects, attributes, and conditions, respectively. A triadic context is read as follows: an object o has attributes $\{a_1, a_2, \ldots, a_i\}$ under conditions $\{c_1, c_2, \ldots, c_j\}$. This approach is well suited to medical data and not just to the adverse drug reaction analysis. For instance, a selection of natural triads arising from the datasets might be (Patient, adverse drug reactions, treatment), (Cancer, evolution, drug) or (Cancer, metastasis location, patient demographic). Considering patient IDs as objects, adverse drug reactions as attributes and certain patient demographics as conditions can offer a view of what patient related factors can be considered risk factors for adverse reactions. The triconcepts for this selection show the distribution of adverse drug reactions relative to the age of the patient. A partial list of the obtained triconcepts is given in Table 1.

Out of the 35 concepts, 14 are combinations of adverse reactions for the 30-40 age interval and 7 are combinations of adverse reactions for the 40-50 age interval, leading to the conclusion that the greatest variety of reactions happen between the ages of 30 and 50 years. The most frequent combination is nausea and vomiting, occurring in 50% of the patients in the 30-40 age interval and 43% of the patients in the 40-50 age interval. 42% of the patients in the 30-40 age

Table 1. The percentages in the table are relative to the value of the condition, in this case, the percentage of patients that fall into the age interval and exhibit the specified drug reactions

	Age interval	Adverse reactions	%
1.	30-40	Nausea, neutropenia	21%
2.	30-40	Nausea, vomiting, nephropathy, edema, hair loss	7%
3.	30-40	Hair loss, skin-flesh-nails	28%
4.	30-40	Nausea, skin-flesh-nails	28%
5.	40-50	Nausea, vomiting, nephropathy	21%
6.	60-70	Nausea, neutropenia	66%
7.	60-70	Nausea	100%

Table 2. Triadic concepts: ID-Intensity of reactions-Age

	Age interval	Intensities	%
1.	30-40	1,2,3	21%
2.	30-40	2,3	28%
3.	30-40	2,4	7%
4.	40-50	1,2	39%
5.	40-50	1	56%
6.	40-50	1,2,3	8%
7.	50-60	3	26%
8.	50-60	1,2,3	21%
9.	60-70	1	100%
10.	60-70	1,2	66%
11.	60-70	1,2,3	33%

interval also suffer hair loss. This combination of reactions (nausea, vomiting and hair loss) also appears in the 40-50 age interval, but is always accompanied by other reactions such as neutropenia or nephropathy. To see the distribution of the intensity for adverse reactions by age group, the concepts are built with patient IDs as objects, age as conditions and the intensity of the reactions as attributes. The results are presented in table 2.

In this case there are only 11 concepts because there are less possible combinations of intensities (intensities are represented by 4 degrees) than there are combinations of adverse reactions. Table 2 shows that all patients over 60 years of age suffer adverse reactions of intensity at least 1, while only 56% of patients in the 40-50 age interval suffer adverse reactions of intensity at least 1. The 30-40 age interval is characterized by a greater percentage of patients that suffer adverse reactions of degrees 1, 2 and 3 (21%) compared to the 40-50 age interval, where only 8% of patients suffer all three adverse reaction degrees. The triadic concepts for this distribution are represented in Figure 3.

Concepts consisting of treatments as objects and adverse reactions as attributes can be used to discover adverse reactions to drug interactions if another

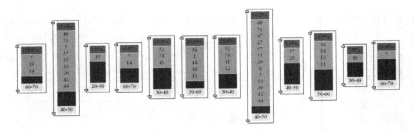

Fig. 3. Distribution of the intensity for adverse reactions by age group

treatment is selected as condition. If there are more treatment options available, selecting a demographic factor as object, the adverse reactions as attributes and the available treatments as conditions can show which treatment yields more adverse reactions for a demographic group and help decide which treatment course to administer.

References

1. Estacio-Moreno, A., Toussaint, Y., Bousquet, C.: Mining for adverse drug events with formal concept analysis. Stud. Health Technol. Inform. 136, 8038 (2008)
2. Lehmann, F., Wille, R.: A Triadic Approach to Formal Concept Analysis. In: Ellis, G., Levinson, R., Rich, W., Sowa, J.F. (eds.) Conceptual Structures: Applications, Implementation and Theory 1995. LNCS (LNAI), vol. 954, pp. 32–43. Springer, Heidelberg (1995)
3. Hotho, A., et al.: Folkrank: A ranking algorithm for folksonomies. In: Sure, Y., Domingo, J. (eds.) Proc. FGIR, pp. 2–5 (2006)
4. Glodeanu, C.V.: Tri-ordinal Factor Analysis. In: Cellier, P., Distel, F., Ganter, B. (eds.) ICFCA 2013. LNCS, vol. 7880, pp. 125–140. Springer, Heidelberg (2013)
5. Belohlavek, R., Glodeanu, C., Vychodil, V.: Optimal factorization of three-way binary data using triadic concepts. Order 30(2), 437–454 (2013)
6. Jaschke, R., Hotho, A., Schmitz, C., Ganter, B., Stumme, G.: TRIAS–An Algorithm for Mining Iceberg Tri-Lattices. In: Sixth IEEE International Conference on Data Mining (ICDM 2006), pp. 907–911 (2006)
7. Wille, R.: Methods of Conceptual Knowledge Processing. In: Missaoui, R., Schmidt, J. (eds.) ICFCA 2006. LNCS (LNAI), vol. 3874, pp. 1–29. Springer, Heidelberg (2006)
8. ToscanaJ homepage, http://toscanaj.sourceforge.net
9. Becker, P., Hereth, J., Stumme, G.: ToscanaJ: An Open Source Tool for Qualitative Data Analysis. In: Proc. Workshop FCAKDD of the 15th European Conference on Artificial Intelligence, ECAI 2002 (July 2002)

Eco-Efficient Packaging Material Selection for Fresh Produce: Industrial Session

Nouredine Tamani[1], Patricio Mosse[2], Madalina Croitoru[1], Patrice Buche[2], Valérie Guillard[2], Carole Guillaume[2], and Nathalie Gontard[2]

[1] LIRMM, University Montpellier 2, France
[2] UMR IATE INRA 2 Place Pierre Viala Montpellier, France

Abstract. Within the framework of the European project EcoBioCap (ECOefficient BIOdegradable Composite Advanced Packaging), we model a real world use case aiming at conceiving the next generation of food packagings. The objective is to select packaging materials according to possibly conflicting requirements expressed by the involved parties (food and packaging industries, health authorities, consumers, waste management authority, etc.). The requirements and user preferences are modeled by several ontological rules provided by the stakeholders expressing their viewpoints and expertise. To deal with these several aspects (CO_2 and O_2 permeance, interaction with the product, sanitary, cost, end of life, etc.) for packaging selection, an argumentation process has been introduced.

1 Introduction

Within the framework of the European project EcoBioCap (www.ecobiocap.eu) about the design of next generation packagings using advanced composite structures based on constituents derived from the food industry, we aim at developing a Decision Support System (DSS) to help parties involved in the packaging design to make rational decisions based on knowledge expressed by the experts of the domain.

The DSS is made of two parts, as depicted in Figure 1:

1. a flexible querying process which is based on a bipolar approach dealing with imprecise data [8] corresponding to the characteristics related to the food product to pack like the optimal permeance, the dimension of the packaging, its shape, etc.,
2. an argumentation process which aims at aggregating several stakeholders (researchers, consumers, food industry, packaging industry, waste management policy, etc.) requirements expressed as simple textual arguments, to enrich the querying process by stakeholders' justified preferences. Each argument supports/opposes a choice justified by the fact that it either meets or not a requirement according to a particular aspect of the packagings.

We implemented of the second part of the DSS, called argumentation system, which aims at aggregating preferences associated with justifications expressed by stakeholders about the characteristics of a packaging. This module

N. Hernandez et al. (Eds.): ICCS 2014, LNAI 8577, pp. 305–310, 2014.
DOI: 10.1007/978-3-319-08389-6_27, © Springer International Publishing Switzerland 2014

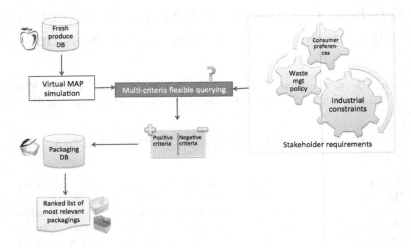

Fig. 1. Global insight of the EcoBioCap DSS

has as inputs stakeholders' arguments supporting or opposing a packaging choice which could be seen as preferences combined with their justifications, and returns consensual preferences which may be candidates to enrich the bipolar querying system.

The DSS consists of two steps: (i) aggregating possibly conflicting needs expressed by the involved several parties (ii) querying a database of packagings with the resulting aggregation obtained at point (i).

In this real case, packagings have to be selected according to several aspects or criteria (permeance, interaction with the packed food, end of life, etc.), highlighted by the expressed stakeholders' arguments. The problem at hand does not simply consist in addressing a multi-criteria optimization problem [4]: the domain experts would need to be able to justify why a certain packaging (or set of possible packagings) are chosen. Argumentation theory in general [9,3,11] is actively pursued in the literature, some approaches even combining argumentation and multi criteria decision making [2].

2   Approach

Stakeholder's set of arguments i is then modeled as concepts, facts and rules to build a partial knowledge bases $\mathcal{K}_{\mathcal{I}_i}$. The union of every stakeholder knowledge base $\mathcal{K} = \bigcup_{i=1,\ldots,n} \mathcal{K}_{\mathcal{I}_i}$ will be used to instantiate the ASPIC [1] argumentation system, as shown on Figure 2.

The main salient points of our work in the EcoBioCap project are the following:

1. A DLR-Lite [7,5] ontology extended to a negation to express stakeholders' arguments about packaging characteristics as combination of concepts (defined

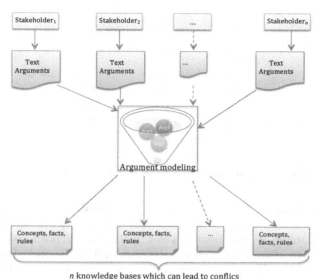

Fig. 2. The global knowledge base of the system

as m-ary relations connected to a database) and inference rules (specified as subsumptions). The language is detailed in the technical report [12],

2. An instantiation of ASPIC argumentation system (AS) with the proposed DLR-Lite logical language. The instantiated ASPIC AS satisfies the rationality postulates [6], please see details in [12],

3. The study of the influence of the modeling rules on the argumentation results. We showed the limitation of the crisp split of the inference rules into *defeasible* and *strict*, and we propose to overcome this limitation a viewpoint approach in which arguments are gathered according to packaging aspects. Each viewpoint delivers subsets of non-conflicting arguments supporting or opposing a kind of packaging according to a single aspect (shelf life parameters, cost, materials, sanitary, end elf life, etc.),

4. The use of the argumentation results for a bipolar querying of the packaging database. Indeed, we can gather the results onto positive and negative collections. We can then deduce automatically such queries from the collections the users formed during the argumentation process.

5. Implementation of the approach. A java GXT/GWT web interface was developed and a open version is accessible on
   http://pfl.grignon.inra.fr/EcoBioCapProduction/.

3 Architecture of the Argumentation System

As illustrated in Figure 3, the proposed argumentation system relies on 5 main modules, described below.

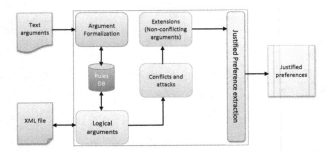

Fig. 3. The architecture of the argumentation system

- *Argument formalization module*: this module implements a user-friendly interface for a semi-automatic translation of text arguments into a formal representation made of concepts and rules (claims and hypothesis). A graphical representation of the expressed rules is also built as the users formalize their text arguments. The formal representation obtained is finally saved in a database for a persistent storage allowing to reload argumentation projects without rebuilding all the arguments and to reuse also the already formatted rules in other projects.
- *Logical arguments*: this module receives as inputs the list of concepts and rules corresponding to text arguments. This list can be the result of the formalization module or given by the user as an XML file. Then, by a derivation process, this module builds all possible arguments according to the logical process defined in ASPIC/ASPIC+ logic-based argumentation frameworks [1,10] and reused in [13,14]. This modules implements also a function to export the argument list into an XML document.
- *Conflicts and attacks*: this module relies on the logical arguments built by the previous module. According to the negation operator, it detects all the conflicts among arguments and models them as attacks with respect to the definition of attacks introduced in [13,14]. The output of this module is an argumentation graph made of arguments (nodes) and attacks (edges).
- *Extensions*: an extension is a subset of non-conflicting (consistent) arguments which defend themselves from attacking arguments. The computation of extensions is made under one semantics (preferred, stable, grounded, etc.) as defined in [9]. This module allows the computation of one or all semantics considered (preferred, stable, grounded, eager, semi-stable). We notice that theoretically we can get empty extensions under any semantics. This situation occurs when a user expresses at least one self-defeated argument, which is not attacked by any other argument, but attacks all the others. This kind of arguments are called contaminating arguments [15]. The current version of the system detects the rules leading to such arguments and discards them before performing the process of extension computations.
- *Extraction of the justified preferences*: the computation of extensions delivers one or several extensions. In the case of several extensions, the system

lets the users selecting the more suitable one according to their objectives. The selected extension is then used to extract corresponding preferences underlying the contained concepts. These preferences are expressed as a list of couples $(attribut, value)$, where $attribute$ stands for a packaging attribute as defined in the packaging database schema of the flexible querying system part of the DSS, and $value$ is the preferred value expressed for the considered attribute.

4 Conclusion

We applied an argumentation approach on a real use case from the industry allowing stakeholders to express their preferences and providing the system with stable concepts and subsumptions of a domain. We have proposed an argumentation system in which each criterion (attribute or aspect) is considered as a viewpoint in which stakeholders express their arguments in homogenous way. Each viewpoint delivers extensions supporting or opposing certain choices according one packaging aspect, which are then used in the querying process. The approach is implemented as freely accessible web application.

References

1. Amgoud, L., Bodenstaff, L., Caminada, M., McBurney, P., Parsons, S., Prakken, H., Veenen, J., Vreeswijk, G.: Final review and report on formal argumentation system. Deliverable d2.6 aspic. Technical report (2006)
2. Amgoud, L., Prade, H.: Using arguments for making and explaining decisions. Artificial Intelligence 173(3-4), 413–436 (2009)
3. Besnard, P., Hunter, A.: Elements of Argumentation. The MIT Press (2008)
4. Bouyssou, D., Dubois, D., Pirlot, M., Prade, H.: Decision-making process – Concepts and Methods. Wiley (2009)
5. Calvanese, D., Giacomo, G.D., Lembo, D., Lenzerini, M., Rosati, R.: Data complexity of query answering in description logics. In: KR, pp. 260–270 (2006)
6. Caminada, M., Amgoud, L.: On the evaluation of argumentation formalisms. Artificial Intelligence 171, 286–310 (2007)
7. Colucci, S., Noia, T.D., Ragone, A., Ruta, M., Straccia, U., Tinelli, E.: Informative Top-k retrieval for advanced skill management. In: De Virgilio, R., et al. (eds.) Semantic Web Information Management, ch. 19, pp. 449–476. Springer, Heidelberg (2010)
8. Destercke, S., Buche, P., Guillard, V.: A flexible bipolar querying approach with imprecise data and guaranteed results. Fuzzy Sets and Systems 169, 51–64 (2011)
9. Dung, P.M.: On the acceptability of arguments and its fundamental role in non-monotonic reasoning, logic programming and n-persons games. Artificial Intelligence 77(2), 321–357 (1995)
10. Prakken, H.: An abstract framework for argumentation with structured arguments. Technical report, Department of Information and Computing Sciences. Utrecht University (2009)

11. Rahwan, I., Simari, G.: Argumentation in Artificial Intelligence. Springer (2009)
12. Tamani, N., Croitoru, M.: Fuzzy argumentation system for decision making. Technical report, INRIA LIRMM (2013), https://drive.google.com/file/d/0BODPgJDRNwbLdE5wdzFQekJocXM/edit?usp=sharing
13. Tamani, N., Croitoru, M., Buche, P.: A viewpoint approach to structured argumentation. In: Bramer, M., Petridis, M. (eds.) The Thirty-third SGAI International Conference on Innovative Techniques and Applications of Artificial Intelligence, pp. 265–271 (2013)
14. Tamani, N., Croitoru, M., Buche, P.: Conflicting Viewpoint Relational Database Querying: An Argumentation Approach. In: Scerri, L., Huhns, B. (eds.) Proceedings of the 2014 International Conference on Autonomous Agents and Multiagent Systems, AAMAS 2014, pp. 1553–1554. International Foundation for Autonomous Agents and Multiagent Systems, Richland (2014)
15. Wu, Y.: Between argument and conclusion. Argument-based approaches to discussion. Inference and Uncertainty. PhD thesis, Université du Luxembourg (2012)

Author Index